FUNCTIONAL EQUATIONS ON GROUPS

FUNCTIONAL EQUATIONS ON GROUPS

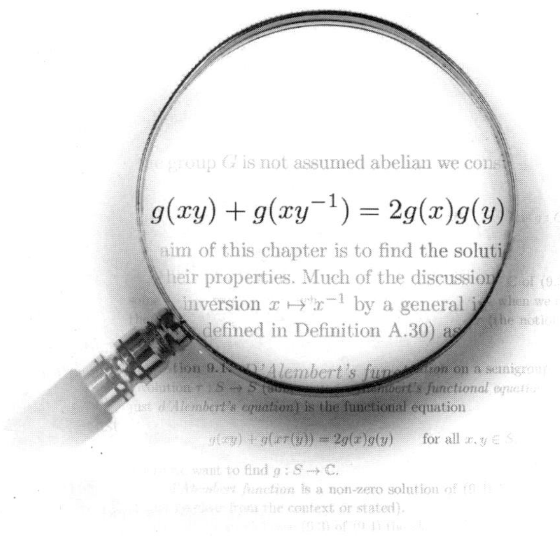

Henrik Stetkær
Aarhus University, Denmark

World Scientific

NEW JERSEY • LONDON • SINGAPORE • BEIJING • SHANGHAI • HONG KONG • TAIPEI • CHENNAI

Published by

World Scientific Publishing Co. Pte. Ltd.
5 Toh Tuck Link, Singapore 596224
USA office: 27 Warren Street, Suite 401-402, Hackensack, NJ 07601
UK office: 57 Shelton Street, Covent Garden, London WC2H 9HE

British Library Cataloguing-in-Publication Data
A catalogue record for this book is available from the British Library.

FUNCTIONAL EQUATIONS ON GROUPS

ISBN 978-981-4513-12-8

Printed in Singapore by World Scientific Printers.

Preface

Our story

The story we wish to tell is about the progress in the last 10–20 years of the theory of a number of trigonometric functional equations. We shall present recent results about these functional equations on groups, in particular nonabelian groups, reveal how they contain earlier results as special cases and apply them to important and interesting groups. Many of the earlier results deal with functions defined on $\mathbb{R}$ or more generally $\mathbb{R}^n$. But not all groups look like $\mathbb{R}^n$ with its algebraic, topological and differentiable structures. Just think of the groups $\mathbb{Z}$ and $\mathrm{SL}(2,\mathbb{C})$. New, interesting and non-obvious phenomena make their appearance in the transition from $\mathbb{R}^n$ to other groups. The theory throws light on the properties of solutions of the functional equations that are valid on all groups or at least on all groups of a special type (like connected, solvable Lie groups). The table of contents gives a schematic survey of which functional equations we study. We illustrate the theory by thoroughly calculating explicit formulas for the solutions of the functional equations on concrete examples of such groups, like the Heisenberg group, the $(ax+b)$-group, $\mathrm{SL}(2,\mathbb{R})$ etc., which have not been mentioned, let alone discussed, in earlier monographs about functional equations.

We wish to tell our story to graduate students and professional mathematicians seeking an accessible and self-contained introduction to those functional equations that generalize classical relations between the trigonometric functions, like the sine addition formula. The present book is meant as

(1) a place where some of the new results of the last 10–20 years are validated and archived in an accessible way.

(2) a source where the reader can find references for further study and get an account of recent and current research in this area of mathematics.
(3) a springboard for research in the area.
(4) a textbook that can be used in a graduate course to give an easily read introduction to important functional equations on groups, because we want to induct a new generation into the recent developments.

Much of the material of the book has appeared in research journals, written for experts, but not in book form. Our exposition is detailed and requires little background material, so that the non-expert reader can follow the arguments. Special technical results are collected in appendices. Combining this with notes and remarks we hope that the book may serve the purposes mentioned above.

I collected the results of the present book, because they are interesting and beautiful extensions of classical results about functional equations and because they interconnect with parts of harmonic analysis. They combine concepts and results from algebra, analysis and topology. I confess that I have had fun playing around with various identities that extend classical ones, and that I appreciate the clever tricks, intricate computations and ingenious methods invented by the mathematicians of the field. Hopefully the reader will enjoy my choice of topics.

The organization of this book

We assume the reader is familiar with linear algebra and point set topology and has taken an introductory course in algebra containing group theory. For some sections and examples a rudimentary knowledge of locally compact groups and their Haar measures is required, but the rest of the book does not depend on them. Apart from that the book is self-contained, except for isolated facts from other subjects that can be found in standard textbooks.

The book is written in the concise style of mathematics literature with precisely formulated statements and rigorous proofs of them, and it is built up with a view to the logical order of the topics. Regrettably the logical order does not always coincide with the pedagogical way of proceeding. We have remedied that by presenting in each chapter a number of examples usually before the theory. Furthermore we have tried to make the various chapters reasonably independent, so that the reader can read the text anywhere in the book without having to go through all of the previous material.

A consultation of one or two previous theorems and/or an appendix should suffice. It has been my intention to make the book suited not just for a course, but also for private study, by including enough details, explanations, motivation and manageable exercises.

Each chapter contains suitably designed exercises meant to fulfill one or more of the following purposes:

(1) enable the reader to enhance his comprehension by answering accessible, simple questions about the theory or by computing illuminating examples.
(2) illustrate the theory by supplying more examples than the ones of the main text. For the sake of reference we often write down the solution formulas.
(3) present and improve upon results in the literature by extending them to non-abelian groups.

Chapters 2 and 3 introduce additive and multiplicative functions, which are two of our building blocks.

Each chapter from Chapter 4 onwards treats its own topic, mainly a specific functional equation, for which we derive explicit formulas for the solutions in terms of our building blocks.

The section Notes and remarks at the end of each chapter points out generalizations and other directions of research than those of the main text. They also give credit for some of the recent developments in the field. Although some of the results of the book are new as far as we know, failure to cite a reference for a given result should not be construed as a claim of originality on our part.

Appendix A contains background material with the basic terminology, examples, definitions and simple consequences of the definitions. It should not be read from start to end, as the reader will surely know most of the material in advance.

The other appendixes contain facts (and sometimes proofs of them) from other parts of mathematics than functional equations, facts that are used in the book and may be time-consuming for a non-expert to localize in the literature. In particular the appendixes contain examples of groups and involutions. These facts are not needed throughout the text, so an appendix need only be consulted by the reader when he encounters a reference to it in the main text.

The material was used successfully in two 1-semester courses for Master and PhD students. The parts discussed in the first course were Chapters 2, 3,

4, 5, 7, 8 and 9 plus the sections of Chapter 11 that deal with abelian groups. Most of the rest of the book was done in the sequel 1-semester course.

Topics we do not discuss

We have not been able to cover all of the recent developments of functional equations on groups. The field is too big. We were furthermore limited by a wish to make the book reasonably short and by our own abilities. So we had to make some choices. Not just of topics, but even of which results to include and which to omit in the chosen topics.

Fields and vector spaces will occur in examples, but we will not systematically discuss functional equations on other algebraic structures than groups. Thus hypergroups are excluded (works by Orosz, Roukbi, Székelyhidi, Zeglami). Even with those restrictions interesting and important topics have been left out: We do not discuss Hyers-Ulam stability of functional equations that many mathematicians have contributed to. There is no iteration theory, and no applications of spectral synthesis on locally compact groups (works by Orosz, Székelyhidi). There are no distribution functions (work by Feldman). There are no operator-valued solutions (works by Chojnacki, Kisyński, Sinopoulos). Special functions do not occur, although many of them are related to groups. We study continuous solutions, and measurable solutions occur only sporadically. Solutions in the sense of distributions are not treated (works by Baker, Deeba and Koh and others). The domain of definition of our solutions is the whole group, which means that we do not discuss restricted and conditional solutions (works by Ger and others).

We are not particularly interested in finite groups with their number theoretical ramifications. Essentially we view finite groups as special instances of compact groups.

We study functional equations for their own sake. As a consequence this book does not contain applications outside of mathematics. It is long even without such applications.

Acknowledgements

I thank professors Thomas M. K. Davison, Bruce Ebanks, Roman Ger, Che Tat Ng and Dilian Yang for many constructive and illuminating discussions. Also the many organizers of the annual International Symposium on Functional Equations deserve thanks. The ISFE's are splendid fora for fruitful discussions with colleagues. Last but not least, I am very much in debt to LaTeX expert Lars Madsen for his invaluable help with practical details and his insistence on quality and consistency of the typesetting.

April 2013
Henrik Stetkær

Contents

Chapter 1

Introduction

1.1 A first glimpse at functional equations

A functional equation is an equation in which the unknown or unknowns are functions. To avoid a too extensive theory differential, difference and integral equations are not counted as parts of functional equations, these theories being huge separate subjects with their own own life and their own special methods.

The present book concentrates on special types of functional equations: Trigonometrical functional equations on groups, i.e., equations that extend and generalize classical relations among the trigonometric and hyperbolic functions. So our point of departure is formulas of elementary trigonometry. To take an example, the function cosine satisfies the identity

$$\cos(x+y)+\cos(x-y)=2\cos x\cos y \qquad \text{for all } x,y\in\mathbb{R}.$$

We seek the functions $g:\mathbb{R}\to\mathbb{C}$ that satisfy the corresponding functional equation (called d'Alembert's functional equation or the cosine equation)

$$g(x+y)+g(x-y)=2g(x)g(y) \qquad \text{for all } x,y\in\mathbb{R}, \tag{1.1}$$

in which we have replaced cos in the identity by g. To solve (1.1) is to find all functions $g:\mathbb{R}\to\mathbb{C}$ for which (1.1) holds. The equation is a functional equation, because its solutions g are functions, not numbers. Incidentally, the cosine equation has other solutions than $g=\cos$, for instance $g=\cosh$.

Another important functional equation that we shall study, is the sine addition equation

$$f(x+y)=f(x)g(y)+g(x)f(y) \qquad \text{for all } x,y\in\mathbb{R}, \tag{1.2}$$

where both $f:\mathbb{R}\to\mathbb{C}$ and $g:\mathbb{R}\to\mathbb{C}$ are unknown functions that we want to determine. So here the functional equation contains two unknown functions, not just one. The ordered pair $f = \sin$, $g = \cos$ is a solution of (1.2), because

$$\sin(x+y) = \sin x \cos y + \cos x \sin y \qquad \text{for all } x, y \in \mathbb{R},$$

but this pair is not the only solution. We find all solutions in Chapter 4.

The distance function (or rather the square of it) $f(x) := \|x\|^2$ on $\mathbb{R}^n$ satisfies the parallelogram identity

$$f(x+y) + f(x-y) = 2f(x) + 2f(y) \qquad \text{for all } x, y \in \mathbb{R}^n. \tag{1.3}$$

The functional equation (1.3) is called the quadratic functional equation.

The examples above show that some of the functional equations are linear, while others involve products, so we study not just linear functional equations.

We are not satisfied with finding the functions on $\mathbb{R}$ and $\mathbb{R}^n$ that satisfy the classical functional equations above: We want a deeper understanding and to see the functional equations in a wider context, so we extend the scope by replacing the domain of definition $\mathbb{R}$ (or $\mathbb{R}^n$) of the functions by a group G, and instead of the classical range space $\mathbb{R}$ we take $\mathbb{C}$ or sometimes even just an abelian group. Thus given a group G we want to describe the solutions $g : G \to \mathbb{C}$ of the cosine equation

$$g(xy) + g(xy^{-1}) = 2g(x)g(y) \qquad \text{for all } x, y \in G,$$

and to find common properties of the solutions on various types of groups (like abelian or compact groups). Similarly for the sine addition formula (1.2) that takes the form

$$f(xy) = f(x)g(y) + g(x)f(y) \qquad \text{for all } x, y \in G,$$

where $f, g : G \to \mathbb{C}$ are the unknowns, and for the quadratic functional equation (1.3) that becomes becomes

$$f(xy) + f(xy^{-1}) = 2f(x) + 2f(y) \qquad \text{for all } x, y \in G$$

with $f : G \to \mathbb{C}$ to be found. We shall study the solutions of these and other functional equations on groups in which the group composition rule and the form of the functional equation are intertwined like in the examples above. The equations are simple to formulate, but not always to solve. The above examples are not the only ones, but are meant to give a first impression and a starting point. It turns out, as could be expected, that the composition of the set of solutions of the functional equations depends much on the structure of the group, for instance of whether it is abelian. The equations have in the special case of the abelian group $G = \mathbb{R}$ attracted the attention of a number of mathematicians during the times. Many of their results shall in satisfactory ways be extended to other groups, so as a by-product of our investigations we come to look at relations between the classical trigonometric functions from a higher point of view. We shall also shed light on how different functional equations on the same group are related.

1.2 Our basic philosophy

What we do is to ask for properties that characterize trigonometrical functions on $\mathbb{R}$ and their extensions to groups: The cosine equation (1.1) expresses a property of the cosine functions $x \mapsto \cos(\lambda x)$, $\lambda \in \mathbb{C}$, and we want to find all solutions of (1.1) to see whether (1.1) characterizes cosine functions, so that the cosine functions are the only non-zero ones with the property (1.1). (This is indeed the case: See Proposition 9.4.) So we turn things around: Our point of departure is the functional equation, not a function like cos motivating it. We want to derive qualitative properties of the solutions from the functional equation, not from knowledge of the individual solutions. In other words, we look for joint properties of the solutions. For instance, what are the properties shared by the solutions of the cosine functional equation?

To solve a functional equation is to express the unknown function or functions in terms of (supposedly known) simple, basic functions. These simple, basic functions are to be our building blocks. To see what this means when we pass from $\mathbb{R}$ to a group G we look at Euler's formulas

$$\cos x = \frac{e^{ix} + e^{-ix}}{2}, \quad \sin x = \frac{e^{ix} - e^{-ix}}{2}.$$

The functions $x \mapsto e^{ix}$ and $x \mapsto e^{-ix}$ are homomorphisms of the group $(\mathbb{R}, +)$ into the multiplicative group $\mathbb{C}^*$ of all non-zero complex numbers. This is an omen of things to come: The solutions of functional equations on groups are expressed in terms of homomorphisms. The functions sin and cos make no sense on an arbitrary group, but homomorphisms do: Our basic functions on a group G will always include the characters, i.e., the homomorphisms of G into $\mathbb{C}^*$. The roles of cosine and sine will be taken by $x \mapsto (\chi(x) + \chi(x^{-1}))/2$ and multiples of $x \mapsto \chi(x) - \chi(x^{-1})$. The characters will be combined with other basic functions like the homomorphisms of G into $(\mathbb{C}, +)$, i.e., the additive functions, and bi-additive maps of $G \times G$ into $\mathbb{C}$. A function like $f(x) := 2x$ is an example of an additive function on $G = \mathbb{R}$. Some of the building blocks may be spherical functions. Whatever the building blocks are, they will always be closely connected to the structure of the underlying group, just like it is in the special case of $G = \mathbb{R}$. The building blocks are formally the same kind of functions on any group: Characters, additive maps, bi-additive maps, representations etc. But such functions vary from group to group according to the nature of the group and should be computed for each group or type of group. Our theory will typically present results that are common for all groups of a certain type, like abelian groups, compact groups or semisimple Lie groups.

Lots of efforts have gone into the study of functional equations on abelian groups, in particular on $\mathbb{R}$ and $\mathbb{C}$. We will not duplicate the previous works on abelian groups here, but just refer to the monographs by Aczél [3], Aczél and Dhombres [6], Székelyhidi [197] and Kannappan [132], because our main interest lies in the study of functional equations on groups that need not be abelian. During the last 10–20 years much progress has been made in this area. Progress has been made not just for classical functional equations like d'Alembert's and Wilson's functional equations, but also for new functional equations, like the symmetrized sine addition formula, that occur naturally after the transition from the abelian to the non-abelian case. Another example: The two functional equations $f(xy) + f(xy^{-1}) = 2f(x)$ and $f(xy) + f(y^{-1}x) = 2f(x)$ are different, although they coincide on abelian groups, where they are Jensen's classical functional equation. We want to complete the picture by studying functional equations on non-abelian groups. That entails that concepts from non-abelian harmonic analysis like representations will enter to supplement characters and additive maps.

Many important groups are topological by their very origin (think of matrix-groups), so for them it is natural to take continuity into account and concentrate on continuous solutions of the functional equations. I do not agree with the sweeping statement that one of my fellow functional equationists has made: "We lead a relentless crusade against all regularity assumptions". Superfluous assumptions shall of course be weeded out, but there is no need to use unprovoked violence to an inoffensive topological group by robbing it of its natural topology and couching in nothing (the discrete topology). Topological groups are important in analysis and topology, so we cannot avoid them, and we do not want to. It is an extra bonus, that the continuous solutions often can be found explicitly, because continuity of a solution frequently entails its differentiability, so that we may apply the theory of differential equations to compute it. So a unifying aspect of our presentation is that the functional equations are defined on topological groups. Our philosophy is that the basic results should depend only on the topology and the group structure. Extra structures, like that of a linear space, will of course be present and be exploited in specific examples, but not in the general theory.

Functional equations occur at mathematical competitions, which gives the spectator the impression that functional equations are singular and isolated from one another, and that each equation needs its own solution trick. You get the same impression at your first encounter with partial differential equations. However, partial differential equations fall into different types: Think of the Laplace equation, the heat equation and the wave equation on domains in $\mathbb{R}^n$ for varying n, each equation with its own characteristic behavior. There are many different functional equations on groups and no common method of solution exists. But functional equations on groups, like differential equations on varying domains of $\mathbb{R}^n$, also fall into certain types. In our comparison with differential equations domains shall be replaced by topological groups. To take an example, the two differently looking functional equations

$$g(x+y)+g(x-y)=2g(x)g(y) \qquad \text{for } x,y\in\mathbb{R},$$

and

$$\begin{aligned} &g(x_1+x_2,y_1+y_2,z_1+z_2+x_1y_2)\\ &\quad +g(x_1-x_2,y_1-y_2,z_1-z_2-(x_1-x_2)y_2)\\ &\qquad =2g(x_1,y_1,z_1)g(x_2,y_2,z_2) \qquad \text{for } x_1,y_1,z_1,x_2,y_2,z_2\in\mathbb{R}, \end{aligned}$$

are both manifestations of d'Alembert's functional equation, the first on the additive group $\mathbb{R}$, the other on the Heisenberg group. Our results in the chapter about d'Alembert's functional equation hold on any group, but they simplify of course on abelian groups and on other special groups. Similarly we treat a number of other important functional equations. Each gets a chapter of its own containing results about its solutions, many results valid on any group and some valid on large classes of groups like abelian groups. So our main project is to analyze functional equations, each one simultaneously on large classes of groups. The present book shows this is feasible.

However, this is not the complete story, because another aspect need to be taken into account: There are relations between various functional equations on the same underlying group. This was observed a hundred years ago by Wilson [209] for functions defined on $\mathbb{R}$ or $\mathbb{C}$. It is apparent from Theorems 4.1 and 4.15, that show the sine and cosine addition formulas are related. Or from Lemma 11.2(e) that leads from Wilson's functional equation to d'Alembert's long functional equation. Thus a study of one functional equation can be instrumental for the investigation of another. As a consequence our presentation does not, actually cannot, consist of isolated chapters, having nothing in common and each dealing with its own particular type of functional equation. By the contrary: The functional equations are intricately intertwined.

1.3 Exercises

Exercise 1.1. Let X be a set. Let $f, g : X \to \mathbb{C}$.

(a) Assume that $f(x)g(y) = f(y)g(x)$, $x, y \in X$. Show that $f = 0$ or g is proportional to f.

(b) Assume that $f(x)g(y) = -f(y)g(x)$, $x, y \in X$. Show that $f = 0$ or $g = 0$.

Exercise 1.2. Let X be a non-empty set. Show that the solutions $\varphi : X \times X \to \mathbb{C}$ of *Sincov's functional equation*

$$\varphi(x, y) + \varphi(y, z) = \varphi(x, z), \qquad x, y, z \in X,$$

are the functions of the form $\varphi(x, y) = \psi(x) - \psi(y)$, where $\psi : X \to \mathbb{C}$ is a complex function on X.

Hint: Fixing $y = x_0 \in X$ we have $\varphi(x, z) = f(x) + g(z)$, where $f(x) := \varphi(x, x_0)$ and $g(z) := \varphi(x_0, z)$.

Can you generalize this, replacing $(\mathbb{C}, +)$ by a possibly non-abelian group?

1.4 Notes and remarks

Exercise 1.2: For the history around Sincov's functional equation consult Gronau [103].

Chapter 2

Around the Additive Cauchy Equation

The main topic of this chapter is the additive Cauchy equation (2.1). It is included, not so much because its theory is nice, interesting and important (it is), but because the additive functions (= the solutions of the additive Cauchy equation) are among the building blocks, out of which we construct solutions of other functional equations. We are in particular interested in the continuous, additive maps.

Two minor related topics of this chapter are bi-additive functions and the symmetrized additive Cauchy equation (2.3).

Our account of the additive Cauchy equation is brief, because it is discussed in details in the literature.

2.1 The additive Cauchy equation

That a map $L : V \to W$ between two vector spaces is linear means that it is homogeneous ($L(\alpha v) = \alpha Lv$ for all scalars α and all $v \in V$) and additive ($L(v+w) = Lv+Lw$ for all $v, w \in V$). The homogeneity does not make sense on a group or semigroup, but the additivity does as a homomorphism. In the following definition we restrict a bit further than to a general homomorphism by requiring the range to be abelian.

Definition 2.1. Let S be a semigroup, and $(H, +)$ an abelian semigroup. A solution of *the additive Cauchy equation*

$$a(xy) = a(x) + a(y), \quad x, y \in S, \tag{2.1}$$

is a map $a : S \to H$ satisfying (2.1). A solution is called an *additive map* or an *additive function* from S to H or an additive map/function on S with values in H.

Examples 2.2.

1. An example of an additive function $a : \mathbb{R} \to \mathbb{C}$ on $S = \mathbb{R}$ with values in $H = \mathbb{C}$ is the linear function $a(x) := \alpha x$, $x \in \mathbb{R}$, where α is any complex number.
2. An example of an additive function $a : \mathbb{N} \to \mathbb{N}$ on $S = \mathbb{N}$ with values in $H = \mathbb{N}$ is $a(n) := 2n$, $n \in \mathbb{N}$.
3. An example of an additive function $a : \mathbb{R}^* \to \mathbb{C}$ on the multiplicative group $S = \mathbb{R}^*$ with values in $H = \mathbb{C}$ is $a(x) := \alpha \log|x|$, $x \in \mathbb{R}^*$, where α is any complex number.

On a locally compact group very little is needed for an additive map to become continuous. Measurability with respect to a Haar measure suffices (see Proposition D.4(a) for the precise statement). Exercise 2.7 says that local integrability suffices on $S = \mathbb{R}$.

We shall not treat the additive Cauchy equation in full generality. We shall discuss the case of $S = V$ where V is a vector space over $\mathbb{R}$ and $H = \mathbb{C}$. Here (2.1) becomes

$$a(x + y) = a(x) + a(y), \quad x, y \in V. \tag{2.2}$$

It is neither required nor a consequence of the additivity that an additive function $a : V \to \mathbb{C}$ is linear, i.e., that furthermore $a(\alpha x) = \alpha a(x)$ for all $\alpha \in \mathbb{R}$ and $x \in V$. In examples we compute the additive functions on some groups that are not vector spaces.

Lemma 2.3. *Let V be a vector space over $\mathbb{R}$ and let $a : V \to \mathbb{C}$ be a solution of* (2.2).

(a) $a(qx) = qa(x)$ *for all $q \in \mathbb{Q}$ and $x \in V$, i.e., a is $\mathbb{Q}$-linear.*

(b) *If V is a topological vector space and a is continuous at a point then a is continuous.*

(c) *If V is a topological vector space and a is continuous, then a is linear.*

Proof. (a) By induction we get that $a(nx) = na(x)$ for all $n \in \mathbb{N}$ and $x \in V$. Taking $x = y = 0$ in (2.2) we get that $a(0) = 0$, so that $a(nx) = na(x)$ for all $n \in \mathbb{N} \cup \{0\}$ and $x \in V$. Taking $y = -x$ in (2.2) we get that $a(-x) = -a(x)$. It follows that $a(nx) = na(x)$ for all $n \in \mathbb{Z}$ and $x \in V$. Now, we get for any $q \in \mathbb{N}$ and $p \in \mathbb{Z}$ that

$$qa\left(\frac{p}{q}x\right) = a\left(q\frac{p}{q}x\right) = a(px) = pa(x), \quad \text{so that} \quad a\left(\frac{p}{q}x\right) = \frac{p}{q}a(x).$$

(b) is Exercise 2.6 or more generally Lemma D.1.

(c) follows from (a) and the fact that the rational numbers are dense in $\mathbb{R}$. □

Corollary 2.4. *The continuous solutions of* (2.2) *on* $V = \mathbb{R}^n$ *are the linear functionals on* $\mathbb{R}^n$. *Each such one consists in taking inner product with a vector* $c \in \mathbb{C}^n$: $a(x) = \langle c, x \rangle$, $x \in \mathbb{R}^n$. *In particular, the continuous solutions* a *of* (2.2) *on* $\mathbb{R}$ *are the functions of the form* $a(x) = cx$, $x \in \mathbb{R}$, *where the constant* c *ranges over* $\mathbb{C}$.

Proof. Left to the reader, who should use that all linear functions $L : \mathbb{R}^n \to \mathbb{C}$ are continuous. □

Thus continuous, additive functions $a : \mathbb{R}^n \to \mathbb{C}$ are extremely well behaved. The situation is the opposite for discontinuous, additive functions: They have pathological properties. In Corollary 2.4 we may replace the requirement of continuity by Lebesgue-measurability, because measurability of a solution implies its continuity (see Proposition D.4(a)), so there is a dichotomy: A solution is either extremely nice (linear) or it is very badly behaved (non-measurable). As a further illustration of the pathological behaviour we shall now prove in an elementary way that the values of any discontinuous, additive function $a : \mathbb{R} \to \mathbb{R}$ have to fluctuate so wildly that its graph is dense in the plane $\mathbb{R}^2$. This phenomenon would not worry us if discontinuous, additive functions did not exist. But they do (Proposition 2.8).

Proposition 2.5. *Let* $a : \mathbb{R} \to \mathbb{R}$ *be an additive function, which is not of the form* $a(x) = cx$, $x \in \mathbb{R}$ *for some constant* $c \in \mathbb{R}$. *Then its graph* $\{(x, a(x)) \mid x \in \mathbb{R}\}$ *is dense in* $\mathbb{R}^2$.

Proof. We shall prove that there to any given point $(x_0, y_0) \in \mathbb{R}^2$ exists a sequence $(x_1, a(x_1)), (x_2, a(x_2)), \ldots, (x_n, a(x_n)), \ldots$ from the graph of a such that $(x_n, a(x_n)) \to (x_0, y_0)$ as $n \to \infty$.

By our hypothesis the graph of a is not contained in a line through the origin. So there exist two points $(\xi, a(\xi))$ and $(\eta, a(\eta))$ such that the vectors $(\xi, a(\xi))$ and $(\eta, a(\eta))$ are not collinear. Thus they are linearly independent, and so they constitute a basis of $\mathbb{R}^2$. We may therefore write $(x_0, y_0) \in \mathbb{R}^2$ in the form

$$(x_0, y_0) = \alpha_0(\xi, a(\xi)) + \beta_0(\eta, a(\eta)) \qquad \text{for some } \alpha_0, \beta_0 \in \mathbb{R}.$$

Take now a sequence $\{\alpha_n\}$ of rational numbers converging to α_0 and a sequence $\{\beta_n\}$ of rational numbers converging to β_0. Then the sum

$\alpha_n(\xi, a(\xi)) + \beta_n(\eta, a(\eta))$ converges to (x_0, y_0), so it suffices to verify that $\alpha_n(\xi, a(\xi)) + \beta_n(\eta, a(\eta))$ belongs to the graph of a. And using Lemma 2.3(a) we find that any linear combination with rational coefficients of elements from the graph is again in the graph, finishing the proof. □

In particular, if an additive function $a : \mathbb{R} \to \mathbb{R}$ is bounded from above or below on an interval (for instance is positive everywhere on that interval) then a is linear. See Exercise 2.18 for other conditions.

Definition 2.6. A *Hamel basis* for a vector space V over a field F is a linearly independent subset $\{v_i \mid i \in I\}$ of V that spans V. So any $v \in V$ can in exactly one way be written as $v = \sum_{i \in I} q_i(v)v_i$, where only finitely many of the coefficients $q_i(v)$ are different from zero. The functions $q_i : V \to F$, $i \in I$, are called *the coefficient functions*. Note that $q_i(v_i) = 1$ and that q_i is additive for all $i \in I$ (due to the uniqueness of the coefficients).

Proposition 2.7. *Every vector space $\neq \{0\}$ has a Hamel basis.*

Proof. We skip the proof. The reader may supply one by using Zorn's lemma. A proof can be found in [115, Chapter X.1]. □

All linear functions $L : \mathbb{R}^n \to \mathbb{C}$ are continuous, but the statement becomes false, if we replace the linearity assumption by the weaker notion of additivity. This is shown by the following Proposition 2.8.

Proposition 2.8. *There exists a discontinuous additive function*

$$a : \mathbb{R} \to \mathbb{R}.$$

Proof. Let v_{i_0} be an element of a Hamel basis $\{v_i \mid i \in I\}$ for $\mathbb{R}$ as a vector space over the field $\mathbb{Q}$. It suffices to show that q_{i_0} is not continuous, because in that case we may take $a = q_{i_0}$. Assume to the contrary that $q_{i_0} : \mathbb{R} \to \mathbb{Q} \subseteq \mathbb{R}$ is continuous. Then $q_{i_0}(\mathbb{R})$ is an interval in $\mathbb{R}$ as the continuous image of a connected set. But no interval consists entirely of rational numbers, except if it is a one point interval. Thus q_{i_0} assumes only one value, so it is constant. In particular $q_{i_0}(0) = q_{i_0}(v_{i_0})$. But $q_{i_0}(0) = 0$ and $q_{i_0}(v_{i_0}) = 1$, so we have arrived at a contradiction. □

Remarks 2.9.

(1) The discontinuous additive function in Proposition 2.8 can be chosen so that its graph in $\mathbb{R}^2$ is connected! See [121, Theorem 5] for a proof of this weird fact.

(2) Our proof of the existence of discontinuous solutions (Proposition 2.8) relies on Zorn's lemma, so it provides no explicit formulas for them. We just know they exist.

(3) This remark is really a digression:

There are other notions of bases for vector spaces than that of a Hamel basis. The notions coincide for finite-dimensional vector spaces, but not for infinite-dimensional. In case of a Hamel basis the expansion of any element in the vector space is a finite sum. This is not so for, e.g., an orthonormal basis of a Hilbert space or more generally for a Schauder basis of a Banach space, so in those cases the complicated question of what is meant by convergence of an infinite sum arises. Orthonormal bases and Schauder bases are closely tied to the underlying norm of the vector space, which the Hamel bases are not; they only pertain to the algebraic structure. Furthermore orthonormal bases and Schauder bases can be explicitly constructed, and not just shown to exist by help of the axiom of choice. To take an example, no one has written down explicit formulas for a Hamel basis for $L^2(\mathbb{R})$, but concrete and useful orthonormal bases abound: The Haar system, the Hermite-functions etc. The monograph [40] is a nice modern treaty on bases and frames. A frame is roughly speaking an overdetermined basis.

We have so far only studied additive maps on the additive group of a vector space. In Exercise 2.9 we find the additive maps on the multiplicative groups $\mathbb{R}^+$ and $\mathbb{R}^*$. In the following four examples we compute the continuous additive functions $a : G \to \mathbb{C}$ on certain non-abelian groups G.

Example 2.10. Let $h : G \to \mathbb{C}$ be a continuous, additive map on the $(ax+b)$-group

$$G := \left\{ \begin{pmatrix} a & b \\ 0 & 1 \end{pmatrix} \middle| \, a > 0, \ b \in \mathbb{R} \right\}$$

from Examples A.17(i). Any additive map vanishes on commutators and so on the commutator subgroup

$$[G, G] = \left\{ \begin{pmatrix} 1 & b \\ 0 & 1 \end{pmatrix} \middle| \, b \in \mathbb{R} \right\}.$$

Thus

$$h \begin{pmatrix} a & b \\ 0 & 1 \end{pmatrix} = h \left(\begin{pmatrix} 1 & b \\ 0 & 1 \end{pmatrix} \begin{pmatrix} a & 0 \\ 0 & 1 \end{pmatrix} \right) = 0 + h \begin{pmatrix} a & 0 \\ 0 & 1 \end{pmatrix} = h \begin{pmatrix} a & 0 \\ 0 & 1 \end{pmatrix},$$

which shows that h is essentially an additive map of $(\mathbb{R}^+, \cdot)$ into $(\mathbb{C}, +)$. Being also continuous it has according to Exercise 2.9 the form

$$h\begin{pmatrix} a & b \\ 0 & 1 \end{pmatrix} = h\begin{pmatrix} a & 0 \\ 0 & 1 \end{pmatrix} = c\log a,$$

where $c \in \mathbb{C}$ is a constant.

Conversely, any function of this form is a continuous, additive function on the $(ax+b)$-group.

Example 2.11. Let $a : H_3 \to \mathbb{C}$ be a continuous additive function on the Heisenberg group H_3 (Examples A.17(a)). Any additive function vanishes on the commutator group $[H_3, H_3]$, so

$$\begin{aligned} a\begin{pmatrix} 1 & x & z \\ 0 & 1 & y \\ 0 & 0 & 1 \end{pmatrix} &= a\left(\begin{pmatrix} 1 & 0 & z \\ 0 & 1 & 0 \\ 0 & 0 & 1 \end{pmatrix}\begin{pmatrix} 1 & 0 & 0 \\ 0 & 1 & y \\ 0 & 0 & 1 \end{pmatrix}\begin{pmatrix} 1 & x & 0 \\ 0 & 1 & 0 \\ 0 & 0 & 1 \end{pmatrix}\right) \\ &= 0 + a\begin{pmatrix} 1 & 0 & 0 \\ 0 & 1 & y \\ 0 & 0 & 1 \end{pmatrix} + a\begin{pmatrix} 1 & x & 0 \\ 0 & 1 & 0 \\ 0 & 0 & 1 \end{pmatrix} \\ &= a\begin{pmatrix} 1 & x & 0 \\ 0 & 1 & 0 \\ 0 & 0 & 1 \end{pmatrix} + a\begin{pmatrix} 1 & 0 & 0 \\ 0 & 1 & y \\ 0 & 0 & 1 \end{pmatrix}. \end{aligned}$$

Since an additive continuous function from $\mathbb{R}$ to $\mathbb{C}$ is linear we get

$$a\begin{pmatrix} 1 & x & z \\ 0 & 1 & y \\ 0 & 0 & 1 \end{pmatrix} = \alpha x + \beta y, \qquad \text{where}$$

$$\alpha = a\begin{pmatrix} 1 & 1 & 0 \\ 0 & 1 & 0 \\ 0 & 0 & 1 \end{pmatrix} \in \mathbb{C} \quad \text{and} \quad \beta = a\begin{pmatrix} 1 & 0 & 0 \\ 0 & 1 & 1 \\ 0 & 0 & 1 \end{pmatrix} \in \mathbb{C}.$$

Conversely, any function on H_3 of this form is a continuous additive function from H_3 to $\mathbb{C}$.

Example 2.12. A Lie group is semisimple if its connected identity component is semisimple, because semisimplicity by definition means that the Lie algebra is semisimple. Examples of semisimple Lie groups are listed in Helgason's monograph [108, Ch. X, 2]. Let us among them mention the two connected groups $\mathrm{SL}(n,\mathbb{R})$ and $\mathrm{SL}(n,\mathbb{C})$ for $n \geq 2$, and the Lorentz group $O(3,1)$ that has four connected components.

If G is a semisimple Lie group with finitely many components, then any additive map $a : G \to \mathbb{C}$ vanishes. We need not assume that the additive map is continuous.

To prove the statement we note that the identity component G_e of G satisfies $G_e = [G_e, G_e]$ ([201, Corollary 3.18.10]), so that $a(G_e) = \{0\}$. Thus a is a function on the finite group G/G_e. And then $a = 0$ (Exercise 2.5(c)).

Example 2.13. Let us consider the group $G := \mathbb{R}^2 \times_s \mathrm{SO}(2)$ of rigid motions of $\mathbb{R}^2$ from Example A.22(c). We can decompose any element $(x,\rho) \in G$ as a product of elements from the two subgroups $\mathbb{R}^2$ and $\mathrm{SO}(2)$ in the following way $(x,\rho) = (x,I)(0,\rho)$.

Let $a : G \to \mathbb{C}$ be an additive function on G. We let $a_1(x) := a(x,I)$ and $a_2(\rho) := a(0,\rho)$ denote the corresponding additive functions on the subgroups $\mathbb{R}^2$ and $\mathrm{SO}(2)$. Then $a(x,\rho) = a((x,0)(0,\rho)) = a(x,0) + a(0,\rho) = a_1(x) + a_2(\rho)$. A small computation shows that a is an additive function if and only if a_1 and a_2 are additive functions such that $a_1(\rho x) = a_1(x)$ for all $\rho \in \mathrm{SO}(2)$ and $x \in \mathbb{R}^2$. Taking $\rho = -I$ we see that $a_1(x) = a_1(-x)$ for all $x \in \mathbb{R}^2$, which implies that $a_1 = 0$. Thus a reduces to $a(x,\rho) = a_2(\rho)$.

If a and hence also a_2 is continuous, then a_2 a continuous additive function on the compact group $\mathrm{SO}(2)$, and so $a_2 = 0$ (Exercise 2.5(b)). We conclude that the only continuous additive function on G is $a = 0$.

2.2 Pexiderization

The functional equations $f(x+y) = g(x) + h(y)$ and $f(x+y) = g(x)h(y)$ and similar ones where f, g and h are the unknown functions, were studied by Pexider [157]. They are generalizations of the corresponding Cauchy equations in which $f = g = h$. To generalize a functional equation like, e.g.,

$$f(xy) + f(xy^{-1}) = 2f(x)f(y),$$

with one unknown function f to

$$f(xy) + g(xy^{-1}) = 2h(x)k(y),$$

with f replaced by several unknown functions (here f, g, h and k), is therefore in colloquial language often called to *Pexiderize* the original functional equation, and the resulting functional equation is called the *Pexiderization* of the original functional equation.

With S and H being abelian semigroups the functional equation

$$f(x+y) = g(x) + h(y), \qquad x, y \in S,$$

where $f, g, h : S \to H$, is often called *Pexider's functional equation*. For an example see Exercise 2.23.

2.3 Bi-additive maps

In our discussion of the quadratic equation (Chapter 13) and of the cocycle equation (Chapter 16) on a group G we meet bi-additive functions $B : G \times G \to \mathbb{C}$. Here we note the following simple results for the case of $G = \mathbb{R}^n$.

Lemma 2.14. *If $B : \mathbb{R}^n \times \mathbb{R}^n \to \mathbb{C}$ is a continuous, bi-additive function, then there exists exactly one matrix $A \in M(n \times n, \mathbb{C})$ such that $B(x, y) = \langle Ax, y\rangle$ for all $x, y \in \mathbb{R}^n$. If B is symmetric, resp. skew-symmetric, then so is the matrix A.*

If B is symmetric, then its continuity is implied by the continuity of the function $f : \mathbb{R}^n \to \mathbb{C}$ defined by $f(x) := B(x, x)$, $x \in \mathbb{R}^n$.

Proof. Since B is continuous and additive in each variable we get from Corollary 2.4 that B is bilinear. It is known from linear algebra that a bilinear form on $\mathbb{R}^n \times \mathbb{R}^n$ has the desired form for a unique matrix $A \in M(n \times n, \mathbb{C})$.

The following formula, valid for all $x, y \in \mathbb{R}^n$ when B is symmetric,

$$4B(x, y) = B(x+y, x+y) - B(x-y, x-y) = f(x+y) - f(x-y),$$

implies the continuity of B, given that of f. □

In some cases there are restrictions on a bi-additive, symmetric map $B : G \times G \to \mathbb{C}$, stemming from an involution $\tau : G \to G$: It should satisfy that $B(\tau(x), y) = -B(x, y)$ for all $x, y \in G$. See for example Theorem 13.6. Going back to the set up of Lemma 2.14 we write down what this restriction means for the corresponding symmetric matrix $A \in M(n \times n, \mathbb{C})$ for various τ:

- If τ is the involution given by $\tau(x) = x$ for all $x \in \mathbb{R}^n$, then the restriction $B(\tau(x), y) = -B(x, y)$ forces $A = 0$.

- If $\tau(x_1, x_2, \ldots, x_n) = (-x_1, x_2, \ldots, x_n)$ for all $(x_1, x_2, \ldots, x_n) \in \mathbb{R}^n$, then the restriction means that the first column vector of A is 0. There are no restrictions on the remaining elements of the matrix A.

2.4 The symmetrized additive Cauchy equation

Definition 2.15. Let S be a semigroup and H an abelian group. The functional equation

$$f(xy) + f(yx) = 2f(x) + 2f(y) \qquad \text{for all } x, y \in S, \tag{2.3}$$

in which $f : S \to H$ is the unknown function, is called *the symmetrized additive Cauchy equation*.

The symmetrized additive Cauchy equation (2.3) is a non-commutative version of the additive Cauchy equation (2.1), because it reduces to (2.1) if S is abelian and H is 2-torsion-free, i.e., $[h \in H \text{ and } 2h = 0] \Rightarrow h = 0$. The last requirement holds if $H = \mathbb{C}$.

Proposition 2.17 below shows that the solutions of the symmetrized additive Cauchy equation on a group are the normalized solutions of Jensen's functional equation, when H is 2-torsion-free.

Definition 2.16. Let G be a group and H an abelian group. The functional equation

$$f(xy) + f(xy^{-1}) = 2f(x) \qquad \text{for all } x, y \in G, \tag{2.4}$$

in which $f : G \to H$ is the unknown function, is called *Jensen's functional equation*. The set of solutions for which $f(e) = 0$ will be denoted $S(G, H)$.

The formulation (2.3) allows the underlying space to be a semigroup, while the group structure in form of the group inversion is present in the formulation of Jensen's functional equation.

Proposition 2.17. *Let $f : G \to H$, where G is a group and H is an abelian group. If $f \in S(G, H)$ then f satisfies* (2.3). *If f satisfies* (2.3) *then $4f \in S(G, H)$.*

Proof. Assume first that $f \in S(G, H)$, which implies that f is odd. We obtain (2.3) by interchanging x and y in Jensen's functional equation and adding the two identities (cf. the proof of Lemma 11.3(d)).

Conversely assume (2.3) holds. Taking $x = y = e$ we get that $2f(e) = 0$. Replacing f by $f - f(e)$ we may from now on assume that $f(e) = 0$. Putting $y = x^{-1}$ in (2.3) we find that $2f$ is odd.

To get an expression like $f(ab)$ on the right hand side we replace x by ab and to simplify matters on the left hand side we replace y by a^{-1}:

$$f(aba^{-1}) + f(b) = 2f(ab) + 2f(a^{-1}) = 2f(ab) - 2f(a).$$

When we here replace b by b^{-1} and add the identities we get, starting with the right hand side, that

$$\begin{aligned}2f(ab) + 2f(ab^{-1}) - 4f(a) &= f(aba^{-1}) + f(b) + f(ab^{-1}a^{-1}) + f(b^{-1})\\ &= f(aba^{-1}) + f((aba^{-1})^{-1}) + f(b) + f(b^{-1}),\end{aligned}$$

which upon multiplication by 2 yields the desired identity, because $2f$ is odd. □

Definition 2.18. Let $\phi : S \to H$ where S is a semigroup and H an abelian group.

(a) The *Cauchy difference* $C\phi : S \times S \to H$ of ϕ is

$$(C\phi)(x, y) := \phi(xy) - \phi(x) - \phi(y) \qquad \text{for } x, y \in S.$$

(b) *The second Cauchy difference* $C^2\phi : S \times S \times S \to H$ of ϕ is for $x, y, z \in S$ defined by

$$\begin{aligned}C^2\phi(x, y, z) &:= C\phi(xy, z) - C\phi(x, z) - C\phi(y, z) &(2.5)\\ &= C\phi(x, yz) - C\phi(x, y) - C\phi(x, z)\\ &= \phi(xyz) - \phi(xy) - \phi(xz) - \phi(yz) + \phi(x) + \phi(y) + \phi(z).\end{aligned}$$

Some authors use the terminology *Cauchy kernel* instead of Cauchy difference. The Cauchy difference is related to additive maps, because its kernel $\{\phi : S \to H \mid C\phi = 0\}$ is the set of additive functions from S to H.

Two immediate observations from the definition are

Lemma 2.19. *Let $\phi : S \to H$ where S is a semigroup and H an abelian group.*

(a) *$C^2\phi = 0$ if and only if $C\phi(-, x)$ is additive for each $x \in G$ or equivalently if and only if $C\phi(x, -)$ is additive for each $x \in G$. In particular, if $C^2\phi = 0$ then $C\phi$ is bi-additive.*

(b) *If ϕ is central then $C^2\phi(x,y,z) = C^2\phi(y,z,x)$ for all $x,y,z \in S$. We express this by saying that $C^2\phi$ is invariant under cyclic permutations.*

Lemma 2.20 notes some properties of the solutions of the symmetrized additive Cauchy equation (2.3) and hence of functions in $S(G,H)$. The annoying factors of 2 may be divided away if $[h \in H$ and $2h = 0] \Rightarrow h = 0$.

Lemma 2.20. *Let S be a semigroup and let H be an abelian group. Let $f : S \to H$ be a solution of the functional equation* (2.3). *For all $x,y,z \in S$:*

(a) $2f(x^n) = n2f(x)$ *for* $n \in \mathbb{Z}$.

(b) *Cf is anti-symmetric in the sense that $Cf(x,y) = -Cf(y,x)$.*

(c) $2C^2 f = 0$.

(d) $2Cf(-,-) : S \times S \to H$ *is bi-additive.*

Proof. (a) is proved by induction on n for $n \geq 0$. For negative n note from (2.3) that $2f(e) = 0$, which makes $2f$ odd.

(b) The identity is just a reformulation of (2.3).

(c) To get an expression for $f(xyz)$ we replace y by yz in (2.3) which gives us $f(xyz)+f(yzx) = 2f(x)+2f(yz)$. To get rid of the term $f(yzx)$ we apply the functional equation once more: $f(yzx)+f(zxy) = 2f(y)+2f(zx)$. That introduces the term $f(zxy)$ that we also compute by help of the functional equation: $f(zxy) + f(xyz) = 2f(z) + 2f(xy)$. Subtracting the middle identity from the sum of the other two we get

$$2f(xyz) = 2\{f(xy) - f(zx) + f(yz) + f(x) - f(y) + f(z)\}.$$

When we apply (2.3) to the term $f(zx)$ to change it to $f(xz)$ we get (c).

(d) Lemma 2.19(a). □

We refer to the thorough discussion of Jensen's equation in Chapter 12 for more information about it and hence about the symmetrized additive Cauchy equation on groups. When S is a group then the solutions $f : S \to \mathbb{C}$ of the equation $C^2 f = 0$ from Lemma 2.20(c) can be found in Exercise 13.10.

2.5 Exercises

Exercise 2.1. Show that the additive maps $a : \mathbb{N} \to \mathbb{C}$ are the maps of the form $a(n) = \alpha n$, $n \in \mathbb{N}$, where $\alpha \in \mathbb{C}$ is a constant.

Exercise 2.2. Prove the following result which is well known for groups:

> **Lemma 2.21.** *Let M be a monoid with neutral element e, and H an abelian group with neutral element 0. If $a : M \to H$ is additive, then $a(e) = 0$.*

Exercise 2.3. Let G be a group and H an abelian group. Let $a : G \to H$ be an additive map. Show that $a(x^n) = na(x)$ for all $x \in G$ and $n \in \mathbb{Z}$.

Exercise 2.4. Let G be a group, $[G, G]$ its commutator subgroup.

(a) Let $a : G \to \mathbb{C}$ additive map. Show that a vanishes identically on $[G, G]$. Show that a is a function on $G/[G, G]$.

(b) Generalize (a) to the following statement: If $a : G \to H$ is a homomorphism from G to an abelian group H, then a vanishes identically on $[G, G]$. What does this tell you about $\chi([G, G])$, if $\chi : G \to \mathbb{C}^*$ is multiplicative (i.e., $\chi(xy) = \chi(x)\chi(y)$ for all $x, y \in G$)?

Exercise 2.5.

(a) Let $a : S \to \mathbb{C}$ be a bounded, additive function on a semigroup S. Show that $a = 0$.

Hint: Prove that $a(x^n) = na(x)$ for all $x \in S$ and $n \in \mathbb{N}$.

(b) Let S be a compact, topological semigroup and $a \in C(S)$ an additive function. Show that $a = 0$.

(c) Let G be a finite group and $a : G \to \mathbb{C}$ an additive function. Show that $a = 0$.

Exercise 2.6. Let $a : \mathbb{R} \to \mathbb{C}$ be additive.

(a) Assume a is continuous at $0 \in \mathbb{R}$. Show that a is continuous at each point $x \in \mathbb{R}$ (cf. Lemma D.1).

(b) Assume a is continuous at some point. Show that a is continuous.

Exercise 2.7. Let $a : \mathbb{R} \to \mathbb{C}$ be additive and locally integrable (i.e., integrable over every bounded interval). Show that a is linear.

Hint: A smart way to proceed can be found in the 1/2 page paper [169]: Let $x, y \in \mathbb{R}$. Derive from the symmetry of the right hand side of

$$\begin{aligned} ya(x) &= \int_0^y a(x)\,dt = \int_0^y (a(x+t) - a(t))\,dt = \cdots \\ &= \int_0^{x+y} a(t)\,dt - \int_0^x a(t)\,dt - \int_0^y a(t)\,dt, \end{aligned}$$

that $ya(x) = xa(y)$. Then finish the proof.

The requirement of local integrability can weakened to Lebesgue-measurability (Exercise 2.18(b)).

Exercise 2.8. Consider the additive semigroup $\mathbb{N}_0^2 := \mathbb{N}_0 \times \mathbb{N}_0$ with the involution $\tau(m,n) := (n,m)$ for $(m,n) \in \mathbb{N}_0^2$. Show that the additive maps $a : \mathbb{N}_0^2 \to \mathbb{C}$ satisfying $a \circ \tau = -a$ are the maps of the form $a(m,n) = \alpha(m-n)$ for $(m,n) \in \mathbb{N}_0^2$, where $\alpha \in \mathbb{C}$ is a constant.

Exercise 2.9. In this exercise we find the continuous, additive functions on the (multiplicative) groups $\mathbb{R}^+$ and $\mathbb{R}^*$. Each of the functional equations in (a), (b) and (c) of this exercise is called the *logarithmic equation.*

(a) Show that the continuous solutions $L : \mathbb{R}^+ \to \mathbb{C}$ of the functional equation $L(xy) = L(x) + L(y)$, $x, y \in \mathbb{R}^+$, are the functions $L(x) = c\log x$, $x \in \mathbb{R}^+$, where the constant c ranges over $\mathbb{C}$.

Hint: Which functional equation does $L \circ \exp : \mathbb{R} \to \mathbb{R}^+$ satisfy?

(b) Find the continuous solutions $L : \mathbb{R}^* \to \mathbb{C}$ of the functional equation $L(xy) = L(x) + L(y)$, $x, y \in \mathbb{R}^*$.

Answer: The functions $L(x) = c\log|x|$, $x \in \mathbb{R}^*$, where the constant c ranges over $\mathbb{C}$.

(c) Find the continuous solutions $L : \mathbb{C}^* \to \mathbb{C}$ of the functional equation $L(xy) = L(x) + L(y)$, $x, y \in \mathbb{C}^*$.

Answer: The functions $L(z) = c\log|z|$, $z \in \mathbb{C}^*$, where the constant c ranges over $\mathbb{C}$.

Hint: The map $(r, e^{i\theta}) \mapsto re^{i\theta}$ is a topological isomorphism of $\mathbb{R}^+ \times \mathbb{T}$ onto $\mathbb{C}^*$.

Exercise 2.10. Let $A > 0$. Find the solutions $f \in C(\mathbb{R})$ of the functional equation

$$f(x+y) = f(x)A^y + f(y)A^x, \qquad x, y \in \mathbb{R}.$$

Hint: Divide by A^{x+y}.

Exercise 2.11. Let $A > 0$. Show that $f_0(x) := A^x - 1$, $x \in \mathbb{R}$, satisfies

$$f_0(x+y) = f_0(x) + f_0(y) + f_0(x)f_0(y), \qquad x, y \in \mathbb{R}.$$

Find the solutions $f \in C(\mathbb{R})$ of

$$f(x+y) = f(x) + f(y) + (A^x - 1)(A^y - 1), \qquad x, y \in \mathbb{R}.$$

Exercise 2.12. Let G be a group, and $c \in \mathbb{C}$ a constant.

(a) Let $a : G \to \mathbb{C}$ be an additive function. Show that the solutions $f : G \to \mathbb{C}$ of

$$f(xy) = f(x) + f(y) + ca(x)a(y), \qquad x, y \in G,$$

are the functions of the form

$$f(x) = \tfrac{c}{2}a(x)^2 + a_1(x), \qquad x \in G,$$

where $a_1 : G \to \mathbb{C}$ is additive.

(b) Let $\chi : G \to \mathbb{C}$ be multiplicative (i.e., $\chi(xy) = \chi(x)\chi(y)$ for all $x, y \in G$). Show that the solutions $f : G \to \mathbb{C}$ of

$$f(xy) = f(x) + f(y) + c(\chi(x) - 1)(\chi(y) - 1), \qquad x, y \in G,$$

are the functions of the form

$$f(x) = c(\chi(x) - 1) + a(x), \qquad x \in G,$$

where $a : G \to \mathbb{C}$ is additive.

Exercise 2.13. Let $a : \mathbb{C} \to \mathbb{C}$ be a continuous, additive function. Show that it has the form $a(z) = \alpha z + \beta \overline{z}$, $z \in \mathbb{C}$, where α and β are complex constants.

Exercise 2.14. Let G be a group and let $\tau : G \to G$ satisfy that $\tau(e) = e$. Let $f : G \to \mathbb{C}$ be a solution of

$$f(xy) - f(x) = -f(\tau(y)) \qquad \text{for all } x, y \in G. \tag{2.6}$$

(a) Show that $f(e) = 0$.

(b) Show that $f(y) = -f(\tau(y))$ for all $y \in G$.

(c) Show that f is additive.

(d) Find the continuous solutions $f : \mathbb{R}^2 \to \mathbb{C}$ of (2.6) with $G = \mathbb{R}^2$ and $\tau\binom{a}{b} = \binom{b}{a}$, $a, b \in \mathbb{R}$.

(e) Find the continuous solutions $f : \mathbb{R}^* \to \mathbb{C}$ of (2.6) with $G = \mathbb{R}^*$ and $\tau(x) := 1/x$ for all $x \in \mathbb{R}^*$ (Exercise 2.9 may be used).

Exercise 2.15. Extend Lemma 2.3 to the following.

Lemma 2.22. *Let V and W be vector spaces over $\mathbb{R}$. Let $a : V \to W$ be additive, i.e., a satisfies* (2.2).

(a) *$a(qx) = qa(x)$ for all $q \in \mathbb{Q}$ and $x \in V$.*

(b) *Let V and W be normed spaces. If a is bounded on the open unit ball of V, then $a : V \to W$ is a linear, continuous operator.*

Exercise 2.16. Let $a : ([0, \infty[, +) \to \mathbb{C}$ be additive.

(a) Show that there exists an additive map $A : \mathbb{R} \to \mathbb{C}$ such that $A = a$ on $[0, \infty[$.

Hint: Define $A(x - y) := a(x) - a(y)$ for $x, y \in [0, \infty[$, and make sure A is well defined by this prescription.

(b) Is there more than one additive extension of a from $[0, \infty[$ to $\mathbb{R}$?

(c) Assume furthermore that $a([0, \infty[) \subseteq [0, \infty[$. Show that there exists a constant $c \geq 0$ such that $a(x) = cx$ for all $x \in [0, \infty[$.

Exercise 2.17. Show that the functions $F : \mathbb{R} \to \mathbb{R}$ that satisfy the functional equation

$$F(x)F(y) = F(\sqrt{x^2 + y^2}), \quad x, y \in \mathbb{R},$$

are the ones of the following three forms: $F = 0$, $F = 1_{\{0\}}$ and $F(x) = e^{A(x^2)}$, $x \in \mathbb{R}$, where $A : \mathbb{R} \to \mathbb{R}$ is any additive function.

What simplifications occur if F is assumed continuous?

Hints on how to proceed:

(a) Note that $F(0) = 0$ or $F(0) = 1$, and that F is even, i.e., that $F(x) = F(|x|)$ for all $x \in \mathbb{R}$. Prove that $F(0) = 0$ implies $F = 0$ which is the first form.

We may thus from now on assume that $F(0) = 1$.

(b) Assume that F has a zero $x_0 \in \mathbb{R}$. Show that $F(x) = 0$ for all $x \in \mathbb{R}$ such that $|x| \geq |x_0|$. Combining this with the fact that $F(y) = 0$ implies $F(y/2) = 0$ infer that $F(x) = 0$ for all $x \neq 0$. Conclude that F has the second form.

We may thus from now on assume that F vanishes nowhere.

(c) Put $x = y$ in the functional equation to get that $F(2y) = F(y)^2$ for all $y \in \mathbb{R}$. Conclude (F is assumed real-valued!) that $F(x) > 0$ for all $x \in \mathbb{R}$, so that we may define $a(x) := \log F(\sqrt{x})$, $x \geq 0$. Observe that $a(x^2) + a(y^2) = a(x^2 + y^2)$ for all $x, y \in \mathbb{R}$, i.e., that $a(x + y) = a(x) + a(y)$ for all $x \geq 0$ and $y \geq 0$.

Finally extend a to an additive function on all of $\mathbb{R}$ (Exercise 2.16).

Exercise 2.18. Let $a : \mathbb{R} \to \mathbb{C}$ be additive.

(a) Assume a is bounded on a set of positive measure. Show that a is linear.

Hint: Steinhaus' theorem (Theorem D.2).

(b) Assume a is Lebesgue measurable. Show that a is linear.

Hint: $\bigcup_{n=1}^{\infty} \{x \in [0,1] \mid |a(x)| < n\} = [0,1]$, so one of the sets in the union on the left hand side has positive measure.

Exercise 2.19. Let $a : \mathbb{R}^n \to \mathbb{C}$ be additive.

(a) Assume that a is bounded on the unit ball $B(0,1) := \{x \in \mathbb{R}^n \mid \|x\| < 1\}$. Show that a is linear and hence continuous

Hint: Apply Proposition 2.5 to radii.

(b) Assume that a is bounded on the unit sphere $S := \{x \in \mathbb{R}^n \mid \|x\| = 1\}$, and that $n \geq 2$. Show that a is linear.

Hint: Show that $B(0,1) \subseteq S + S$ (this is in reality a 2-dimensional problem), and apply (a).

(c) Assume that a is bounded on a set of positive measure. Show that a is linear.

Hint: Steinhaus' theorem.

Exercise 2.20. Let $a : \mathbb{R} \to \mathbb{R}$ be a real-valued, additive function such that furthermore $a(xy) = a(x)a(y)$ for all $x, y \in \mathbb{R}$ (i.e., a is also multiplicative).

Show that $a = 0$ or $a(x) = x$ for all $x \in \mathbb{R}$.

Hint: Combine Proposition 2.5 with the fact that $a(x^2) = a(x)^2 \geq 0$ for all $x \in \mathbb{R}$.

An interesting application is that the only automorphism of the field $\mathbb{R}$ is the identity in contrast to the field $\mathbb{C}$: Exercise 3.11 shows there are exactly two continuous automorphisms of $\mathbb{C}$. It is known there exists a nowhere continuous automorphism of $\mathbb{C}$ ([6, Theorem 5.7]) or [211]).

Exercise 2.21. Let $a : \mathbb{R} \to \mathbb{R}$ be an additive function such that $a(1) = 1$ and $a(x)a(\frac{1}{x}) = 1$ for all $x \neq 0$. Show that a is continuous and find a.
Hint: Apply a to the identity

$$\frac{2}{x^2 - 1} = \frac{1}{x - 1} - \frac{1}{x + 1}, \qquad x \in \mathbb{R} \setminus \{\pm 1\},$$

and deduce that $a(x^2) = a(x)^2 \geq 0$.

Exercise 2.22. In this exercise we find that $(\mathbb{R}, +)$ possesses discontinuous involutions, i.e., there exist discontinuous, additive maps $a : \mathbb{R} \to \mathbb{R}$ such that $a \circ a = \mathrm{id}$.

Let $\{v_0, v_1, v_i \mid i \in I\}$ be a basis of $\mathbb{R}$ as a vector space over the field $\mathbb{Q}$ (in other words a Hamel basis). Define a on the basis by

$$a(v) = \begin{cases} v_1 & \text{if } v = v_0 \\ v_0 & \text{if } v = v_1 \\ v_i & \text{if } i \in I \end{cases}$$

and extend a to an additive map $a : \mathbb{R} \to \mathbb{R}$.

(a) Show that $a \circ a = \mathrm{id}$.

(b) Show that a is discontinuous.
Hint: If a is continuous, then $a(v)/v$ is constant.

Exercise 2.23.

(a) Show that the solutions $f, g, h : \mathbb{R} \to \mathbb{C}$ of *Pexider's functional equation*

$$f(x + y) = g(x) + h(y), \qquad x, y \in \mathbb{R},$$

are the triples $f = a + c_1$, $g = a + c_2$, $h = a + c_1 - c_2$, where $a : \mathbb{R} \to \mathbb{C}$ is additive and $c_1, c_2 \in \mathbb{C}$.

(b) Find all continuous solutions $f, g, h : \mathbb{R} \to \mathbb{C}$ of Pexider's functional equation.

Exercise 2.24. Find all continuous solutions $f, g, h : \mathbb{R}^* \to \mathbb{C}$ of the functional equation

$$f(xy) = g(x) + h(y), \qquad x, y \in \mathbb{R}^*.$$

Hint: Exercise 2.9(b).

Exercise 2.25. Let $B : \mathbb{R} \times \mathbb{R} \to \mathbb{C}$ be bi-additive and continuous. Show that there is a constant $c \in \mathbb{C}$ such that $B(x, y) = cxy$ for all $(x, y) \in \mathbb{R} \times \mathbb{R}$.

Exercise 2.26. Let X be a Hausdorff topological vector space over $\mathbb{R}$ or $\mathbb{C}$.

(a) Show that the only separately continuous, skew-symmetric, bi-additive function $\Psi : \mathbb{R} \times \mathbb{R} \to X$ is 0.

Hint: Show that $\Psi(x, y) = xy\Psi(1, 1)$. What is $\Psi(1, 1)$?

(b) Show that the only separately continuous, skew-symmetric bi-homomorphism $\Psi : \mathbb{R}^+ \times \mathbb{R}^+ \to X$ is 0.

Exercise 2.27. Let G be the $(ax + b)$-group from Examples A.17(i). Show that the continuous bi-additive functions $B : G \times G \to \mathbb{C}$ are the functions of the form

$$B\left(\begin{pmatrix} a_1 & b_1 \\ 0 & 1 \end{pmatrix}, \begin{pmatrix} a_2 & b_2 \\ 0 & 1 \end{pmatrix}\right) = c(\log a_1)(\log a_2), \text{ where } c \in \mathbb{C} \text{ is a constant.}$$

Exercise 2.28. Let G be the Heisenberg group from Examples A.17(a). Show that the continuous bi-additive functions $B : G \times G \to \mathbb{C}$ are the functions of the form

$$B\left(\begin{pmatrix} 1 & x_1 & z_1 \\ 0 & 1 & y_1 \\ 0 & 0 & 1 \end{pmatrix}, \begin{pmatrix} 1 & x_2 & z_2 \\ 0 & 1 & y_2 \\ 0 & 0 & 1 \end{pmatrix}\right) = \left\langle A \begin{pmatrix} x_1 \\ y_1 \end{pmatrix}, \begin{pmatrix} x_2 \\ y_2 \end{pmatrix} \right\rangle,$$

where $A \in M(2 \times 2, \mathbb{C})$.

Exercise 2.29. Let $Q : S \times S \to \mathbb{C}$ be a symmetric, bi-additive function on a topological semigroup S. Show that Q is continuous on $S \times S$ if and only if the function $x \mapsto Q(x, x)$ is continuous on S.

Exercise 2.30. Let $\Psi \in C(\mathbb{R}^2 \times \mathbb{R}^2)$ be bi-additive and skew-symmetric. Show that there exists a constant $c \in \mathbb{C}$ such that

$$\Psi\left(\begin{pmatrix} x_1 \\ x_2 \end{pmatrix}, \begin{pmatrix} y_1 \\ y_2 \end{pmatrix}\right) = c \det \begin{pmatrix} x_1 & y_1 \\ x_2 & y_2 \end{pmatrix} \qquad \text{for all } \begin{pmatrix} x_1 \\ x_2 \end{pmatrix}, \begin{pmatrix} y_1 \\ y_2 \end{pmatrix} \in \mathbb{R}^2.$$

Exercise 2.31. Let $f : G \to H$, where G is a group and $(H, +)$ is an abelian group.

(a) Let f be a solution of Jensen's equation (2.4) such that $f(e) = 0$. Show that f is a semi-homomorphism (Definition 2.23).

(b) Let f be a semi-homomorphism. Show that f is a solution of Jensen's functional equation (2.4).

Exercise 2.32. Let G be a group and H an abelian group. That a map $F : G \to H$ has $CF = 0$ means that F is additive. Show that if $C^2 f = 0$ for $f : G \to H$ then $Cf : G \times G \to H$ is bi-additive.

Exercise 2.33. Let G be a group and H an abelian group.

(a) Let $f : G \to H$. Show that if $C^2 f = 0$ then $Cf : G \times G \to H$ is bi-additive.

Hint: That a map $F : G \to H$ has $CF = 0$ means that F is additive.

(b) Assume that $\mathrm{Hom}(G, H) = \{0\}$. Show that the kernel of the second Cauchy difference is zero, i.e., that $\{f : G \to H \mid C^2 f = 0\} = \{0\}$.

(c) Is the assumption $\mathrm{Hom}(G, H) = \{0\}$ satisfied if $G = \mathbb{Q}$ and $H = \mathbb{Z}$? What about $G = \mathrm{SL}(n, \mathbb{R})$, where $n \geq 2$, and H arbitrary?

Exercise 2.34. Let G be a group and H an abelian group.

(a) Let $f : G \to H$ satisfy $C^2 f = 0$. Show that

$$f(x^n) = nf(x) + \frac{n(n-1)}{2} Cf(x, x) \qquad \text{for } x \in G \text{ and } n \in \mathbb{Z}.$$

In the computations you might exploit that $Cf : G \times G \to H$ is bi-additive.

(b) Show that the solutions $f : \mathbb{Z} \to H$ of $C^2 f = 0$ are the maps of the form

$$f(n) = nh_0 + \frac{n(n-1)}{2} h_1, \qquad n \in \mathbb{Z},$$

where $h_0, h_1 \in H$.

2.6 Notes and remarks

We refer the reader to [6] for the history and the vast literature about the additive Cauchy equation, and to [135] for a systematic exposition of the theory of additive functions on $\mathbb{R}^n$.

Both Jensen's functional (2.4) and the symmetrized additive Cauchy equation (2.3) are closely related to the algebraic notion of a semi-homomorphism.

Definition 2.23. A map $f : G \to G'$ from one group to another is a *semi-homomorphism*, if $f(xyx) = f(x)f(y)f(x)$ for all $x, y \in G$.

$G' = H$ is abelian in this chapter, so the definition of a semi-homomorphism $f : G \to H$ reduces here to $f(xyx) = 2f(x) + f(y)$ for all $x, y \in G$. The relation alluded to above is for Jensen's equation spelled out in Exercise 2.31. For the symmetrized additive Cauchy equation it is the following:

If f is a solution of (2.3) then $2f$ is a semi-homomorphism (take $z = x$ in Lemma 2.20(c)). Conversely, if f is a semi-homomorphism, then $f(x^{-1}) = -f(x)$ and then we get (2.3) as follows:

$$\begin{aligned} f(xy) &= f(xyx^{-1}y^{-1}yx) = 2f(x) + f(yx^{-1}y^{-1}y) \\ &= 2f(x) + 2f(y) + f(x^{-1}y^{-1}) = 2f(x) + 2f(y) + f((yx)^{-1}) \\ &= 2f(x) + 2f(y) - f(yx). \end{aligned}$$

About products of additive functions Halter-Koch, Reich and Schwaiger [105] deduced: Suppose $a_1, a_2, \dots, a_n : G \to \mathbb{C}$ are additive functions on a group G and $a_1(x)a_2(x)\cdots a_n(x) = 0$ for all $x \in G$. Then there is a $j \in \{1, 2, \dots, n\}$ such that $a_j = 0$.

Proposition 2.8 is due to Hamel [106].

Section 2.2: For more about the classical theory of Pexiderized equations consult [135, Ch. XIII §3], [6, Section 4.3 and Ch. 15] or [132, Section 1.6.2].

Definition 2.15: (2.3) is the special case of $g = 1$ of the symmetrized sine addition formula (6.1) that we study in Chapter 6. See also (11.5).

Proposition 2.17 is due to Peter de Place Friis (unpublished).

Lemma 2.20: See the comment to Proposition 12.5 in Section 12.10 Notes and remarks.

Exercises 2.10 and 2.11 are from Boswell [28].

Exercise 2.16 can be generalized to subsemigroups of non-abelian groups. See Aczél, Baker, Djoković, Kannappan and Radó [4].

Exercise 2.17 is adapted from the William Lowell Putnam Competition 1947.

Exercise 2.19: See the comment to Exercise A.9.

Exercise 2.21 was one of the longlisted problems of the International Mathematical Olympiad 1989.

Exercise 2.23: The solutions can be found in [6, Section 4.3] and in [101, Lemma 2] for more general set ups, in which $\mathbb{R}$ and $\mathbb{C}$ are replaced by semigroups.

Exercise 2.33: The result is due to C. T. Ng (private communication) who remarked that it is true even with 2 replaced by $k \in \mathbb{N}$.

Exercise 2.34 is taken from C. T. Ng's lecture at the 50th International Symposium on Functional Equations, Hungary, June 2012.

Chapter 3

The Multiplicative Cauchy Equation

In this chapter we derive some basic properties of group characters, compute the continuous characters of important topological groups and prove Artin's result about linear independence of characters (Corollary 3.20).

Regularity properties of characters like measurability and continuity are discussed in the context of homomorphisms between locally compact groups in Appendix D.

3.1 Group characters

As stated in the introduction we will express the solutions of functional equations on a group in terms of certain fundamental building blocks that are related to structure of the group. We met some of these building blocks, the additive functions, in Chapter 2. Other building blocks are the group characters that we study in this chapter. They are ingredients of the solution formulas of, e.g., the sine addition and d'Alembert's functional equations, but are important in other fields as well. The continuous, unitary characters on a locally compact, abelian group like $\mathbb{R}$ are the building stones for the Fourier analysis on the group: Any square integrable function on the group is according to Fourier's inversion formula a superposition of continuous, unitary characters. Non-unitary exponential functions play an important role in other parts of mathematics, for instance in the theory of partial differential equations with constant coefficients. In our set up solutions are not assumed bounded. Therefore we cannot restrict ourselves to unitary = bounded characters.

Definition 3.1. A *group character* on G or for brevity just a *character* on G is a homomorphism of G into $\mathbb{C}^*$.

A character χ is said to be *unitary*, if $|\chi(x)| = 1$ for all $x \in G$.

Definition 3.2. Let G be a group. The *multiplicative Cauchy equation* for a function $\chi : G \to \mathbb{C}$ is the functional equation

$$\chi(xy) = \chi(x)\chi(y), \qquad x, y \in G. \tag{3.1}$$

It is in the literature also known as the *Cauchy exponential equation*.

The non-zero solutions $\chi : G \to \mathbb{C}$ of (3.1) are the characters on G (see Lemma 3.4(a)). The formula (3.1) expresses that χ is a multiplicative function on G (cf. Definition A.26).

Example 3.3. The function $\chi(x) := e^{ix}$, $x \in \mathbb{R}$, is a continuous unitary character on $(\mathbb{R}, +)$. $\chi_\lambda(x) := e^{\lambda x}$, $x \in \mathbb{R}$, is a continuous character on $(\mathbb{R}, +)$ for any $\lambda \in \mathbb{C}$, which is unitary if and only if λ is purely imaginary. According to Example 3.7 below we get all continuous characters on $(\mathbb{R}, +)$ when λ ranges over $\mathbb{C}$.

The basic properties of characters on groups expressed in Lemma 3.4 will often be used without explicit mentioning. We recall the notations $\check{\chi}(x) = \chi(x^{-1})$ and $\overline{\chi}(x) = \overline{\chi(x)}$ for any x.

Lemma 3.4. *Let G be a group, and let $\chi : G \to \mathbb{C}$ be a non-zero multiplicative function.*

(a) *χ is a character on G. In particular χ does not vanish at any point of G, and $\chi(e) = 1$.*

(b) *If χ is bounded, then $|\chi(x)| = 1$ for all $x \in G$, so χ is a unitary character. If χ is unitary, then $\check{\chi} = \overline{\chi}$.*

(c) *χ is identically 1 on the commutator subgroup $[G, G]$.*

Proof. (a) Lemma A.28 implies that $\chi(e) = 1$. Let $x \in G$ be arbitrary. From $1 = \chi(e) = \chi(xx^{-1}) = \chi(x)\chi(x^{-1})$ we see that $\chi(x) \neq 0$, so $\chi(x) \in \mathbb{C}^*$.

(b) Lemma A.27 provides the inequality $|\chi(x)| \leq 1$. To get the reverse inequality we use that $|\chi(x^{-1})| \leq 1$:

$$|\chi(x)| = \frac{1}{|\chi(x^{-1})|} \geq \frac{1}{1} = 1.$$

The rest of (b) and (c) are left to the reader. □

Warning: According to Lemma 3.4(a) a non-zero multiplicative function on a group vanishes nowhere. This is not so on a monoid. As a counter-example take $M = ([0, \infty[, +)$ and $\chi = \mathbf{1}_{\{0\}}$.

Lemma 3.4(b) applies to finite groups, and more generally to continuous characters on compact groups where the boundedness condition is automatically satisfied: Any continuous complex function on a compact set is bounded.

The next lemma explains why certain degenerate cases singled out by the condition $\check{\chi} = \chi$, sometimes simplify.

Lemma 3.5. *Let χ be a character on a group G such that $\check{\chi} = \chi$. If G is generated by its squares, in particular if G is 2-divisible, then $\chi = 1$.*

Examples 3.6.

(a) The group $\mathbb{R}$ is generated by its squares, so Lemma 3.5 applies. Hence if χ is a character on $\mathbb{R}$ such that $\check{\chi} = \chi$ then $\chi = 1$.

(b) The quaternion group $Q_8 = \{\pm 1, \pm i, \pm j, \pm k\}$ from Example A.17(h) is not generated by its squares. The map $\chi : Q_8 \to \mathbb{C}^*$ given by $\chi(\pm 1) = \chi(\pm i) = 1$, $\chi(\pm j) = \chi(\pm k) = -1$ is a character on Q_8 such that $\check{\chi} = \chi$, but $\chi \neq 1$. Actually any character χ on Q_8 satisfies $\check{\chi} = \chi$ (see Example 3.15 for a list of all the characters of Q_8). So the assumption in Lemma 3.5 that G is generated by its squares cannot be deleted.

3.2 Continuous characters on selected groups

In this section we compute the continuous characters of some important topological groups. Exponential functions will play prominent roles. As mentioned above the characters will come forth later as building blocks of solutions of functional equations.

Examples 3.7.

(a) For any $\lambda \in \mathbb{C}$ the function

$$\chi_\lambda(x) := e^{\lambda x}, \qquad x \in \mathbb{R}, \tag{3.2}$$

is a continuous character on $(\mathbb{R}, +)$. Conversely, for any continuous character χ on $(\mathbb{R}, +)$ there exists exactly one $\lambda \in \mathbb{C}$ such that $\chi = \chi_\lambda$.

The unitary, continuous characters are the ones for which $\lambda \in i\mathbb{R}$.

(b) More generally, the continuous characters on $(\mathbb{R}^n,+)$, $n \in \mathbb{N}$, are the functions of the form

$$\chi_\lambda(x) := e^{\langle\lambda,x\rangle}, \qquad x \in \mathbb{R}^n, \tag{3.3}$$

where λ ranges over $\mathbb{C}^n$. Here we use for $\lambda = (\lambda_1, \lambda_2, \ldots, \lambda_n) \in \mathbb{C}^n$ and $x = (x_1, x_2, \ldots, x_n) \in \mathbb{R}^n$ the notation $\langle\lambda, x\rangle := \lambda_1 x_1 + \lambda_2 x_2 + \cdots + \lambda_n x_n$.

The unitary, continuous characters are the ones for which $\lambda_k \in i\mathbb{R}$ for $k = 1, 2, \ldots, n$.

Proof. (a) It is clear that χ_λ is a continuous character for any $\lambda \in \mathbb{C}$, so it is left to show that any continuous character $\chi : \mathbb{R} \to \mathbb{C}^*$ has this form for a unique $\lambda \in \mathbb{C}$. The uniqueness of λ is trivial, so left is the existence. We first show that χ is differentiable: Since $\chi(0) = 1$ there is an $\varepsilon > 0$ such that $\int_0^\varepsilon \chi(y)\,dy \neq 0$. Integrating the identity $\chi(x+y) = \chi(x)\chi(y)$ over $[0,\varepsilon]$ with respect to y we get after a change of variables that

$$\int_x^{x+\varepsilon} \chi(u)\,du = \chi(x) \int_0^\varepsilon \chi(y)\,dy.$$

The left hand side is differentiable as a function of $x \in \mathbb{R}$ because χ is continuous (the fundamental theorem of calculus). Hence so is the right hand side. But the constant $\int_0^\varepsilon \chi(y)\,dy \neq 0$, so χ is differentiable.

Differentiating $\chi(x+y) = \chi(x)\chi(y)$ with respect to y at $y = 0$ we find χ satisfies the differential equation $\chi' = \chi'(0)\chi$, implying that χ has the form

$$\chi(x) = Ae^{\chi'(0)x}, \qquad x \in \mathbb{R},$$

where $A \in \mathbb{C}$ is a constant. But $\chi(0) = 1$, so $A = 1$.

(b) It is clear that χ_λ is a continuous character for any $\lambda \in \mathbb{C}^n$, so it is left to show that any continuous character $\chi : \mathbb{R}^n \to \mathbb{C}^*$ has this form. Now,

$$\begin{aligned}\chi(x_1, x_2, \ldots, x_n) &= \chi((x_1, 0, \ldots, 0) + (0, x_2, \ldots, 0) + \cdots + (0, 0, \ldots, x_n)) \\ &= \chi(x_1, 0, \ldots, 0)\,\chi(0, x_2, \ldots, 0) \cdots \chi(0, 0, \ldots, x_n).\end{aligned}$$

Here $x_1 \mapsto \chi(x_1, 0, \ldots, 0)$ is a continuous character on $\mathbb{R}$, so by (a) there exists a complex constant λ_1 such that $\chi(x_1, 0, \ldots, 0) = \exp \lambda_1 x_1$ for all $x_1 \in \mathbb{R}$. Similarly for the other factors. So we find complex constants $\lambda_1, \lambda_2, \ldots, \lambda_n$ such that

$$\begin{aligned}\chi(x_1, x_2, \ldots, x_n) &= e^{\lambda_1 x_1} e^{\lambda_2 x_2} \cdots e^{\lambda_n x_n} = e^{\lambda_1 x_1 + \lambda_2 x_2 + \cdots + \lambda_n x_n} \\ &= e^{\langle(\lambda_1, \lambda_2, \ldots, \lambda_n), (x_1, x_2, \ldots, x_n)\rangle}.\end{aligned}$$

□

To find the continuous characters on $\mathbb{R}$ we showed that the functional equation combined with the assumption of continuity implies that the solution in question of the functional equation (i.e., the character) is differentiable. By differentiating the functional equation we obtained a differential equation that we subsequently solved. This way of proceeding is one of the standard methods for solving functional equations on $\mathbb{R}^n$ under certain regularity conditions (like continuity).

The proof of the result in Example 3.7(b) can be generalized to

Proposition 3.8. *Let $G_1, G_2, \ldots, G_n$ be topological groups. The continuous characters on $G_1 \times G_2 \times \cdots \times G_n$ are the ones of the form $\chi_1 \otimes \chi_2 \otimes \cdots \otimes \chi_n$, where χ_i is a continuous character on G_i for $i = 1, 2, \ldots, n$.*

The meaning of the notation $\chi_1 \otimes \chi_2 \otimes \cdots \otimes \chi_n$ can be found in Section A.2.

Examples 3.9.

(a) The continuous characters on the multiplicative group $(\mathbb{R}^+, \cdot)$ are the functions

$$\phi_\lambda(t) := t^\lambda = e^{\lambda \log t}, \qquad t \in \mathbb{R}^+, \tag{3.4}$$

where λ ranges over $\mathbb{C}$.

(b) The continuous characters on the multiplicative group $(\mathbb{R}^*, \cdot)$ are the functions

$$\phi_\lambda(t) := |t|^\lambda \quad \text{and} \quad \psi_\lambda(t) := |t|^\lambda \operatorname{sgn}(t), \qquad t \in \mathbb{R}^*,$$

where λ ranges over $\mathbb{C}$.

(c) There are only the following maps $M \in C(\mathbb{R})$ such that $M(st) = M(s)M(t)$ for all $s, t \in \mathbb{R}$. In the formulas $\alpha > 0$ and $\beta \in \mathbb{R}$:

$$M = 0, \quad M = 1,$$

$$M(t) = \begin{cases} |t|^{\alpha+i\beta} & \text{if } t \neq 0 \\ 0 & \text{if } t = 0 \end{cases}$$

and

$$M(t) = \begin{cases} |t|^{\alpha+i\beta} \operatorname{sgn}(t) & \text{if } t \neq 0 \\ 0 & \text{if } t = 0\,. \end{cases}$$

Proof. (a) It is easy to check that functions of the form (3.4) are continuous characters, so it is left to show that any continuous character ϕ has the form (3.4). Now, $\Phi(x) := \phi(\exp x)$, $x \in \mathbb{R}$, is a continuous character on

$(\mathbb{R}, +)$, so according to Example 3.7(a) there exists a $\lambda \in \mathbb{C}$ such that $\Phi(x) = \exp(\lambda x) = (\exp x)^\lambda$. But this says that $\phi(t) = t^\lambda$ for all $t \in \mathbb{R}^+$.

(b) follows from (a). We leave the details to the reader except for noting that $\chi(-1)^2 = \chi(1) = 1$, so that either $\chi(-1) = 1$ or $\chi(-1) = -1$.

(c) Left to the reader. □

Example 3.10. The continuous characters on the circle group $\mathbb{T} := \{z \in \mathbb{C} \mid |z| = 1\}$ are the functions of the form

$$\chi_n(z) = z^n, \qquad z \in \mathbb{T}, \tag{3.5}$$

where n ranges over the integers.

Proof. It is easy to check that functions of the form (3.5) are continuous characters, so it is left to show that any continuous character χ has the form (3.5). The composite map $x \mapsto e^{ix} \mapsto \chi(e^{ix})$ is a continuous character on $(\mathbb{R}, +)$. According to Example 3.7(a) there exists a $\lambda \in \mathbb{C}$ such that

$$\chi(e^{ix}) = e^{i\lambda x} \qquad \text{for all } x \in \mathbb{R}.$$

Taking $x = 2\pi$ we get that $1 = e^{i\lambda 2\pi}$, which means that λ is an integer, say $\lambda = n \in \mathbb{Z}$. Thus

$$\chi(e^{ix}) = e^{inx} = (e^{ix})^n \qquad \text{for all } x \in \mathbb{R}.$$ □

Example 3.11. The characters of the group $(\mathbb{Z}, +)$ of integers are the functions $\chi_\alpha(n) := \alpha^n$, $n \in \mathbb{Z}$, where α ranges over $\mathbb{C}^*$.

Indeed, $\chi(n) = \chi(1)^n$ for any character χ, so as α we may take $\chi(1)$.

Example 3.12. We shall here exhibit the continuous characters on the multiplicative group $\mathbb{C}^*$: The map $(r, e^{i\theta}) \mapsto re^{i\theta}$ is a topological isomorphism of $\mathbb{R}^+ \times \mathbb{T}$ onto $\mathbb{C}^*$, so from the Examples 3.9 and 3.10 we find that the continuous characters on $\mathbb{C}^*$ are the functions $\{\chi_{\lambda,n} \mid \lambda \in \mathbb{C},\ n \in \mathbb{Z}\}$ defined by

$$\chi_{\lambda,n}(re^{i\theta}) := r^\lambda e^{in\theta},\ r > 0,\ \theta \in \mathbb{R}, \qquad \text{i.e., } \chi(z) = |z|^{\lambda - n} z^n \text{ for } z \in \mathbb{C}^*. \tag{3.6}$$

Example 3.13. Let χ be a character on the $(ax + b)$-group G from Examples A.17(i). Let us for brevity write

$$(a, b) := \begin{pmatrix} a & b \\ 0 & 1 \end{pmatrix} \qquad \text{for } a > 0 \text{ and } b \in \mathbb{R}.$$

Any character is 1 on the commutator group $[G,G]$, which here is $[G,G] = \{(1,b) \mid b \in \mathbb{R}\}$, so $\chi(1,b) = 1$ for all $b \in \mathbb{R}$. Now $\chi(a,b) = \chi((1,b)(a,0)) = \chi(1,b)\chi(a,0) = \chi(a,0)$. Assume furthermore that χ is continuous. The map $a \mapsto \chi(a,0)$ is then a continuous character on $(\mathbb{R}^+,\cdot)$, so according to Example 3.9 there exists a $\lambda \in \mathbb{C}$ such that

$$\chi(a,b) = a^\lambda \qquad \text{for all } (a,b) \in G.$$

Conversely, any map of this form is a continuous character on G.

Example 3.14. The continuous characters on the Heisenberg group H_3 from Examples A.17(a) are the functions

$$\begin{pmatrix} 1 & x & z \\ 0 & 1 & y \\ 0 & 0 & 1 \end{pmatrix} \mapsto e^{ax+by}, \tag{3.7}$$

where a and b range over $\mathbb{C}$.

Proof. It is easy to check that functions of the form (3.7) are continuous characters, so it is left to show that any continuous character χ on H_3 has the form (3.7). Let us for brevity write

$$(x,y,z) = \begin{pmatrix} 1 & x & z \\ 0 & 1 & y \\ 0 & 0 & 1 \end{pmatrix} \qquad \text{for } x,y,z \in \mathbb{R}.$$

Any character is 1 on the commutator group $[H_3,H_3]$, which here is $[H_3,H_3] = \{(0,0,z) \,|\, z \in \mathbb{R}\}$, so $\chi(0,0,z) = 1$ for all $z \in \mathbb{R}$. Thus $\chi(x,y,z) = \chi((0,0,z)(x,y,0)) = \chi(0,0,z)\chi(x,y,0) = \chi(x,y,0) = \chi(0,y,0)\chi(x,0,0)$. The maps $y \mapsto \chi(0,y,0)$ and $x \mapsto \chi(x,0,0)$ are continuous characters on $(\mathbb{R},+)$, so according to Example 3.7(a) there exist $\lambda_1, \lambda_2 \in \mathbb{C}$ such that

$$\chi(x,y,z) = e^{\lambda_2 y}e^{\lambda_1 x} = e^{\lambda_1 x+\lambda_2 y} \qquad \text{for all } (x,y,z) \in H_3. \qquad \square$$

Example 3.15. We shall in this example compute all characters on the quaternion group $Q_8 = \{\pm 1, \pm i, \pm j, \pm k\}$ from Example A.17(h). Let $\chi : Q_8 \to \mathbb{C}^*$ be a character on Q_8. Since characters are identically equal to 1 on the commutator group $[Q_8,Q_8] = \{\pm 1\}$, we get that $\chi(1) = \chi(-1) = 1$. It follows that $\chi(q) = \chi(-q)$ for all $q \in Q_8$.

Now, $\chi(i)^2 = \chi(i^2) = \chi(-1) = 1$, so $\chi(i) \in \{\pm 1\}$. By the same argument $\chi(j) \in \{\pm 1\}$ and $\chi(k) \in \{\pm 1\}$. At least one of $\chi(i)$, $\chi(j)$ and $\chi(k)$

must be $+1$: They cannot all be -1, because that will give us a contradiction by $-1 = \chi(k) = \chi(ij) = \chi(i)\chi(j) = (-1)(-1) = 1$.

There are now 4 possibilities:

(i) $\chi = 1$.

(ii) $\chi(\pm i) = 1$. Then $1 = \chi(i) = \chi(jk) = \chi(j)\chi(k)$, so either $\chi(j) = \chi(k) = 1$ or $\chi(j) = \chi(k) = -1$. The first instance is the possibility (i), so left is $\chi(\pm j) = \chi(\pm k) = -1$.

(iii) $\chi(\pm j) = 1$, and $\chi(\pm i) = \chi(\pm k) = -1$. This can be derived like (ii) was.

(iv) $\chi(\pm k) = 1$, and $\chi(\pm i) = \chi(\pm j) = -1$. This can also be derived like (ii) was.

It is simple to verify that the four functions from the possibilities are characters.

Example 3.16. The only character on SO(3) (continuous or not) is the constant 1. See [67, Section 4.8.4, p. 232] for a proof. A more sophisticated way of seeing this is to refer to the fact that SO(3) is a connected, semisimple Lie group so that $[\mathrm{SO}(3), \mathrm{SO}(3)] = \mathrm{SO}(3)$ (see [201, Corollary 3.18.10]).

Example 3.17. Let us consider the group $G := \mathbb{R}^2 \times_s \mathrm{SO}(2)$ of rigid motions of $\mathbb{R}^2$ from Example A.22(c). We can decompose any element $(x, \rho) \in G$ as a product of elements from the two subgroups $\mathbb{R}^2$ and SO(2) in the following way $(x, \rho) = (x, I)(0, \rho)$.

Let $\chi : G \to \mathbb{C}^*$ be a character on G. We let $\alpha(x) := \chi(x, I)$ and $\beta(\rho) := \chi(0, \rho)$ denote the corresponding characters on the subgroups $\mathbb{R}^2$ and SO(2). Then $\chi(x, \rho) = \chi((x, 0)(0, \rho)) = \chi(x, 0)\chi(0, \rho) = \alpha(x)\beta(\rho)$, so $\chi = \alpha \otimes \beta$. A small computation shows that $\chi = \alpha \otimes \beta$ is a character if and only if α and β are characters such that $\alpha(\rho x) = \alpha(x)$ for all $\rho \in \mathrm{SO}(2)$ and $x \in \mathbb{R}^2$. Taking $\rho = -I$ we see that $\alpha(2x) = 1$ for all $x \in \mathbb{R}^2$, so that $\alpha = 1$. Thus χ reduces to $\chi = 1 \otimes \beta$.

If χ and hence also β is continuous we get from Example 3.10 that the continuous characters on G are the functions χ_n, $n \in \mathbb{Z}$, defined by

$$\chi_n\left(x, \begin{pmatrix} \cos\theta & -\sin\theta \\ \sin\theta & \cos\theta \end{pmatrix}\right) = e^{in\theta} \qquad \text{for } x \in \mathbb{R}^2 \text{ and } \theta \in \mathbb{R}.$$

3.3 Linear independence of multiplicative functions

In this section we prove extensions of Artin's result (Corollary 3.20) that characters are linearly independent. Artin's result will be very useful for us at many later occasions, because it means that a function can be written as a linear combination of characters in at most one way. From an advanced point of view the proper place of Artin's result is in the framework of non-commutative harmonic analysis as described in Theorem E.13.

The set of exponential functions $\{e^{\lambda x} \mid \lambda \in \mathbb{C}\}$ is a linearly independent subset of the vector space $C(\mathbb{R})$. So, if $\lambda_1, \lambda_2, \ldots, \lambda_n$ are different complex numbers and $a_1, a_2, \ldots, a_n$ are complex numbers such that

$$a_1 e^{\lambda_1 x} + a_2 e^{\lambda_2 x} + \cdots + a_n e^{\lambda_n x} = 0 \qquad \text{for all } x \in \mathbb{R},$$

then $a_1 = a_2 = \cdots = a_n = 0$. The reader should try to prove this fact on his own, because he will then appreciate the simplicity of the proof below of Theorem 3.18.

Theorem 3.18. *Let S be a semigroup and $n \in \mathbb{N}$. Let $\chi_1, \chi_2, \ldots, \chi_n : S \to \mathbb{C}$ be n different multiplicative functions, and let $a_1, a_2, \ldots, a_n \in \mathbb{C}$. Define $f := a_1\chi_1 + a_2\chi_2 + \cdots + a_n\chi_n : S \to \mathbb{C}$.*

(a) *If $f = 0$, then $a_1\chi_1 = a_2\chi_2 = \cdots = a_n\chi_n = 0$. In other words, the multiplicative functions form a direct sum.*

(b) *The set of non-zero multiplicative functions is a linearly independent subset of the complex-valued functions on S.*

(c) *If f is bounded, then each of the functions $a_1\chi_1$, $a_2\chi_2$, $\ldots$, $a_n\chi_n$ is also bounded.*

(d) *If S is a topological semigroup and f is continuous, then each of the functions $a_1\chi_1$, $a_2\chi_2$, $\ldots$, $a_n\chi_n$ is also continuous.*

(e) *If S is a topological semigroup and f is Borel measurable, then each of the functions $a_1\chi_1$, $a_2\chi_2$, $\ldots$, $a_n\chi_n$ is also Borel measurable.*

(f) *If S is a locally compact group and f is measurable with respect to the left Haar measure on S, then the functions $a_1\chi_1$, $a_2\chi_2$, $\ldots$, $a_n\chi_n$ are continuous.*

Proof. (b) is an immediate consequence of (a). The statements (a), (c), (d) and (e) can all be proved by induction on n. The proofs are quite similar, so we just treat (d) and leave the proofs of the remaining statements to the reader:

The statement (d) is trivially true if $n = 1$. To get the induction step we assume that it is true for an $n \geq 1$ and prove it for $n + 1$. So

$$f(x) = a_1\chi_1(x) + a_2\chi_2(x) + \cdots + a_n\chi_n(x) + a_{n+1}\chi_{n+1}(x) \tag{3.8}$$

is a continuous function of $x \in S$. Let $x, y \in S$ be arbitrary. Evaluating (3.8) at the point $xy \in S$ we get that

$$\begin{aligned} f(xy) = a_1\chi_1(x)\chi_1(y) + a_2\chi_2(x)\chi_2(y) + \cdots + a_n\chi_n(x)\chi_n(y) \\ + a_{n+1}\chi_{n+1}(x)\chi_{n+1}(y). \end{aligned} \tag{3.9}$$

When we multiply (3.8) by $\chi_{n+1}(y)$ and subtract the result from (3.9) the last terms on the right hand sides cancel one another and we get that

$$\begin{aligned} f(xy) - f(x)\chi_{n+1}(y) &= a_1\chi_1(x)[\chi_1(y) - \chi_{n+1}(y)] \\ &+ a_2\chi_2(x)[\chi_2(y) - \chi_{n+1}(y)] + \cdots + a_n\chi_n(x)[\chi_n(y) - \chi_{n+1}(y)]. \end{aligned}$$

The left hand side is for each fixed $y \in S$ a continuous function of $x \in S$, so by our induction hypothesis each of the terms on the right hand side is continuous. In particular the last one $x \mapsto [\chi_n(y) - \chi_{n+1}(y)]a_n\chi_n(x)$. Now we use that $\chi_n \neq \chi_{n+1}$, so there exists a $y_0 \in S$ such that $\chi_n(y_0) - \chi_{n+1}(y_0) \neq 0$. Taking $y = y_0$ it follows that $a_n\chi_n$ is continuous. Going back to (3.8) we may move $a_n\chi_n$ from the right hand side to the left hand side and still preserve the continuity of the left hand side. Having done that there are only n terms on the right hand side. By our induction hypothesis each term on the right hand side is continuous.

(f) Proceeding as above we find that the functions $a_1\chi_1, a_2\chi_2, \ldots, a_n\chi_n$ are measurable with respect to the left Haar measure. But then they are continuous by Proposition D.4(b). □

The next uniqueness result will be useful at several places, for instance in our discussion of abelian pre-d'Alembert and d'Alembert functions.

Corollary 3.19. *Let S be a semigroup. Let $\chi_1, \chi_2, \chi_3, \chi_4 : S \to \mathbb{C}$ be multiplicative functions. If $\chi_1 + \chi_2 = \chi_3 + \chi_4$, then there are only the following two possibilities: $\chi_1 = \chi_3$ and $\chi_2 = \chi_4$, or $\chi_1 = \chi_4$ and $\chi_2 = \chi_3$.*

So a decomposition $\chi_1 + \chi_2$ with just two terms is unique up to interchange of χ_1 and χ_2.

Proof. When $\chi_1 + \chi_2 = \chi_3 + \chi_4$ we have the linear combination $\chi_1 + \chi_2 + (-1)\chi_3 + (-1)\chi_4 = 0$. Here $\chi_1, \chi_2, \chi_3, \chi_4$ cannot be 4 different functions, because in that case they are all 0 by Theorem 3.18(a) and so they are equal and not different. So two of them coincide. The only interesting case is that of $\chi_1 = \chi_2$, where we get the linear combination $2\chi_1 + (-1)\chi_3 + (-1)\chi_4 = 0$. Again, two of χ_1, χ_3, χ_4 must coincide by Theorem 3.18(a). We leave the rest of the proof to the reader. □

Corollary 3.20 (Artin). *The set of characters on a group G is a linearly independent subset of the vector space of all complex-valued functions on G.*

Proof. The corollary is a corollary of Theorem 3.18(b). □

As we saw above in Theorem 3.18, the method of the proof of Artin's result is so versatile that it has several consequences. Thus we have got a proof with results, not just a result with a proof!

3.4 The symmetrized multiplicative Cauchy equation

If f is a function on a semigroup S then the expression $f(xy)$ is not symmetric in the variables $x, y \in S$: x comes before y, so in general the expression changes value if x and y are interchanged. In various connections $f(xy)$ is therefore replaced by the symmetric expression $(f(xy) + f(yx))/2$. When we do this in the multiplicative Cauchy equation we obtain the symmetrized Cauchy equation (3.10) below. This new equation actually plays a role in the literature:

In his paper [218] on the spectrum of a Banach algebra Żelazko asked the question: When is a linear functional f on a real or complex algebra A multiplicative, i.e., when is $f(xy) = f(x)f(y)$ for all $x, y \in A$? He answered the question by showing that a sufficient condition is that $f(xy + yx) = 2f(x)f(y)$ for all $x, y \in A$. The necessity of this condition is of course easy to verify.

In [161] Radó answered the following question posed by I. Corovei: Let G be a group, K a division ring of characteristic $\neq 2$, and $f : G \to K$ a map for which $f(xy) + f(yx) = 2f(x)f(y)$ for all $x, y \in G$. Is it then true that $f(xy) = f(x)f(y)$ for all $x, y \in G$? The answer is yes.

At first glance the two results seem to be quite different, the domain of definition of the function f being an algebra in the first case and a group in the second one. On the other hand in both cases the domain of definition of f is a semi-group, and in both cases f is a solution of the equation $f(xy) + f(yx) = 2f(x)f(y)$ where x and y range over the domain of definition of f.

Theorem 3.21 encompasses the two results for complex-valued functions. The proof of it is short, elementary and self-contained.

Theorem 3.21. *Let $f : S \to \mathbb{C}$ be a function on a semi-group S satisfying the* symmetrized multiplicative Cauchy equation

$$\frac{f(xy) + f(yx)}{2} = f(x)f(y) \qquad \text{for all } x, y \in S. \tag{3.10}$$

Then $f : S \to \mathbb{C}$ is a multiplicative function.

Proof. Putting $y = x$ in (3.10) we get $f(x^2) = f(x)^2$, which we use without mentioning below. We get for any $u, v \in S$, using (3.10) repeatedly, that

$$\begin{aligned} 2f(vu^2v) &= f(v(u^2v)) + f(u^2v^2) + f((vu^2)v) + f(v^2u^2) \\ &\quad - f(u^2v^2) - f(v^2u^2) \\ &= 2f(u^2v)f(v) + 2f(vu^2)f(v) - 2f(u^2)f(v^2) \\ &= 4f(u^2)f(v)f(v) - 2f(u)^2f(v)^2 = 2f(u)^2f(v)^2. \end{aligned}$$

Using this identity and (3.10) we derive the theorem as follows:

$$\begin{aligned} [f(xy) - f(yx)]^2 &= [f(xy) + f(yx)]^2 - 4f(xy)f(yx) \\ &= [2f(x)f(y)]^2 - 2\{f(xy^2x) + f(yx^2y)\} \\ &= 4f(x)^2f(y)^2 - 2f(x)^2f(y^2) - 2f(y)^2f(x^2) \\ &= 0. \end{aligned}$$

We see that $f(xy) = f(yx)$, that implies the theorem. □

We cannot obtain an analogous result to Theorem 3.21 by replacing the multiplication by addition + on the right hand side of (3.10): Doing so the equation becomes according to Lemma 2.17 Jensen's functional equation with $f(e) = 0$, and that equation has on some groups more solutions than the additive functions (see Example 12.4).

3.5 Exercises

Exercise 3.1. Let $f : \mathbb{R} \to \mathbb{R}$, $f \neq 0$, satisfy $f(x + y) = f(x)f(y)$ for all $x, y \in \mathbb{R}$. Show that there is an additive function $a : \mathbb{R} \to \mathbb{R}$ such that $f(x) = \exp a(x)$ for all $x \in \mathbb{R}$.

Hint: Note first that $f(x) \neq 0$ for all $x \in \mathbb{R}$. Show next that $f(x) > 0$ for all $x \in \mathbb{R}$. Finally take the logarithm of the functional equation.

Replace the domain of definition $\mathbb{R}$ of f by a 2-divisible group, and derive a more general result. What about a group which is generated by its squares?

Exercise 3.2. Find the solutions $f \in C(\mathbb{R})$ of

$$f(x+y) = e^{2xy} f(x)f(y), \quad x, y \in \mathbb{R}.$$

Hint: $2xy = (x+y)^2 - x^2 - y^2$.

Exercise 3.3. Write the character $\chi : G \to \mathbb{C}^*$ on the group G in the form $\chi = g + if$, where $g, f : G \to \mathbb{R}$. So $g = \operatorname{Re}\chi$ and $f = \operatorname{Im}\chi$.

Show that $g(e) = 1$ and $f(e) = 0$. Derive the addition formulas

$$\begin{aligned} g(xy) &= g(x)g(y) - f(x)f(y) \qquad \text{for all } x, y \in G, \\ f(xy) &= f(x)g(y) + f(y)g(x) \qquad \text{for all } x, y \in G, \end{aligned}$$

and the general version $(g(x) + if(x))^n = g(x^n) + if(x^n)$, $x \in G$, $n \in \mathbb{Z}$, of de Moivre's formula.

Exercise 3.4. Let χ and σ be characters of a group such that $\operatorname{Re}\chi = \operatorname{Re}\sigma$. Show that $\chi = \sigma$ or $\chi = \overline{\sigma}$.

Exercise 3.5. Let the group G be generated by its squares. We recall the notation $\check{F}(x) := F(x^{-1})$ for any function F on G. Show that the only multiplicative functions $m : G \to \mathbb{C}$ such that $\check{m} = m$ are $m = 0$ and $m = 1$. What are the additive functions $a : G \to \mathbb{C}$ such that $\check{a} = a$?

Exercise 3.6. Let S be a semigroup. Let $a : S \to \mathbb{C}$ be additive and $m : S \to \mathbb{C}$ be multiplicative. Assume there is a constant $c \in \mathbb{C}$ such that $m = a + c$. Show that $a = 0$. Show that $m = 1$ or $m = 0$.

Exercise 3.7. Discuss the following equations, taken from Jensen's 1878 paper [117, p. 150–151]. The unknown function f is a continuous, complex-valued function defined on an interval on the real line.

(a) $f(x+y) = f(x) + f(y) + f(x)f(y)$.

Hint: Find a functional equation that $f + 1$ satisfies.

(b)

$$f(x+y) = \frac{f(x)f(y)}{f(x)+f(y)}.$$

Hint: Consider $1/f(x+y)$.

(c)

$$f(x+y) = \frac{f(x)f(y)}{1+f(x)+f(y)}.$$

(d)

$$f(x+y) = \frac{f(x)+f(y)}{1+c^2 f(x)f(y)},$$

where $c \in \mathbb{C}$ is a given constant.

Hint: Reformulate the expression

$$\frac{1+cf(x+y)}{1-cf(x+y)}.$$

Exercise 3.8. Let $\alpha > 0$. Show that the solutions $f \in C(]-\alpha,\infty[) \setminus \{0\}$ of the functional equation

$$f(x+y+xy/\alpha) = f(x)f(y), \qquad x > -\alpha,\ y > -\alpha,$$

are $f_\lambda(x) := (1+x/\alpha)^\lambda$, where λ ranges over $\mathbb{C}$.

Hint: Let $\Phi :]-\alpha,\infty[\to \mathbb{R}^+$ be defined by $\Phi(x) := 1 + x/\alpha$ for $x > -\alpha$, and apply Example 3.9(a) to $F := f \circ \Phi^{-1} : \mathbb{R}^+ \to \mathbb{C}$.

Exercise 3.9. Let S be a semigroup. Let $w, g : S \to \mathbb{C}$ satisfy the symmetrized sine addition formula, i.e.,

$$\frac{w(xy)+w(yx)}{2} = w(x)g(y) + w(y)g(x) \qquad \text{for all } x, y \in S.$$

Assume furthermore that there exists a $\lambda \in \mathbb{C}$ such that

$$\frac{g(xy)+g(yx)}{2} = g(x)g(y) + \lambda^2 w(x)w(y) \qquad \text{for all } x, y \in S.$$

Show that $g + \lambda w : S \to \mathbb{C}$ is multiplicative.

Exercise 3.10. Find for given $\alpha \in \mathbb{C}$ the general solution $f, g \in C(\mathbb{R})$ of the following system of (coupled) functional equations

$$\begin{cases} f(x+y) = f(x)g(y) + f(y)g(x), \\ g(x+y) = g(x)g(y) + \alpha^2 f(x)f(y), \end{cases}$$

where x and y range over $\mathbb{R}$. Maybe you recognize the sine and cosine addition formulas.

Hint: Show that $g + \alpha f$ and $g - \alpha f$ are multiplicative.

Exercise 3.11. Let $f \in C(\mathbb{C})$ be a non-zero, additive function satisfying $f(zw) = f(z)f(w)$ for all $z, w \in \mathbb{C}$. Show that either $f(z) = z$ for all $z \in \mathbb{C}$ or $f(z) = \overline{z}$ for all $z \in \mathbb{C}$.

Hint: Exercise 2.13.

What are the continuous automorphisms of the field of complex numbers? See in this connection Exercise 2.20.

Exercise 3.12. Show that the multiplicative functions M on the monoid $(\mathbb{C}, \cdot)$ are $M = 0$, $M = 1$ and

$$M(z) = \begin{cases} m(|z|)\Gamma(e^{i\theta}) & \text{for } z = |z|e^{i\theta} \neq 0 \\ 0 & \text{for } z = 0, \end{cases}$$

where $m : (\mathbb{R}^+, \cdot) \to \mathbb{C}^*$ and $\Gamma : \mathbb{T} \to \mathbb{C}^*$ are arbitrary characters.

Exercise 3.13. Find all multiplicative functions on $\mathrm{GL}(n, \mathbb{C})$, where $n \geq 2$, as follows:

(a) Let $P_j : \mathbb{C}^n \to \mathbb{C}^n$ be the matrix that switches the standard basis vectors e_1 and e_j. So $P_j(e_1) = e_j$, $P_j(e_j) = e_1$ and $P_j(e_k) = e_k$ for all $k \neq 1, j$. Show that $P_j^2 = I$ so that $P_j^{-1} = P_j$, and show that

$$P_j \begin{pmatrix} d_1 & 0 & \dots & 0 & \dots & 0 \\ 0 & d_2 & \dots & 0 & \dots & 0 \\ \vdots & \vdots & \ddots & \vdots & \vdots & \vdots \\ 0 & 0 & \dots & d_j & \dots & 0 \\ \vdots & \vdots & \vdots & \vdots & \ddots & \vdots \\ 0 & 0 & \dots & 0 & \dots & d_n \end{pmatrix} P_j^{-1}$$

$$= \begin{pmatrix} d_j & 0 & \dots & 0 & \dots & 0 \\ 0 & d_2 & \dots & 0 & \dots & 0 \\ \vdots & \vdots & \ddots & \vdots & \vdots & \vdots \\ 0 & 0 & \dots & d_1 & \dots & 0 \\ \vdots & \vdots & \vdots & \vdots & \ddots & \vdots \\ 0 & 0 & \dots & 0 & \dots & d_n \end{pmatrix}.$$

(b) Let $\chi : \mathrm{GL}(n,\mathbb{C}) \to \mathbb{C}$ be a multiplicative function. Show that

$$\chi\begin{pmatrix} d_1 & 0 & \dots & 0 \\ 0 & d_2 & \dots & 0 \\ \vdots & \vdots & \ddots & \vdots \\ 0 & 0 & \dots & d_n \end{pmatrix} = \chi\begin{pmatrix} d_1 d_2 \dots d_n & 0 & \dots & 0 \\ 0 & 1 & \dots & 0 \\ \vdots & \vdots & \ddots & \vdots \\ 0 & 0 & \dots & 1 \end{pmatrix}.$$

Hint: Use the multiplicativity.

(c) Define $\chi' : \mathbb{C}^* \to \mathbb{C}$ by

$$\chi'(z) := \chi\begin{pmatrix} z & 0 & \dots & 0 \\ 0 & 1 & \dots & 0 \\ \vdots & \vdots & \ddots & \vdots \\ 0 & 0 & \dots & 1 \end{pmatrix} \qquad \text{for } z \in \mathbb{C}^*.$$

Show that χ' is multiplicative.

(d) Show that $\chi(x) = \chi'(\det x)$ for all $x \in \mathrm{GL}(n,\mathbb{C})$.

Hint: By the polar decomposition any invertible complex matrix x may be written as a product of two normal (and hence diagonalizable) matrices.

Exercise 3.14. Find all continuous, multiplicative functions on $\mathrm{GL}(n,\mathbb{R})$, where $n \geq 2$.
Answer: There are two types of solutions χ: $\chi(x) = |\det x|^\lambda$, and $\chi(x) = |\det x|^\lambda \operatorname{sgn}(\det x)$. In both cases $\lambda \in \mathbb{C}$ is a constant.

Hint: You may use the decomposition of Example A.22(d), combined with Example 3.9(b) and the fact that $[\mathrm{SL}(n,\mathbb{R}), \mathrm{SL}(n,\mathbb{R})] = \mathrm{SL}(n,\mathbb{R})$ (Example A.17(d)).

Exercise 3.15. Let $\alpha \in \mathbb{C}$.

(a) Let S be a semigroup. Find the solutions $f : S \to \mathbb{C}$ of

$$f(xy) = f(x) + f(y) + \alpha f(x)f(y), \qquad x, y \in S,$$

in terms of additive and multiplicative maps.

Hint: Show that $F(x) := 1 + \alpha f(x)$ is multiplicative.

(b) Find the solutions $f \in C(\mathbb{R})$ of the functional equation

$$f(x + y) = f(x) + f(y) + \alpha f(x)f(y), \qquad x, y \in \mathbb{R}.$$

Exercise 3.16. For any given $\alpha > 0$ find the solutions $f \in C(\mathbb{R})$ of the functional equation

$$f(x+y) = \alpha^{xy} f(x) f(y), \qquad x, y \in \mathbb{R}.$$

Hint: Introduce the function g by $f(x) = \alpha^{x^2/2} g(x)$.

Exercise 3.17. Let m be a continuous, non-zero, multiplicative function on the semigroup $S = (\mathbb{R}^+, +)$. Show that there exists a $\lambda \in \mathbb{C}$ such that $m(x) = e^{\lambda x}$ for all $x \in \mathbb{R}^+$.

Hint: Copy the procedure of the proof in Example 3.7(a).

Exercise 3.18. Consider the abelian group $\mathbb{Z}_N$. For $k = 0, 1, \ldots, N-1$ we define $\chi_k : \mathbb{Z}_N \to \mathbb{C}$ by

$$\chi_k(l + nN) := \exp\left(\frac{2\pi i}{N} kl\right) \qquad \text{for } l, n \in \mathbb{Z}.$$

Show that $\chi_0, \chi_2, \ldots, \chi_{N-1}$ are the characters of $\mathbb{Z}_N$.

Exercise 3.19. Find the solutions $f, g, h \in C(\mathbb{R})$ of $f(x+y) = g(x)h(y)$, $x, y \in \mathbb{R}$ (the Pexiderized version of the multiplicative Cauchy equation $f(x+y) = f(x)f(y)$).

Exercise 3.20. Let M be a monoid. Consider the Pexiderized version $f(xy) = g(x)h(y)$, $x, y \in M$, of the multiplicative Cauchy equation (3.1). So $f, g, h : M \to \mathbb{C}$. Show that

(a) $f = 0$ if and only if either $g = 0$ or/and $h = 0$.

(b) If $f \neq 0$, then there exist a multiplicative function $\chi : M \to \mathbb{C}$ for which $\chi(e) = 1$ and constants $c_1, c_2 \in \mathbb{C} \setminus \{0\}$ such that $g = c_1\chi$, $h = c_2\chi$ and $f = c_1 c_2 \chi$.

Exercise 3.21. Let V be a topological vector space over $\mathbb{R}$. Let χ be a continuous unitary character on V, where V is viewed as a topological abelian group under $+$. Show that there exists exactly one continuous linear functional $\lambda : V \to \mathbb{R}$ on V such that $\chi(v) = \exp(i\lambda(v))$ for all $v \in V$.

Hints on how to proceed:

(a) (Existence and uniqueness of λ) According to Example 3.7(a) there exists for any $v \in V$ exactly one $\lambda(v) \in \mathbb{R}$ such that $\chi(tv) = \exp(i\lambda(v)t)$ for all $t \in \mathbb{R}$.

(b) (Linearity of λ) Show that $\lambda : V \to \mathbb{R}$ is linear.

It is left to prove that $\lambda : V \to \mathbb{R}$ is continuous. This constitutes the rest of the proof.

(c) Note that $\{v \in V \| \chi(v) - 1| < 1\}$ is a neighborhood of 0 in V and choose a balanced neighborhood U of 0 such that $U \subseteq \{v \in V \mid |\chi(v) - 1| < 1\}$ (that U is balanced means by definition that $u \in U$, $|\lambda| \leq 1 \Rightarrow \lambda u \in U$. See [111, B.6]). Observe that $\chi(u) \neq -1$ for any $u \in U$.

(d) Let us for any $u \in U$ write

$$\chi(u) = e^{i \operatorname{Arg} \chi(u)},$$

where Arg denotes the principal branch of the argument function $\arg z$. So $-\pi < \operatorname{Arg} z < \pi$ for all $z \in \mathbb{C} \setminus]-\infty, 0]$.

Show that $u \mapsto \operatorname{Arg} \chi(u)$ is continuous at $0 \in V$.

(e) Let $u \in U$. Then $\chi(u) = e^{i\operatorname{Arg}\chi(u))} = e^{i\lambda(u)}$, so $\lambda(u) - \operatorname{Arg}\chi(u) = 2\pi n$ for some $n = n(u) \in \mathbb{Z}$. Show by contradiction that $n = 0$: Assuming $n \neq 0$ derive that $|\lambda(u)| > \pi$ and obtain a contradiction by considering the value χ assumes at

$$\frac{\pi}{\lambda(u)} u \in U.$$

Conclude that $\lambda(u) = \operatorname{Arg}\chi(u)$.

(f) Show that λ is continuous (Lemma D.1).

Exercise 3.22. Let V be a topological vector space over $\mathbb{R}$. Let $\chi : V \to \mathbb{C}^*$ be a continuous character on V, where V is viewed as a topological abelian group under $+$. In contrast to Exercise 3.21 χ is no longer assumed unitary.

Show that there exists exactly one continuous $\mathbb{R}$-linear functional $a : V \to \mathbb{C}$ on V such that $\chi(v) = e^{a(v)}$ for all $v \in V$.

Exercise 3.23. Find the solutions $f \in C(\mathbb{R})$ of the functional equation $f(xy) = f(x)f(y)$, $x, y \in \mathbb{R}$. Note that the solutions are required to be continuous on all of $\mathbb{R}$, not just on $\mathbb{R}^* = \mathbb{R} \setminus \{0\}$.

Hint: Example 3.9.

Exercise 3.24. The functional equation

$$f(x + y + xy) = f(x) + f(y) + f(x)f(y), \quad x, y \in P := \mathbb{R} \setminus \{-1\}, \quad (3.11)$$

with $f : P \to \mathbb{C}$ being the unknown, is in the literature known as *Pompeiu's functional equation.* Via an abstract set up we explain how *the circle* *operation* $x \circ y := x + y + xy$ from Pompeiu's functional equation is conjugate to the multiplication in $\mathbb{R}$. The conjugation amounts to a change of coordinates that warps the ordinary multiplication into the circle operation. From this point of view Pompeiu's functional equation is the multiplicative Cauchy equation on $(\mathbb{R}^*, \cdot)$ in camouflage, so we may use our results about the multiplicative Cauchy equation to solve it. We start with the abstract set up.

(a) Let X be a set and $\Psi : X \to \mathbb{C}$ a bijection. We transfer the multiplication in $\mathbb{C}$ to X by defining the circle composition $(x, y) \mapsto x \circ y$ from $X \times X$ to X by

$$x \circ y := \Psi^{-1}(\Psi(x)\Psi(y)) \qquad \text{for } x, y \in X.$$

Show that $(X, \circ)$ is a monoid with neutral element $\Psi^{-1}(1)$.

(b) Show that $P := \Psi^{-1}(\mathbb{R}^*) \subseteq X$ is a group with respect to the circle composition $\circ$, that its neutral element is $\Psi^{-1}(1)$ and that the restriction $\Phi := \Psi|_P : P \to \mathbb{R}^*$ is an isomorphism between the groups P and $\mathbb{R}^*$.

(c) Let $F : P \to \mathbb{C}$. Show that F is multiplicative on $(P, \circ)$ if and only if there is a multiplicative function $\chi : \mathbb{R}^* \to \mathbb{C}$ such that $F = \chi \circ \Phi$. (Here and below $\circ$ also denotes composition of maps!) So F must be the transfer of a multiplicative function on $\mathbb{R}^*$ to P by help of Φ.

(d) Let $A : P \to \mathbb{C}$. Show that A is additive on $(P, \circ)$ if and only if there is an additive map $a : \mathbb{R}^* \to \mathbb{C}$ such that $A = a \circ \Phi$.

(e) Let $f : P \to \mathbb{C}$. Show that

$$f(x \circ y) = f(x) \circ f(y) \qquad \text{for all } x, y \in P,$$

if and only if there is a multiplicative function $\chi : \mathbb{R}^* \to \mathbb{C}$ such that $f = \Psi^{-1} \circ \chi \circ \Phi$.

We leave now the general set up and specialize to the concrete case of $X = \mathbb{C}$ and Ψ translation by 1, i.e., $\Psi(z) := z + 1$, $z \in \mathbb{C}$.

(f) Derive the formula $x \circ y = x + y + xy$, $x, y \in \mathbb{C}$, for the circle operation in this concrete case. Show that 0 is the neutral element for the circle operation. Show that the group P is $(\mathbb{R} \setminus \{-1\}, \circ)$ and that Φ is a topological isomorphism of the group P onto $(\mathbb{R}^*, \cdot)$.

(g) Show that the continuous multiplicative functions $F : P \to \mathbb{C}$ on the group $(P, \circ)$ are $F = 0$ and the functions of the forms $F(w) = |w+1|^\lambda$ and $F(w) = |w+1|^\lambda \operatorname{sgn}(w+1)$, where λ ranges over $\mathbb{C}$.

Hint: $F \circ \Phi^{-1} : \mathbb{R}^* \to \mathbb{C}$ is a continuous multiplicative function on the group $(\mathbb{R}^*, \cdot)$, so we can apply Example 3.9(b).

(h) (The Pompeiu functional equation) Find all continuous functions $f : P \to \mathbb{C}$ such that $f(x \circ y) = f(x) \circ f(y)$ for all $x, y \in P$, i.e., all continuous solutions of Pompeiu's functional equation (3.11).

Answer: $f = -1$ and the functions of the forms $f(w) = |w+1|^\lambda - 1$ and $f(w) = |w+1|^\lambda \operatorname{sgn}(w+1) - 1$, where λ ranges over $\mathbb{C}$.

(i) Find all continuous endomorphisms of the group P.

(j) Find the continuous additive maps $A : P \to \mathbb{C}$ on the group $(P, \circ)$ (use Exercise 2.9(b)).

In the theory of operator algebras (see for example [164, p. 16]) you meet the circle operation $(x, y) \mapsto x + y - xy$ with a minus instead of a plus in front of xy. It is conjugate to the circle operation $\circ$ of this exercise, the conjugation being multiplication by -1, because $x+y-xy = -[(-x)\circ(-y)]$.

Exercise 3.25. *Lobachevsky's functional equation* for $f : \mathbb{R} \to \mathbb{R}$ is

$$f(x)^2 = f(x+y)f(x-y), \qquad x, y \in \mathbb{R}.$$

It says that f maps three successive terms of any arithmetic progression to three terms of a geometric progression. It is known that its continuous solutions are multiples of exponential functions. Prove the following generalization:

The solutions $f : G \to \mathbb{C}$ of $f(x)^2 = f(xy)f(y^{-1}x)$, $x, y \in G$, are the multiples of the characters of G, when G is a 2-divisible group.

Hints: Show that $f(e) = 0$ implies $f = 0$, so we may assume that $f(e) \neq 0$. Note that $\chi := f/f(e)$ also satisfies the functional equation, so that we may assume that $f(e) = 1$. Show that $\chi(x)^2 = \chi(x^2)$. Given $u, v \in G$ find $x, y \in G$ such that $x^2 = uv$, $y^{-1}x = u$ and $xy = u$.

3.6 Notes and remarks

Sometimes multiplicative and additive maps are combined. For example, Heuvers and Ng [110] study on a division algebra $\mathcal{R}$ the functional equation

$$F(x) + m(x)G(x^{-1}) = 0 \qquad \text{for all } x \in \mathcal{R},\ x \neq 0,$$

where $F, G : \mathcal{R} \to \mathcal{R}$ are additive and $m : \mathcal{R} \to \mathcal{R}$ is multiplicative.

For other considerations of the interplay between solutions of the additive and the multiplicative Cauchy equations see Ger [100].

Example 3.7(a): The differentiability of χ is immediate from Proposition D.7, but the simple, elementary trick in the text suffices here.

Corollary 3.20: Our proof is from E. Artin [18, Theorem II.12, p. 34–35]. The idea is so simple that the procedure can be used in other connections. We do it in Exercise 10.3.

Theorem 3.21: A shorter answer of Corovei's question than Radó's was later given by Kannappan [128]. Consult Stetkær [190] if you are interested in more general mappings than complex-valued functions. Corovei [52] studies the functional equation $f(xy) + f(yx) = f(x)f(y) + f(y)f(x)$ where $f : S \to R$ is a function from a monoid S to a ring R of characteristic different from 2.

Exercise 3.8 is from Thielman [199]. For other aspects of the case of $\alpha = 1$ see Exercise 3.24.

Exercise 3.9 is part of the proof of Proposition 6.5.

Exercise 3.10 is taken from Jensen [117, p. 154].

Exercise 3.13 is the content of the paper Hosszú [113].

Exercise 3.15(b) is in the case of $\alpha = 1$ in Jensen [117, p. 150–151].

Exercises 3.21 and 3.22: For similar results see Hewitt and Ross [111, 24.43] and Pontrjagin [158, Theorem 63, p. 228].

Exercise 3.24: Kannappan [132, Section 13.1.2] and also Kannappan and Sahoo [167, Ch. 12] discuss generalizations of Pompeiu's functional equation on $\mathbb{R}$ and provide references to the literature. For matrix-valued solutions of the equation in (e) see McKiernan [145].

Chapter 4

Addition and Subtraction Formulas

4.1 Introduction

The trigonometrical functions sin, cos and their hyperbolic counterparts cosh and sinh stem from elementary geometry, although some mathematicians prefer to define them analytically via power series or differential equations. They satisfy various addition and subtraction formulas like the sine and cosine addition formulas

$$\sin(x+y) = \sin x \cos y + \sin y \cos x, \qquad x, y \in \mathbb{R}, \tag{4.1}$$

$$\cos(x+y) = \cos x \cos y - \sin x \sin y, \qquad x, y \in \mathbb{R}. \tag{4.2}$$

Viewing the addition and subtraction formulas as functional equations and solving these we shall in this chapter see to what extent the formulas determine the trigonometrical and hyperbolic functions. We work in a rather abstract setting, where the underlying space need not be $\mathbb{R}$, but can be any group. Some of the results hold even on semigroups. We solve the functional equations in the sense that we express the solutions in terms of our building blocks: characters and additive functions.

A common feature, that holds not just for the functional equations of this chapter, but for most of those of the present book, is that their solutions fall into a small number of types. It is illustrated for example by Corollary 4.17. We find it is satisfactory to obtain conclusions like: This particular functional equation has exactly these three (or another small number) types of solutions. That the feature holds is shown by experience, so it is not a mathematical theorem about functional equations.

Theorem 4.1 is the most important result of the present chapter, because it has applications to functional equations elsewhere.

The sine and cosine addition formulas are special cases of Levi-Civita's functional equation that we discuss in Chapter 5.

4.2 The sine addition formula

In this section we examine the sine addition formula (4.1) in a rather abstract setting, where the underlying space need not even be an abelian group. We merely use the associative property $x(yz) = (xy)z$, i.e., that the underlying space is a semigroup.

We shall in Theorem 4.1(a) in particular see that solutions of the generalized sine addition formula (4.3) automatically also satisfy a version of the cosine addition formula (4.2). A converse result holds for the general cosine addition formula (4.17) (see Theorem 4.15). So you cannot have the sine addition formula without the cosine addition formula.

It is easy to get characters on $\mathbb{R}$ from the two identities (4.1) and (4.2) above: Multiplying the first one by i and adding it to the second one we get after a small reduction that

$$\cos(x+y) + i\sin(x+y) = (\cos x + i\sin x)(\cos y + i\sin y).$$

This identity expresses a very well known property of the complex exponential function $\theta \mapsto e^{i\theta} = \cos\theta + i\sin\theta$ on the real line, namely that it is multiplicative, but the procedure above works in more generality and is used in the proof of Theorem 4.1(b).

We will state and solve extensions to groups of the trigonometrical addition and subtraction formulas above, in which the plus/minus is replaced by an involutive automorphism σ. In that way we obtain a formulation that simultaneously covers both the plus and the minus cases, but we do not require that $\sigma = \pm I$. It is remarkable that the group need not be abelian. It need not even be a group: It suffices that it is a semigroup.

We shall first discuss the simpler case of $\sigma = I$ (Theorem 4.1), because it is interesting in itself, because it has applications to function algebras, and finally because it will be useful in the treatment of other functional equations like d'Alembert's functional equation.

The sine addition formula (4.3) is our first functional equation apart from the ones defining our building blocks. It illustrates an interesting common aspect of functional equations, namely that one equation may determine more than just one unknown function: (4.3) contains two unknown functions, f and g.

The functional equation (4.3) will be called the *sine addition formula.*

Theorem 4.1. *Let S be a topological semigroup. Let $f, g \in C(S)$ satisfy the sine addition formula*

$$f(xy) = f(x)g(y) + f(y)g(x) \qquad \text{for all } x, y \in S. \tag{4.3}$$

Assume furthermore that $f \neq 0$.

(a) *There exists a constant $\alpha \in \mathbb{C}$ such that*

$$g(xy) = g(x)g(y) + \alpha^2 f(x)f(y) \qquad \text{for all } x, y \in S. \tag{4.4}$$

(b) *$\chi_1 := g + \alpha f : S \to \mathbb{C}$ and $\chi_2 := g - \alpha f : S \to \mathbb{C}$ are continuous multiplicative functions such that*

$$g = \frac{\chi_1 + \chi_2}{2}. \tag{4.5}$$

Given g then χ_1 and χ_2 are the only multiplicative functions satisfying (4.5), except that they may be interchanged.

(c) *Assume that $\chi_1 \neq \chi_2$. Then $\alpha \neq 0$ and $f = \frac{1}{2\alpha}(\chi_1 - \chi_2)$.*

(d) *Assume that $\chi_1 = \chi_2$. Then $g = \chi_1 = \chi_2$. With $\chi := g$ the sine addition formula (4.3) becomes to*

$$f(xy) = f(x)\chi(y) + f(y)\chi(x) \qquad \text{for all } x, y \in S, \tag{4.6}$$

where $\chi : S \to \mathbb{C}$ is a continuous, multiplicative function. Furthermore:

If S is a monoid, then $\chi(e) = 1$ and $f(e) = 0$, where e denotes the neutral element of S.

If S is a group, then χ is a continuous character on S, and there exists an additive function $a \in C(S)$ such that $f = \chi a$.

(e) *Both f and g are abelian functions.*

(f) *Define $d(x) := 2g(x)^2 - g(x^2)$, $x \in S$. Then $d = \chi_1\chi_2$, so $d : S \to \mathbb{C}$ is multiplicative.*

Proof. (a) For any $x, y, z \in S$ we find from (4.3) that

$$\begin{aligned} f(xyz) &= f(x(yz)) = f(x)g(yz) + f(yz)g(x) \\ &= f(x)g(yz) + f(y)g(x)g(z) + f(z)g(x)g(y). \end{aligned}$$

Similarly we find that

$$\begin{aligned} f(xyz) &= f((xy)z)) = f(xy)g(z) + f(z)g(xy) \\ &= f(x)g(y)g(z) + f(y)g(x)g(z) + f(z)g(xy). \end{aligned}$$

Comparing these two identities for $f(xyz)$ we find that

$$f(x)[g(yz) - g(y)g(z)] = f(z)[g(xy) - g(x)g(y)]. \tag{4.7}$$

Defining $\Phi : S \times S \to \mathbb{C}$ by $\Phi(u,v) := g(uv) - g(u)g(v)$, $u, v \in S$, the identity (4.7) reads

$$f(x)\Phi(y,z) = f(z)\Phi(x,y), \tag{4.8}$$

which shows that the expression $f(x)\Phi(y,z)$ is unchanged under cyclic permutations of x, y, z. Applying (4.8) twice we get

$$f(z)^2\Phi(x,y) = f(z)f(x)\Phi(y,z) = f(x)f(z)\Phi(y,z) = f(x)f(y)\Phi(z,z),$$

which according to the definition of Φ means that

$$f(z)^2[g(xy) - g(x)g(y)] = f(x)f(y)\Phi(z,z) \qquad \text{for all } x, y, z \in S. \tag{4.9}$$

By assumption $f \neq 0$ so there exists a $z_0 \in S$ such that $f(z_0) \neq 0$. Putting $z = z_0$ in (4.9) and dividing through by $f(z_0)^2$ we get that

$$g(xy) - g(x)g(y) = cf(x)f(y), \qquad \text{where } c := \frac{\Phi(z_0, z_0)}{f(z_0)^2},$$

which becomes (a) when we choose $\alpha \in \mathbb{C}$ such that $\alpha^2 = c$.

(b) Multiplying (4.3) by $\pm\alpha$ and adding the result to the identity (4.4) we find that

$$(g + \alpha f)(xy) = (g + \alpha f)(x)(g + \alpha f)(y),$$

and

$$(g - \alpha f)(xy) = (g - \alpha f)(x)(g - \alpha f)(y).$$

Thus both $\chi_1 := g + \alpha f$ and $\chi_2 := g - \alpha f$ are multiplicative functions from S to $\mathbb{C}$, and $g = (\chi_1 + \chi_2)/2$.

The essential uniqueness of χ_1 and χ_2 is immediate from Lemma 3.19.

(c) follows from the definitions $\chi_1 := g + \alpha f$ and $\chi_2 := g - \alpha f$, because $\alpha \neq 0$ when $\chi_1 \neq \chi_2$.

(d) If S is a monoid, then we see from the functional equation (4.3), that $\chi = g \neq 0$. Indeed, $g = 0$ would entail $f = 0$, contradicting our assumption.

And a non-zero multiplicative function χ on a monoid has $\chi(e) = 1$. To get $f(e) = 0$ put $x = y = e$ in (4.6).

If S is a group we get by Lemma 3.4(a) that χ is a character on S. The functional equation (4.3) becomes

$$f(xy) = f(x)\chi(y) + f(y)\chi(x).$$

Dividing through by $\chi(xy) = \chi(x)\chi(y)$ we see that we as a may take the function $a := f/\chi$.

(e) That f and g are central follows from (4.3) and (4.4). The statement about Kannappan's condition is trivial for g due to the formula (4.5). For f it is trivial in case (c). Left is the case (d) where f satisfies (4.6), but this is easy to treat when we use that f is central.

(f) Substitute the formula (4.5) for g into the definition of d. □

The formulas of Theorem 4.1 define solutions of the sine addition formula (4.3) on any semigroup. No restrictions are necessary:

Lemma 4.2. *Let S be a semigroup, and let $\chi_1, \chi_2, \chi : S \to \mathbb{C}$ be multiplicative functions.*

(a) *The pair $f = c\,(\chi_1 - \chi_2)$, $g = \frac{1}{2}(\chi_1 + \chi_2)$ is a solution of the sine addition formula* (4.3) *for any $c \in \mathbb{C}$.*

(b) *The pair $f = a\chi$, $g = \chi$ is a solution of* (4.3) *for any additive function $a : S \to \mathbb{C}$.*

Proof. This is done by simple computations that we leave to the reader. □

In spite of the fact that nothing whatsoever is assumed about commutativity in Theorem 4.1 (S need not be abelian, and f and g need not even be central) it turns out that f and g are abelian (Theorem 4.1(e)). This striking consequence will benefit us in our treatment of other functional equations.

Remarks 4.3. Here is a discussion of the results of Theorem 4.1.

(a) A constant α must be present in the formula (4.4) because f is only determined up to a multiplicative scalar by (4.3).

(b) If G is a locally compact group, then we may in Theorem 4.1 and Corollary 4.4 replace the assumption about continuity by measurability, because measurable solutions are continuous. More precisely:

If $f \neq 0$ and g are measurable with respect to a (left or right) Haar measure on G and satisfy the sine addition formula (4.3), then they are continuous: Going through the proof of Theorem 4.1 we see that the multiplicative functions $\chi_1 := g+\alpha f$ and $\chi_2 := g-\alpha f$ are measurable. Each of them is either 0 or a character (Lemma 3.4(a)), so they are continuous by Proposition D.4(b). It follows that $g = (\chi_1+\chi_2)/2 \in C(G)$. If $\alpha \neq 0$ then also $f = (\chi_1 - \chi_2)/(2\alpha) \in C(G)$. If $\alpha = 0$, then the additive map $a := f/\chi$ from Theorem 4.1(d) is measurable and so continuous according to Proposition D.4(c). Now $f = \chi a \in C(G)$.

(c) It is allowed that $\chi_1 = 0$ or $\chi_2 = 0$ in Theorem 4.1, even when S has a neutral element. $\chi_1 = 0$ or $\chi_2 = 0$ will of course simplify the expressions: If $\chi_2 = 0$ in (c) then the solution pair takes the form $f = c\chi$, $g = \chi/2$, where $\chi := \chi_1 : S \to \mathbb{C}$ is a continuous multiplicative function such that $\chi \neq 0$, and where $c := 1/(2\alpha) \in \mathbb{C}$ is a non-zero constant. We get the same form if $\chi_1 = 0$.

(d) For a general semigroup S we cannot conclude in Theorem 4.1(d) that $g = \chi \neq 0$, as the following counter-example shows: Take $S := (]1,\infty[,+)$ and $f \in C(]1,\infty[) \setminus \{0\}$ such that $f(x) = 0$ for all $x > 2$.

(e) For a general monoid S we cannot conclude in Theorem 4.1(d) that $g = \chi$ vanishes nowhere and that f has the form $f = \chi a$ for some function $a : S \to \mathbb{C}$, as we can for a group. Here is a counter-example: Take $S := (\mathbb{N} \cup \{0\},+)$, $f := 1_{\{1\}}$ and $g = \chi := 1_{\{0\}}$.

(f) We shall encounter the function $d : S \to \mathbb{C}$ from Theorem 4.1(f) at various places later. See also the discussion after Definition B.6.

Theorem 4.1 simplifies in the important case of S being a group, because the mess mentioned in Remarks 4.3(d) and (e) disappears:

Corollary 4.4. *Let G be a topological group. The solutions $g, f \in C(G)$ of the sine addition formula (4.3) are the following:*

(a) *$f = 0$ and g arbitrary in $C(G)$.*

(b) *$g = \chi/2$ and $f = c\chi$, where $\chi \in C(G)$ is a character on G and $c \in \mathbb{C} \setminus \{0\}$ a constant.*

(c) *There exist two different characters $\chi_1, \chi_2 \in C(G)$ on G and a constant $c \in \mathbb{C} \setminus \{0\}$ such that*

$$g = \frac{\chi_1 + \chi_2}{2} \quad \textit{and} \quad f = c\,(\chi_1 - \chi_2).$$

(d) *There exist a character* $\chi \in C(G)$ *on* G *and an additive function* $a \in C(G)$, $a \neq 0$, *such that* $g = \chi$ *and* $f = \chi a$.

Proof. We leave it to the reader to verify that the pairs listed in the various cases are solutions. Thus it is left to show that any solution $g, f \in C(G)$ is in one of the cases.

If $f = 0$ we deal with case (a), so we may during the rest of the proof assume that $f \neq 0$. Since $g = 0$ implies that $f = 0$ (use (4.3)) we may assume that also $g \neq 0$.

From Theorem 4.1 we find that $g = (\chi_1 + \chi_2)/2$ where χ_1 and χ_2 are continuous multiplicative functions on G. They cannot both be 0, because then $g = 0$, so at least one of them does not vanish.

Assume at first that one of the two multiplicative functions vanishes, say $\chi_2 = 0$, while $\chi := \chi_1 \neq 0$. As a non-zero multiplicative function χ is a character (Lemma 3.4(a)). Since $\chi_1 \neq \chi_2$ we are in case (c) of Theorem 4.1 which implies that we deal with case (b) of the present corollary.

We have left the case where none of the two multiplicative functions vanishes. They are then both characters. If they are different we get from Theorem 4.1(c) that we are in case (c) of the present corollary. If they are equal then we get from Theorem 4.1(d) that we are in case (d) of the present corollary. □

Example 4.5. In this example we list all continuous solutions of the sine addition formula (4.3) on $\mathbb{R}$, i.e., all pairs of functions $f, g \in C(\mathbb{R})$ satisfying

$$f(x+y) = f(x)g(y) + f(y)g(x) \qquad \text{for all } x, y \in \mathbb{R}.$$

By help of Corollary 4.4, combined with Example 3.7(a) and Corollary 2.4, we find that the solutions are the pairs indicated below:

(a) $f = 0$, while $g \in C(\mathbb{R})$ is arbitrary.

(b) There exist $\lambda, c \in \mathbb{C}$ such that

$$g(x) = \tfrac{1}{2}e^{\lambda x} \quad \text{and} \quad f(x) = ce^{\lambda x} \qquad \text{for all } x \in \mathbb{R}.$$

(c) There exist $\lambda_1, \lambda_2, c \in \mathbb{C}$ such that

$$g(x) = \frac{e^{\lambda_1 x} + e^{\lambda_2 x}}{2} \quad \text{and} \quad f(x) = c(e^{\lambda_1 x} - e^{\lambda_2 x}) \qquad \text{for all } x \in \mathbb{R}.$$

(d) There exist $\lambda, c \in \mathbb{C}$ such that

$$g(x) = e^{\lambda x} \quad \text{and} \quad f(x) = cxe^{\lambda x} \qquad \text{for all } x \in \mathbb{R}.$$

You might wonder why linear functions like $a(x) = x$, $x \in \mathbb{R}$, [case (d) with $\lambda = 0$ and $c = 1$] enter. After all a linear function is quite different in nature from the exponential functions in (b) and (c). An explanation is here that a is a pointwise limit of functions f from (c): If we in (c) take $\lambda_2 = 0$ and $c = 1/\lambda_1$ we arrive at the solution

$$g_{\lambda_1}(x) = \frac{e^{\lambda_1 x} + 1}{2} \quad \text{and} \quad f_{\lambda_1}(x) = \frac{e^{\lambda_1 x} - 1}{\lambda_1} \qquad \text{for all } x \in \mathbb{R}.$$

Letting $\lambda_1 \to 0$ we get that $f_{\lambda_1}(x) \to a(x)$ for each $x \in \mathbb{R}$.

4.3 A connection to function algebras

In subsection 4.3 we show how Theorem 4.1 applies to function algebras. The results will not be used later in the book.

We have introduced and motivated the sine addition formula (4.3) from its origin in trigonometry. But that is not the only possible approach, because it occurs in other parts of mathematics. For example in the theory of algebras of functions. The point of view in the theory of function algebras is totally different from ours: The sine addition formula is viewed as a form of Leibniz' rule of differentiation. Also the terminology is different.

For the definition of the concept of an algebra $\mathcal{A}$ and the associated set $\widehat{\mathcal{A}}$ of non-zero multiplicative linear functionals on $\mathcal{A}$ see Section A.4.

Definition 4.6. Let $\mathcal{A}$ be a complex algebra, and let $f, g : \mathcal{A} \to \mathbb{C}$ be linear functionals on $\mathcal{A}$.

(i) (f, g) is a *derivation pair*, if $f(xy) = f(x)g(y) + f(y)g(x)$ for all $x, y \in \mathcal{A}$, i.e., if the pair satisfies the sine addition formula with respect to the multiplication of $\mathcal{A}$.

(ii) f is said to be a *point derivation* on $\mathcal{A}$ at g, if (f, g) is a derivation pair and $g \in \widehat{\mathcal{A}}$.

Theorem 4.1 reveals that the structure of the set of derivation pairs is determined already by the multiplicative semigroup of the algebra, not its linear structure. This fundamental property was apparently first observed by Glaeser [102], who studied derivations of a commutative algebra over a field of characteristic $\neq 0$.

In this section we present three illustrative and basic examples of algebras of functions, and we demonstrate the power of Theorem 4.1 by using it to give a simple and transparent proof of a result used in the theory of function algebras (Theorem 4.11).

Examples 4.8 and 4.9 below motivate the terminology point derivation from Definition 4.6(ii): In these examples any point derivation is differentiation of functions at a point. In this connection it might also be noted that any point derivation f vanishes on scalars just like ordinary differentiation does: $f(\alpha \cdot 1) = 0$ for all $\alpha \in \mathbb{C}$ (Exercise 4.5(a)).

Example 4.7. If T is a compact Hausdorff space then $\mathcal{A} := C(T)$ is a function algebra. $\widehat{C(T)}$ consists of the point evaluations $\mathrm{ev}_t(x) := x(t)$, $x \in C(T)$, where t ranges over T (see [34, Theorem 1.3.1]). $C(T)$ has no non-zero point derivations (see [34, pp. 64–65] or Exercise 4.5).

Example 4.8. Let $D := \{z \in \mathbb{C} \mid |z| < 1\}$. An example of a function algebra is the *disk algebra* $H(D)$ of the functions $\phi \in C(\overline{D})$ which are holomorphic in D. $\widehat{H(D)}$ consists of the point evaluations $\mathrm{ev}_z(x) := x(z)$, $x \in H(D)$, where z ranges over $\overline{D}$ (see [34, pp. 35–36]). If $z \in D$ then $D_z(x) := x'(z)$, $x \in H(D)$, is a point derivation on $H(D)$ at ev_z, and any point derivation on $H(D)$ at ev_z is a scalar multiple of D_z (for the latter statement see Exercise 4.6). The point derivations corresponding to points on the boundary of D are 0 (see [34, p. 65]).

Example 4.9. Let Ω be plane region (an open, connected, non-empty subset of $\mathbb{C}$). The algebra $H(\Omega)$ of holomorphic functions defined on Ω is a function algebra. It is known that $\widehat{H(\Omega)}$ consists of the point evaluations $\mathrm{ev}_z(x) := x(z)$, $x \in H(\Omega)$, where z ranges over Ω (see [217, Section 2]). If $z \in \Omega$ then $D_z(x) := x'(z)$, $x \in H(\Omega)$, is a point derivation on $H(\Omega)$ at ev_z, and any point derivation on $H(\Omega)$ at ev_z is a scalar multiple of D_z.

The main theorem of [217] is a rediscovery of Glaeser's result [102] about characterization of derivation pairs. It is an immediate corollary of Theorem 4.1, taking Remark 4.3(c) into account. We can even delete the assumption of [217] that the algebra is commutative:

Theorem 4.10. *Let $\mathcal{A}$ be a complex algebra. Any derivation pair f, g on $\mathcal{A}$ with $f \neq 0$ is of one of the forms*

1. $f = c\chi$ *and* $g = \chi/2$,
2. $f = c(\chi_1 - \chi_2)$ *and* $g = (\chi_1 + \chi_2)/2$,
3. f *is a point derivation at* χ *and* $g = \chi$,

where $c \in \mathbb{C} \setminus \{0\}$, $\chi \in \widehat{\mathcal{A}}$ *and* $\chi_1, \chi_2 \in \widehat{\mathcal{A}}$ *are such that* $\chi_1 \neq \chi_2$.

Another result that comes out easily from Theorem 4.1 is Theorem 1.6.1 of Browder's monography [34]. This theorem characterizes those linear functionals f on a commutative Banach algebra $\mathcal{A}$ the kernels of which are subalgebras (not just subspaces). We do not need Browder's assumption about the algebra being a Banach-algebra, and its commutativity may be replaced by f being central with respect to the multiplication in $\mathcal{A}$, so we extend his result to a purely algebraic setting as follows:

Theorem 4.11. *Let $\mathcal{A}$ be a complex algebra with unit 1, and let $f : \mathcal{A} \to \mathbb{C}$ be a non-zero linear functional on $\mathcal{A}$ with the properties that $\ker f$ is a subalgebra of $\mathcal{A}$ and that $f(xy) = f(yx)$ for all $x, y \in \mathcal{A}$. Then there are two cases:*

(a) $f(1) \neq 0$. *Here f is a scalar multiple of an element in $\widehat{\mathcal{A}}$.*

(b) $f(1) = 0$. *Here either there exist $\chi_1, \chi_2 \in \widehat{\mathcal{A}}$ such that f is a scalar multiple of $\chi_1 - \chi_2$ or there exists a $\chi \in \widehat{\mathcal{A}}$ such that f is a point derivation on $\mathcal{A}$ at χ.*

Proof. (a) Possibly replacing f by a scalar multiple of f we may assume that $f(1) = 1$. Then $x - f(x)1 \in \ker f$ for all $x \in \mathcal{A}$. Now, $\ker f$ being an algebra, we get that $(x - f(x)1)(y - f(y)1) \in \ker f$ for all $x, y \in \mathcal{A}$, so that

$$\begin{aligned} 0 &= f([x - f(x)1][y - f(y)1]) = f(xy - f(x)y - f(y)x + f(x)f(y)1) \\ &= f(xy) - f(x)f(y) - f(y)f(x) + f(x)f(y) = f(xy) - f(x)f(y), \end{aligned}$$

that proves (a).

(b) Since $f \neq 0$ we may pick $e_0 \in \mathcal{A}$ such that $f(e_0) = 1$. We only need e_0 to define $e := e_0 - \frac{1}{2}f(e_0^2)1 \in \mathcal{A}$ with the properties $f(e) = 1$ and $f(e^2) = 0$.

For any $x, y \in \mathcal{A}$ we have $x - f(x)e \in \ker f$ and $y - f(y)e \in \ker f$, so, $\ker f$ being an algebra, we get

$$\begin{aligned} 0 &= f([x - f(x)e][y - f(y)e]) = f(xy - f(x)ey - f(y)xe + f(x)f(y)e^2) \\ &= f(xy) - f(x)f(ey) - f(y)f(xe). \end{aligned}$$

Since f is central, we get

$$f(xy) = f(x)g(y) + f(y)g(x) \qquad \text{for all } x, y \in \mathcal{A}, \tag{4.10}$$

where $g(x) = f(xe) = f(ex)$ for $x \in \mathcal{A}$, which means that we have arrived at the sine addition formula with an $f \neq 0$. According to Theorem 4.1

there exists a constant $\alpha \in \mathbb{C}$ such that the functions $\chi_1 := g + \alpha f$ and $\chi_2 := g - \alpha f$ are multiplicative. They are of course linear, since g and f are linear. Furthermore $\chi_1(1) = g(1) + \alpha f(1) = f(e) + \alpha \cdot 0 = 1$. Similarly we find $\chi_2(1) = 1$. Thus χ_1 and χ_2 are not identically 0, so $\chi_1, \chi_2 \in \widehat{\mathcal{A}}$.

If $\alpha \neq 0$, then $f = (\chi_1 - \chi_2)/(2\alpha)$ as desired.

If $\alpha = 0$, then $\chi_1 \in \widehat{\mathcal{A}}$, and we see from (4.10) that f is a point derivation on $\mathcal{A}$ at χ_1. □

4.4 The sine subtraction formula

The trigonometrical identity $\sin(x - y) = \sin x \cos y - \sin y \cos x$ inspires us to derive the following result:

Theorem 4.12 (The sine subtraction formula). *Let G be a topological group, and $\sigma : G \to G$ a continuous homomorphism such that $\sigma \circ \sigma = I$, where I denotes the identity map.*

The solutions $f, g \in C(G)$ of the sine subtraction formula

$$f(x\sigma(y)) = f(x)g(y) - f(y)g(x) \qquad \text{for all } x, y \in G, \tag{4.11}$$

are the following pairs of functions, where $\chi \in C(G)$ denotes a continuous character and $c, c_1 \in \mathbb{C}$ denote constants:

(i) *$f = 0$ and g arbitrary in $C(G)$.*

(ii)
$$f = c_1 \frac{\chi - \chi \circ \sigma}{2} \quad \text{and} \quad g = \frac{\chi + \chi \circ \sigma}{2} + c\frac{\chi - \chi \circ \sigma}{2}, \tag{4.12}$$
where $\chi \neq \chi \circ \sigma$.

(iii) *$f = \chi a$ and $g = \chi(1 + ca)$, where $\chi = \chi \circ \sigma$ and where $a \in C(G)$ is an additive map such that $a \circ \sigma = -a$.*

Proof. Verifying that the stated pairs of functions constitute solutions consists of simple computations that we leave to the reader. To see the converse, i.e., that any solution $f, g \in C(G)$ of (4.11) is contained in one of these three cases, we proceed as follows to reduce the equation to the sine addition formula (4.3) so that we can apply Corollary 4.4:

Let $g_o := (g - g \circ \sigma)/2$, respectively $g_e := (g + g \circ \sigma)/2$, denote the odd, respectively even, part of g with respect to σ.

Since the right hand side of (4.11) changes sign when x and y are interchanged we see that $f(x\sigma(y)) = -f(y\sigma(x))$. Putting $y = e$ we get $f \circ \sigma = -f$, i.e., f is odd with respect to σ. Thus $f(x\sigma(y)) = -f(y\sigma(x)) = (f \circ \sigma)(y\sigma(x)) = f(\sigma(y)x)$, so f is furthermore central. Using that f is central on the left hand side of (4.11) (i.e., that $f(x\sigma(y)) = f(\sigma(y)x)$) we get from the right hand sides that

$$f(x)[g(y) - g(\sigma(y))] = f(y)[g(x) - g(\sigma(x))],$$

or equivalently

$$f(x)g_o(y) = f(y)g_o(x) \qquad \text{for all } x, y \in G. \tag{4.13}$$

Substituting $g = g_e + g_o$ into (4.11) we find by help of (22) that

$$f(x\sigma(y)) = f(x)g_e(y) - f(y)g_e(x) \qquad \text{for all } x, y \in G, \tag{4.14}$$

or equivalently, f being odd, that

$$f(xy) = f(x)g_e(y) + f(y)g_e(x) \qquad \text{for all } x, y \in G. \tag{4.15}$$

If $f = 0$ we deal with case (i). So during the rest of the proof we will assume that $f \neq 0$. From (22) we then read that $g_o = cf$ for some constant $c \in \mathbb{C}$, so we may concentrate on the pair (f, g_e).

The identity (4.15) reveals that the pair (f, g_e) is a solution of the sine addition formula (4.3), so we know from Corollary 4.4 that there are only the following 4 possibilities:

(a) $f = 0$ and g_e arbitrary. This does not apply, because $f \neq 0$ by assumption.

(b) $g_e = \chi/2$ and $f = c\chi$ for some $\chi : G \to \mathbb{C}$. Here f and g_e are proportional. But then the odd function f is also even, which implies that $f = 0$. Also this case does not apply, because $f \neq 0$.

(c) There exist two different continuous characters χ_1 and χ_2 on G and a constant $c_1 \in \mathbb{C}$ such that

$$g_e = \frac{\chi_1 + \chi_2}{2} \quad \text{and} \quad f = c_1\,(\chi_1 - \chi_2).$$

g_e being even and $f \neq 0$ being odd with respect to σ mean that $\chi_1 \circ \sigma + \chi_2 \circ \sigma = \chi_1 + \chi_2$ and $\chi_1 \circ \sigma - \chi_2 \circ \sigma = \chi_2 - \chi_1$, from which we infer that $\chi_2 = \chi_1 \circ \sigma$, so that

$$g_e = \frac{\chi_1 + \chi_1 \circ \sigma}{2} \quad \text{and} \quad f = c_1(\chi_1 - \chi_1 \circ \sigma).$$

Since $f \neq 0$ we have $\chi_1 \neq \chi_1 \circ \sigma$. Taking $g_o = cf$ from above into account we see that we deal with case (ii).

(d) $g_e = \chi$ and $f = \chi a$, where $\chi \in C(G)$ is a character and $a \in C(G)$ is an additive map.

From g_e being even with respect to σ we see that $\chi = \chi \circ \sigma$, and from f being odd with respect to σ we see then that $a \circ \sigma = -a$. Taking $g_o = cf$ from above into account we conclude that we deal with case (iii). □

Corollary 4.13. *Let $(G, +)$ be a topological abelian group. The solutions $f, g \in C(G)$ of the functional equation*

$$f(x-y) = f(x)g(y) - f(y)g(x) \qquad \text{for all } x, y \in G,$$

are the following pairs of functions, where $\chi \in C(G)$ denotes a character and $c, c_1 \in \mathbb{C}$ denote constants:

(i) *$f = 0$ and g arbitrary in $C(G)$.*

(ii)
$$f = c_1 \frac{\chi - \check{\chi}}{2} \qquad \text{and} \qquad g = \frac{\chi + \check{\chi}}{2} + c\frac{\chi - \check{\chi}}{2},$$
where $\chi \neq \check{\chi}$.

(iii) *$f = \chi a$ and $g = \chi(1 + ca)$, where $\chi = \check{\chi}$ and where $a \in C(G)$ is an additive map.*

The case (iii) of Corollary 4.13 simplifies even further if G is 2-divisible, because then $\chi = \check{\chi}$ implies that $\chi = 1$ (Lemma 3.5). Let us as an example take $G = \mathbb{R}$. Here the continuous additive maps can be found in Corollary 2.4 and the continuous characters in Example 3.7(a).

Example 4.14. The solutions $f, g \in C(\mathbb{R})$ of the functional equation

$$f(x-y) = f(x)g(y) - f(y)g(x) \qquad \text{for all } x, y \in \mathbb{R},$$

are the following pairs of functions, where $\lambda, \alpha \in \mathbb{C} \setminus \{0\}$ and $\beta \in \mathbb{C}$ denote constants:

(i) $f = 0$ and g arbitrary in $C(\mathbb{R})$.

(ii) $f(x) = \alpha \sin(\lambda x)$ and $g(x) = \cos(\lambda x) + \beta \sin(\lambda x)$ for $x \in \mathbb{R}$.

(iii) $f(x) = \alpha x$ and $g(x) = 1 + \beta x$ for $x \in \mathbb{R}$.

4.5 The cosine addition and subtraction formulas

Throughout Section 4.5 we let G be a topological group, not necessarily abelian, and $\sigma : G \to G$ a continuous automorphism such that $\sigma \circ \sigma = I$. The discrete topology on G is allowed.

We will solve the functional equation

$$g(x\sigma(y)) = g(x)g(y) - f(x)f(y), \qquad x, y \in G, \tag{4.16}$$

where $g, f \in C(G)$ are unknown functions, by comparing it to the special instance of $\sigma = I$, which is known as *the cosine addition formula*:

$$g(xy) = g(x)g(y) - f(x)f(y), \qquad x, y \in G, \tag{4.17}$$

that we solve first (Theorem 4.15). The equations are generalizations of the addition and subtraction formulas for the trigonometric functions cos and cosh, like

$$\cos(x + y) = \cos x \cos y - \sin x \sin y, \qquad x, y \in \mathbb{R},$$
$$\cos(x - y) = \cos x \cos y + \sin x \sin y, \qquad x, y \in \mathbb{R}.$$

The minus-signs on the right hand sides of (4.16) and (4.17) are not essential: We can change them to plus-signs by replacing f by if.

Taking $\sigma(x) = -x$ in (4.16) on an abelian group and replacing f by if gives us *the cosine subtraction formula*

$$g(x - y) = g(x)g(y) + f(x)f(y), \qquad x, y \in G, \tag{4.18}$$

that we solve for any 2-divisible abelian group in Corollary 4.17.

Theorem 4.15. *If $g, f \in C(G)$ is a solution of (4.17), then there exists a constant $\alpha \in \mathbb{C}$ such that the following version of the sine addition formula holds:*

$$f(xy) = f(x)g(y) + f(y)g(x) + \alpha f(x)f(y) \qquad \text{for all } x, y \in G. \tag{4.19}$$

The solutions $g, f \in C(G)$ of (4.17) are the following where $m, M \in C(G)$ denote two different continuous characters.

(a) $g = 0$ *and* $f = 0$.

(b) $g = \frac{1}{1-c^2} m$, $f = \frac{c}{1-c^2} m$, *where* $c \in \mathbb{C} \setminus \{\pm 1\}$.

(c) $g = \frac{\lambda M + \frac{1}{\lambda} m}{\lambda + \frac{1}{\lambda}}$ *and* $f = \frac{M - m}{(\lambda + \frac{1}{\lambda})i}$, *where* $\lambda \in \mathbb{C} \setminus \{0, i, -i\}$.

(d) $g = m(1+a)$, $f = ma$, *where* $a \in C(G)$ *is additive.*

(e) $g = m(1+a)$, $f = -ma$, *where* $a \in C(G)$ *is additive.*

Proof. We will first derive (4.19) from (4.17): We find that

$$\begin{aligned} g(x(yz)) &= g(x)g(yz) - f(x)f(yz) \\ &= g(x)g(y)g(z) - g(x)f(y)f(z) - f(x)f(yz) \end{aligned}$$

and similarly

$$g((xy)z) = g(x)g(y)g(z) - f(x)f(y)g(z) - f(xy)f(z).$$

Using that $g(x(yz)) = g((xy)z)$ we find that

$$f(x)[f(yz) - g(y)f(z) - f(y)g(z)] = f(z)[f(xy) - f(x)g(y) - g(x)f(y)],$$

or with the abbreviation $\Phi(u,v) := f(uv) - f(u)g(v) - g(u)f(v)$ that

$$f(x)\Phi(y,z) = f(z)\Phi(x,y) \qquad \text{for all } x, y \in G. \tag{4.20}$$

The identity (4.19) is trivially true if $f = 0$, so we assume $f \neq 0$, i.e., that there is a $z_0 \in G$ such that $f(z_0) \neq 0$. Then $\Phi(x,y) = f(x)\phi(y)$, where $\phi(y) := \Phi(y, z_0)/f(z_0)$. Substituting this back into (4.20) we get $f(x)f(y)\phi(z) = f(z)f(x)\phi(y)$, from which we infer that $\phi = \alpha f$ where $\alpha = \phi(z_0)/f(z_0)$, and thus that $\Phi(x,y) = f(x)\phi(y) = \alpha f(x)f(y)$. Going back to the definition of Φ we get (4.19).

We leave it to the reader to verify that the pairs stated in (a)–(e) are solutions of (4.17). It is thus left to show that any solution $g, f \in C(G)$ of (4.17) falls into one of the categories (a)–(e).

If $g = 0$, then $f = 0$. This is case (a). From now on we assume that $g \neq 0$.

If g and f are linearly dependent, them there exists a constant $c \in \mathbb{C}$ such that $f = cg$. Substituting this into (4.17) we find that $g(xy) = (1-c^2)g(x)g(y)$. Note that $c \neq \pm 1$, because $g \neq 0$. It follows that $(1-c^2)g \neq 0$ is multiplicative, so $m := (1-c^2)g$ is a character (Lemma 3.4(a)). Since $f = cg$ we are in case (b).

From now on we may assume that g and f are linearly independent. Combining (4.17) and (4.19) we find for any $\lambda \in \mathbb{C}$ after some reduction that

$$(g - \lambda f)(xy) = (g - \lambda f)(x)(g - \lambda f)(y) - (\lambda^2 + \alpha\lambda + 1)f(x)f(y). \tag{4.21}$$

Let λ_1 and λ_2 be the two roots of the polynomial $z^2 + \alpha z + 1$. Now, (4.21) shows that $m := g - \lambda_1 f$ and $M := g - \lambda_2 f$ are continuous multiplicative functions of G into $\mathbb{C}$. Furthermore they are non-zero, because g and f are linearly independent. Hence they are characters (Lemma 3.4(a)). Note that $\lambda_1\lambda_2 = 1$. In particular $\lambda_1 \neq 0$ and $\lambda_2 \neq 0$.

If $\lambda_1 \neq \lambda_2$ then $M \neq m$. In that case we find

$$g = \frac{\lambda_1 M - \lambda_2 m}{\lambda_1 - \lambda_2} \quad \text{and} \quad f = \frac{M - m}{\lambda_1 - \lambda_2}.$$

Using that $\lambda_1\lambda_2 = 1$ and writing $\lambda_1 = i\lambda$ we obtain the formulas from (c) for g and f.

If $\lambda_1 = \lambda_2$ then we see from $\lambda_1\lambda_2 = 1$ that either $\lambda_1 = 1$ or $\lambda_1 = -1$.

Let us consider the first case: Since $\lambda_1 = 1$ is a root of the polynomial $z^2 + \alpha z + 1$ we get that $\alpha = -2$. Noting that $m = g - \lambda_1 f = g - f$ we get from (4.19) that

$$\begin{aligned} f(xy) &= f(x)g(y) + f(y)g(x) + \alpha f(x)f(y) \\ &= f(x)g(y) + f(y)g(x) - 2f(x)f(y) \\ &= f(x)[g(y) - f(y)] + f(y)[g(x) - f(x)] = f(x)m(y) + f(y)m(x). \end{aligned}$$

m is a non-zero homomorphism, so it never vanishes (Lemma 3.4(a)). We see that $a := f/m$ is additive, which means that we are in case (d).

If $\lambda_1 = -1$ then similar arguments show that we are in case (e). □

Theorem 4.16. *The solutions $g, f \in C(G)$ of the cosine addition formula* (4.16) *are the following where $m, M : G \to \mathbb{C}$ denote two different continuous characters such that $M \circ \sigma = M$ and $m \circ \sigma = m$, $\chi : G \to \mathbb{C}^*$ is a continuous character, and $a \in C(G)$ is a continuous additive function such that $a \circ \sigma = a$:*

(a) $g = 0$ *and* $f = 0$.

(b) $g = \frac{1}{1-c^2}\, m$, $f = \frac{c}{1-c^2}\, m$, *where* $c \in \mathbb{C} \setminus \{\pm 1\}$.

(c) $g = \frac{\lambda M + \frac{1}{\lambda} m}{\lambda + \frac{1}{\lambda}}$ *and* $f = \frac{M - m}{(\lambda + \frac{1}{\lambda})i}$, *where* $\lambda \in \mathbb{C} \setminus \{0, i, -i\}$.

Given f of this form there are only two choices of the corresponding g. They arise from one another by a replacement of λ by $1/\lambda$:

$$g = \frac{\lambda M + \frac{1}{\lambda} m}{\lambda + \frac{1}{\lambda}} \quad \textit{and} \quad g = \frac{\frac{1}{\lambda} M + \lambda m}{\lambda + \frac{1}{\lambda}}.$$

(d) $g = \frac{\chi + \chi \circ \sigma}{2}$ *and* $f = \frac{\chi - \chi \circ \sigma}{2}$.

(e) $g = m(1+a)$, $f = ma$.

(f) $g = m(1-a)$, $f = ma$.

Proof. We start by proving that if the pair $\{g, f\} \in C(G)^2$ is a solution of (4.16), then

(i) $g \circ \sigma = g$, and

(ii) $f \circ \sigma = f$ or $f \circ \sigma = -f$. In the first case $\{g, f\}$ is a solution of (4.17), and in the second case $\{g, if\}$ is a solution of (4.17).

(i) Since the right hand side of (4.16) is invariant under interchange of x and y we get that $g(x\sigma(y)) = g(y\sigma(x))$. Taking $y = e$ we get that $g \circ \sigma = g$.

(ii) is trivially true if $f = 0$, so we may from now on assume that $f \neq 0$. Using (i) we get

$$\begin{aligned} g(x)g(y) - f(x)f(y) &= g(x\sigma(y)) = (g \circ \sigma)(x\sigma(y)) = g(\sigma(x)\sigma(\sigma(y))) \\ &= g(\sigma(x))g(\sigma(y)) - f(\sigma(x))f(\sigma(y)) = g(x)g(y) - f(\sigma(x))f(\sigma(y)), \end{aligned}$$

so $f(x)f(y) = f(\sigma(x))f(\sigma(y))$. Choosing $y_0 \in G$ such that $f(\sigma(y_0)) \neq 0$ we obtain

$$f \circ \sigma = \alpha f, \qquad \text{where } \alpha = \frac{f(y_0)}{f(\sigma(y_0))} \in \mathbb{C}.$$

Since $\sigma \circ \sigma = I$ we get that $\alpha^2 = 1$, so that $\alpha = \pm 1$.

If $\alpha = 1$ so that $f \circ \sigma = f$ we find that

$$\begin{aligned} g(xy) &= g(x\sigma(\sigma(y))) = g(x)g(\sigma(y)) - f(x)f(\sigma(y)) \\ &= g(x)g(y) - f(x)f(y), \end{aligned}$$

and if $\alpha = -1$ so that $f \circ \sigma = -f$ we find in a similar way that

$$g(xy) = g(x)g(y) + f(x)f(y) = g(x)g(y) - if(x)if(y).$$

We leave the rest of the proof to the reader: Use Theorem 4.15 and perhaps the result of Exercise 3.6. The essential uniqueness of g in (c) follows from Artin's result (Corollary 3.20). □

The following group version of the cosine subtraction formula (4.18) can be found in Kannappan's paper [131].

Corollary 4.17. *Let G be a 2-divisible, abelian topological group. The solutions $g, f \in C(G)$ of the functional equation*

$$g(x-y) = g(x)g(y) + f(x)f(y), \qquad x, y \in G, \tag{4.22}$$

are the following:

(a) $g = f = 0$.

(b) *There is a $c \in \mathbb{C} \setminus \{\pm 1\}$ such that*

$$g = \frac{1}{1-c^2}, \quad f = \frac{ic}{1-c^2}.$$

(c) *There is a continuous character $\chi : G \to \mathbb{C}^*$ such that*

$$g = \frac{\chi + \check{\chi}}{2}, \quad f = \frac{\chi - \check{\chi}}{2i}.$$

Proof. Combine Theorem 4.16 and Lemma 3.5. □

Example 4.18. The solutions $g, f \in C(\mathbb{R})$ of the functional equation

$$g(x-y) = g(x)g(y) + f(x)f(y), \qquad x, y \in \mathbb{R},$$

are the following:

(a) $g = f = 0$.

(b) There is a $c \in \mathbb{C} \setminus \{\pm 1\}$ such that

$$g = \frac{1}{1-c^2}, \quad f = \frac{ic}{1-c^2}.$$

(c) There is a constant $\lambda \in \mathbb{C} \setminus \{0\}$ such that $g(x) = \cos(\lambda x)$ and $f(x) = \sin(\lambda x)$ for $x \in \mathbb{R}$.

4.6 Exercises

Exercise 4.1. Find all continuous solutions $f, g \in C(\mathbb{T})$ of the sine addition formula (4.3) on the circle group $\mathbb{T}$.

Hint: Example 3.10 and Exercise 2.5 come in handy.

Exercise 4.2. Show that point (d) of Corollary 4.4 does not occur, if G is compact.

Exercise 4.3. Find the solutions $f \in C(\mathbb{R}^+)$ of the functional equation

$$f(xy) = f(x)\frac{y}{2} + f(y)\frac{x}{2}, \qquad x, y \in \mathbb{R}^+.$$

Same question for $f \in C(\mathbb{R}^*)$.

Hint: Example 3.9.

Exercise 4.4. The addition formula $\sin(x+y) = \sin x \cos y + \cos x \sin y$ can also be written in the form

$$\sin(x+y) = \sin x \sin(\tfrac{\pi}{2} + y) + \sin(\tfrac{\pi}{2} + x)\sin y, \qquad x, y \in \mathbb{R},$$

which contains only the function sine. The solutions $f \in C(\mathbb{R})$ of the corresponding functional equation

$$f(x+y) = f(x)f(\tfrac{\pi}{2} + y) + f(\tfrac{\pi}{2} + x)f(y), \qquad x, y \in \mathbb{R}, \tag{4.23}$$

were found by Neder [147]. See [132, Theorem 3.58] for all solutions, continuous or not.

We will in this exercise replace $\mathbb{R}$ by a topological group G. More precisely we ask the reader to prove the following proposition and reflect on what it says about the solutions of the original functional equation (4.23):

Proposition 4.19. *Let G be a topological group and $z_0 \in G$ a fixed element. The solutions $f \in C(G)$ of the functional equation*

$$f(xy) = f(x)f(z_0 y) + f(z_0 x)f(y), \qquad x, y \in G,$$

are the following:

(a) $f = 0$.

(b) *There exists a character $\chi \in C(G)$ such that $f = \frac{1}{2\chi(z_0)}\chi$.*

(c) *There exist two characters $\chi_1, \chi_2 \in C(G)$ such that*

$$\chi_1(z_0) = -\chi_2(z_0) \quad \text{and} \quad f = \frac{1}{2\chi_1(z_0)}(\chi_1 - \chi_2).$$

Hint: Apply Corollary 4.4 with $g(x) := f(z_0 x)$, $x \in G$. Corollary 3.20 and Exercise 3.6 will probably be needed as well.

Exercise 4.5. Let $\mathcal{A}$ be a complex algebra with unit 1. Let f be a point derivation on $\mathcal{A}$ at $g \in \widehat{\mathcal{A}}$.

(a) Show that $f(1) = 0$ and conclude that $f(\alpha \cdot 1) = 0$ for all $\alpha \in \mathbb{C}$.

(b) Let $x \in \mathcal{A}$. Assume that there exists a constant $\alpha \in \mathbb{C}$ such that $x - \alpha \cdot 1 = x_1 x_2$, where $x_1, x_2 \in \mathcal{A}$ and $g(x_1) = g(x_2) = 0$. Show that $f(x) = 0$.

We continue by treating the special case of $\mathcal{A} = C(T)$ where T is a compact Hausdorff space (see Example 4.7).

(c) Let $t_0 \in T$, and let $x \in C(T)$. Assuming $x(t_0) = 0$ show x may be written as $x = x_1 x_2$ for some $x_1, x_2 \in C(T)$ with $x_1(t_0) = x_2(t_0) = 0$.

(d) Show that there are no non-zero point derivations on $C(T)$.

The result of this exercise is a special case of a result from the theory of function algebras ([34, Corollary 1.6.7]) about non-existence of point derivations.

Exercise 4.6. We use the notation from Example 4.8. Let $z \in D$ and let $f : H(D) \to \mathbb{C}$ be a point evaluation at ev_z. Show that there is a constant $c \in \mathbb{C}$ such that $f(x) = cx'(z)$ for all $x \in H(D)$.

Hint: $x \in H(D)$ may be written in the form $x(w) = x(z) + (w - z)y(w)$, $w \in D$, where $y \in H(D)$ and $y(z) = x'(z)$.

Exercise 4.7. Show the converse of Theorem 4.11.

Exercise 4.8. Show that there are exactly two characters $\chi : \mathbb{Z} \to \mathbb{C}^*$ such that $\chi(-n) = \chi(n)$ for all $n \in \mathbb{Z}$.

Hint: Example 3.11.

In the light of this what does Theorem 4.16 say about the solutions of (4.22) on $G = \mathbb{Z}$?

Exercise 4.9. In Kannappan's paper [131, Theorem 1] the equation (4.22) is solved on any abelian, 2-divisible group G. Kannappan shows that its non-constant solutions satisfy

$$g(x + y) = g(x)g(y) - f(x)f(y), \qquad x, y \in G,$$
$$f(x + y) = f(x)g(y) + g(x)f(y), \qquad x, y \in G,$$

and

$$f(x-y) = f(x)g(y) - g(x)f(y), \qquad x, y \in G.$$

Verify Kannappan's result about the non-constant solutions.

Exercise 4.10. How does Example 4.18 simplify if you require the solutions to be real-valued and not just complex-valued?

Exercise 4.11. Let $d \in \mathbb{C} \setminus \{0\}$. Show that the solutions $g : \mathbb{R} \to \mathbb{C}$ of the following special case of the cosine addition formula

$$g(x+y) = g(x)g(y) - d \sin x \sin y, \qquad x, y \in \mathbb{R},$$

are

$$g(x) = \cos x \pm \sqrt{d-1} \sin x, \qquad x \in \mathbb{R}.$$

Exercise 4.12. Let G be an abelian group, $\mathcal{A}$ a $*$-algebra and $f, g : G \to \mathcal{A}$. Let $K \in \mathbb{C}$. Asssume that the pair (f, g) for all $x, y \in G$ is a solution of the following system:

$$f(x-y) = f(x)g(y)^* - g(x)f(y)^*,$$
$$g(x-y) = g(x)g(y)^* - K^2 f(x)f(y)^*.$$

(a) Let $m := g + Kf$ and $n := g - Kf$. Show for all $x, y \in G$ that

$$m(x-y) = m(x)[g(y) - \overline{K}f(y)]^*,$$

and

$$n(x-y) = n(x)[g(y) + \overline{K}f(y)]^*.$$

(b) Assuming $K \in \mathbb{R}$ show that $m(x+y) = m(x)m(y)$ for all $x, y \in G$ and that

$$Kf(x) = \frac{m(x) - m(-x)^*}{2} \quad \text{and} \quad g(x) = \frac{m(x) + m(-x)^*}{2}$$

for all $x \in G$.

(c) Assuming $K \in i\mathbb{R}$ show that $m(x+y) = m(x)m(y)$, $n(x+y) = n(x)n(y)$, $m(x)^* = m(-x)$ and $n(x) = n(-x)^*$ for all $x, y \in G$ and that

$$Kf(x) = \frac{m(x) - n(x)}{2} \quad \text{and} \quad g(x) = \frac{m(x) + n(x)}{2}$$

for all $x \in G$.

Exercise 4.13. Let K be a compact subgroup of a locally compact group G. Let dk denote the normalized Haar measure on K. Study the solutions $f, g \in C(G)$ of the functional equation

$$\int_K f(xkyk^{-1})\,dk = f(x)g(y) + g(x)f(y) \qquad \text{for all } x, y \in G,$$

in terms of functions $\phi \in C(G) \setminus \{0\}$ satisfying

$$\int_K \phi(xkyk^{-1})\,dk = \phi(x)\phi(y) \qquad \text{for all } x, y \in G.$$

4.7 Notes and remarks

Chung, Kannappan and Ng [45, 46] and also Friis and Stetkær [89] deal with equations like

$$\frac{f(x+y) + f(x+\sigma(y))}{2} = f(x)g(y) + f(y)g(x), \qquad x, y \in G,$$

where $\sigma : G \to G$ is an involution on an abelian group G.

The sine and cosine addition formulas have been studied on discrete polynomial hypergroups. See Orosz [153].

Theorem 4.1:

- For references to early works on the sine addition formula (4.3) see [2, Section 3.2.3]. Theorem 4.1 (although formulated for algebras) goes at least back to Glaeser [102], and it has been rediscovered several times. As mentioned in the text by Zalcman [217]. Vincze [204] proved it for abelian groups. It can in another formulation be found in Chung, Kannappan and Ng [45, Lemma 5] for groups. A proof of Theorem 4.1 on a monoid based on Levi-Civita's functional equation can be found in Exercise 5.12.
- It is in Theorem 4.1 straightforward to replace $\mathbb{C}$ by a quadratically closed, commutative field of characteristic different from 2. That is done by for example Sinopoulos [180, p. 258]. The quadratical closedness is needed to ensure the existence of a scalar α such that $\alpha^2 = c$.
- In Theorem 4.1 we saw that the solutions of the sine addition formula satisfy the pair of equations

$$f(xy) = f(x)g(y) + f(y)g(x) \qquad \text{for all } x, y \in S,$$
$$g(xy) = g(x)g(y) + \alpha^2 f(x)f(y) \qquad \text{for all } x, y \in S,$$

where $\alpha \in \mathbb{C}$. In deriving the last equation from the first one we used that the functions were complex-valued, because we divided by $f(z_0)$. If the functions assume their values in an algebra Fechner [80, Corollary 2.3] takes the pair of equations as point of departure and solves it.

Theorem 4.10: We have in Section 4.3 discussed the sine addition formula or more precisely derivation pairs on algebras of functions like $H(\Omega)$, where Ω is a plane region. In connection with Theorem 4.10 see Gaur and Nandakumar [96].

Theorems 4.12 and 4.16 have been generalized by Elqorachi and Redouani in [77]. They consider functional equations of the form

$$\int_G f(xt\sigma(y))\,d\mu(t) = f(x)g(y) \pm f(y)g(x), \qquad x, y \in G,$$
$$\int_G g(xt\sigma(y))\,d\mu(t) = g(x)g(y) \pm f(y)f(x), \qquad x, y \in G,$$

where G is a locally compact group, $\sigma : G \to G$ is a continuous homomorphism such that $\sigma \circ \sigma = I$, and μ is a σ-invariant complex bounded measure on G. Fechner [80, Theorem 3.1] solves under certain conditions the system

$$g(x-y) = g(x)f(y) - f(x)g(y),$$
$$f(x-y) = f(x)f(y) - K^2 g(x)g(y), \qquad x, y \in G,$$

where G is an abelian group, $\mathcal{A}$ a complex algebra, $f, g : G \to \mathcal{A}$ and $K \in \mathbb{C} \setminus \{0\}$.

Corollary 4.17: Kannappan and Nandakumar [133] found the linear solutions of the cosine addition formula on $H(\Omega)$ (see also Kannappan [132, Section 3.4.12]).

Exercise 4.11 extends a problem by Butler [36] for real-valued functions, that was solved by Ebanks [69] and by Rassias [162].

Exercise 4.12 is Fechner [84, Theorem 11].

The formulas of Poulsen and Stetkær [159, Propositions II.4 and III.2] are not quite correct: They contain not just the solutions, but also pairs of functions that are not solutions.

Chapter 5

Levi-Civita's Functional Equation

5.1 Introduction

A *solution of Levi-Civita's functional equation* on a semigroup S is an ordered set of functions $f, g_1, \dots, g_N, h_1, \dots, h_N : S \to \mathbb{C}$ satisfying *Levi-Civita's functional equation*

$$f(xy) = \sum_{l=1}^{N} g_l(x)h_l(y) = g(x)^t h(y), \qquad x, y \in S, \tag{5.1}$$

where on the right hand side we have introduced the vector valued functions

$$g := (g_1, g_2, \dots, g_N)^t : S \to \mathbb{C}^N \quad \text{and} \quad h := (h_1, h_2, \dots, h_N)^t : S \to \mathbb{C}^N.$$

The expression on the right hand side of Levi-Civita's functional equation has a special form, being a finite sum of terms in each of which the variables x and y are separated. Nevertheless (5.1) encompasses interesting and important equations like the Cauchy equations and the sine and cosine addition formulas, so it is a natural sequel to the topics of Chapter 4.

In this chapter we describe the structure of any solution of Levi-Civita's functional equation by help of the close connection between the solutions of (5.1) and the matrix-coefficients of N-dimensional subrepresentations of the right regular representation R of S. The most important result of the present chapter is Theorem 5.2 that describes this connection. The connection is easy to obtain if $N = 1$ and S is a group: Here the non-zero solutions of (5.1) are multiples a character (Exercise 3.20), and a character is a 1-dimensional subrepresentation of R. Theorem 5.2 can be used to replace the earlier ad hoc proofs from, say, our discussion of the sine addition formula, by arguments that reveal more clearly why the various types of solutions arise (for an example see Exercise 5.12).

5.2 Structure of the solutions

Let M be a monoid. We shall solve Levi-Civita's functional equation (5.1) under the additional assumption that $g_1, \ldots, g_N$ are linearly independent in the vector space of functions $\mathbb{C}^M$ and that so are $h_1, \ldots, h_N$. If not we may not be able to conclude much about the functions on the right hand side. To take an example we cannot infer anything about h_2 from the equation $f(xy) = g_1(x)h_1(y) + 0 \cdot h_2(y)$. If one or both of the N-tuples $g_1, \ldots, g_N$ or $h_1, \ldots, h_N$ are linearly dependent, then the right hand side of (5.1) may be reduced to a sum of $N-1$ or fewer terms: If, say, $h_N = \sum_{l=1}^{N-1} \alpha_l h_l$ for some constants $\alpha_1, \ldots, \alpha_{N-1}$ then

$$\sum_{l=1}^{N} g_l(x)h_l(y) = \sum_{l=1}^{N-1} [g_l(x) + \alpha_l g_N(x)]h_l(y).$$

This reduction process may be repeated until we get linearly independent sets of functions or 0 on the right hand side.

Our main result about Levi-Civita's functional equation is the following Theorem 5.2 about the structure of its solutions. Like in the previous chapters it turns out that homomorphisms play an important role. Now in the guise of matrix-coefficients of representations, not just characters and non-zero multiplicative functions (= matrix-coefficients of 1-dimensional representations).

It is expedient to introduce the right regular representation R from Appendix E.3:

Definition 5.1. Let S be a semigroup. For any $y \in S$ we define $R(y) : \mathbb{C}^S \to \mathbb{C}^S$ by $[R(y)F](x) := F(xy)$, $x \in S$, for $F \in \mathbb{C}^S$. $R(y)$ is a linear operator for each $y \in S$.

We recall that a subset V of the functions on S is said to be invariant under R if $R(y)(V) \subseteq V$ for all $y \in S$. The space of polynomials on $S = \mathbb{R}$ is an example of an R-invariant subspace. Exercise 5.2 gives other examples.

Theorem 5.2. *Let M be a monoid with unit element $e \in M$ and let $N \in \mathbb{N}$. Let $\{f, g, h\}$ be a solution of* (5.1) *such that both $g_1, \ldots, g_N$ and $h_1, \ldots, h_N$ are linearly independent. Let $V := \operatorname{span}\{R(y)f \mid y \in M\}$.*

(a) *$V = \operatorname{span}\{g_1, g_2, \ldots, g_N\}$, so the right regular representation R leaves $\operatorname{span}\{g_1, \ldots, g_N\}$ invariant. (For (a) we need only that $h_1, \ldots, h_N$ are linearly independent, not that $g_1, \ldots, g_N$ are linearly independent.)*

(b) *Let* $\rho(x) = \{\rho_{ij}(x)\}_{i,j=1}^N$ *for* $x \in M$ *denote the matrix of* $R(x) : V \to V$ *with respect to the basis* $\{g_1, \dots, g_N\}$ *of* V*, i.e.,* $R(x)g_i = \sum_{j=1}^N \rho_{ji}(x)g_j$*. Then* ρ *is a representation of* M *on* $\mathbb{C}^N$ *with* $h(e) \in \mathbb{C}^N$ *as a cyclic vector.*

(c) *We have the solution formulas*

$$f(x) = g(e)^t \rho(x) h(e), \quad g(x) = \rho(x)^t g(e), \quad h(x) = \rho(x)h(e) \quad (5.2)$$

for all $x \in M$*. In particular* f *and the components of* g *and* h *are matrix-coefficients of the* N*-dimensional representation* ρ*.*

(d) $C(\rho) = \operatorname{span}\{y \mapsto f(x_0 y y_0) \mid x_0, y_0 \in M\}$*, where* $C(\rho) := \operatorname{span}\{\rho_{ij} \mid i,j = 1,2,\dots,N\}$ *denotes the space of matrix-coefficients of* ρ*.*

Proof. (a) To see the inclusion $\subseteq$ note that $R(y)f = \sum_{l=1}^N h_l(y) g_l$ by (5.1). The reverse inclusion is immediate from Corollary C.6(b), when we recall that $h_1, \dots, h_N$ are linearly independent.

(b) Since $R(xy) = R(x)R(y)$ and $R(e) = I$ we get from the correspondence between operators and matrices that $\rho(xy) = \rho(x)\rho(y)$ and $\rho(e) = I$, which shows that ρ is a representation of M on $\mathbb{C}^N$.

Let $\Phi : V \to \mathbb{C}^N$ denote the isomorphism

$$\Phi\left(\sum_{l=1}^N \xi_l g_l\right) := (\xi_1, \xi_2, \dots, \xi_N)^t, \qquad (\xi_1, \xi_2, \dots, \xi_N)^t \in \mathbb{C}^N$$

of V onto $\mathbb{C}^N$. Then $\rho(x) = \Phi \circ R(x) \circ \Phi^{-1}$ for all $x \in M$. This shows ρ is equivalent to $R|_V$ under $\Phi : V \to \mathbb{C}^N$. $f = \sum_{l=1}^N h_l(e) g_l \in V$ is cyclic for $R|_V$, hence $\Phi(f) = (h_1(e), h_2(e), \dots, h_N(e))^t = h(e) \in \mathbb{C}^N$ is cyclic for ρ.

(c) By definition of ρ we get $g_i(yx) = \sum_{j=1}^N \rho_{ji}(x) g_j(y)$. Taking $y = e$ we get that $g_i(x) = \sum_{j=1}^N \rho_{ji}(x) g_j(e)$ and thus $g(x) = \rho(x)^t g(e)$ as desired.

Putting $y = e$ in equation (5.1) we find that $f(x) = g(x)^t h(e) = (\rho(x)^t g(e))^t h(e) = g(e)^t \rho(x) h(e)$ as desired.

For any $x, y \in M$ we compute that

$$\begin{aligned} g(x)^t \rho(y) h(e) &= [\rho(x)^t g(e)]^t \rho(y) h(e) = g(e)^t \rho(x)\rho(y) h(e) \\ &= g(e)^t \rho(xy) h(e) = f(xy) = g(x)^t h(y). \end{aligned}$$

Lemma C.2(c) tells us that $\rho(y)h(e) = h(y)$, since $g_1, g_2, \dots, g_N$ are linearly independent.

(d) $W := \operatorname{span}\{y \mapsto f(x_0 y y_0) \mid x_0, y_0 \in M\}$ is contained in $C(\rho)$ because

$$f(x_0 x y_0) = g(e)^t \rho(x_0 x y_0) h(e) = [\rho(x_0)^t g(e)]^t \rho(x) [\rho(y_0) h(e)].$$

To get the converse inclusion we note first that $\dim W \leq \dim C(\rho) \leq N^2 < \infty$, so that W is finite dimensional. Then $W = C(R|W)$ by Proposition E.16(b). Finally, we get using Lemma E.3 that $C(\rho) = C(R|V) \subseteq C(R|W) = W$. □

A partial converse of Theorem 5.2 holds:

Proposition 5.3. *Let S be a semigroup and let $N \in \mathbb{N}$. Let $\rho : S \to \mathrm{Mat}(N \times N, \mathbb{C})$ satisfy $\rho(xy) = \rho(x)\rho(y)$ for all $x, y \in S$. Then for any $\eta, \zeta \in \mathbb{C}^N$ the following triple of functions on S*

$$f(x) := \eta^t \rho(x)\zeta, \; g(x) := \rho(x)^t \eta, \; h(x) := \rho(x)\zeta, \qquad x \in S,$$

is a solution of Levi-Civita's functional equation (5.1).

Our underlying semigroup is not assumed abelian in the definition of the Levi-Civita functional equation. However, sometimes Kannappan's condition (see Section B) is satisfied, so that the functions are abelian:

Lemma 5.4. *Let $f, g_1, \ldots, g_N, h_1, \ldots, h_N : M \to \mathbb{C}$ be a solution of Levi-Civita's functional equation* (5.1) *on a monoid M.*

(a) *If $h = (h_1, \ldots, h_N)^t$ or $g = (g_1, \ldots, g_N)^t$ is central, then f is abelian.*

(b) *Conversely, assume f is abelian. If both $\{g_1, \ldots, g_N\}$ and $\{h_1, \ldots, h_N\}$ are linearly independent, then each of the functions $g_1, \ldots, g_N, h_1, \ldots, h_N$ is abelian. In this case $\rho(x)\rho(y) = \rho(y)\rho(x)$ for all $x, y \in M$.*

Proof. Immediate from the definition of Kannappan's condition and Corollary C.6. For the last statement note that $R(xy) = R(yx) : V \to V$ in the notation of Theorem 5.2. □

Theorem 5.2 exhibits the general form of the solutions with no ties between the functions except for the equation. But given a Levi-Civita's functional equation of a special form like the sine addition formula (4.3) (where $f = g_1$ etc), Theorem 5.2 is not the final word about its solutions. There will be a possible linear dependence between the functions on the right hand side of (5.1) to take into account. Even having done so, further investigations are needed, because Theorem 5.2 just sets up some guidelines about how to proceed, and explicit formulas for the representation ρ may be hard to obtain. Of course, the very fact that $\rho(xy) = \rho(x)\rho(y)$ provides us with useful relations between the functions on the right hand side of (5.1).

These relations depend on the form of the special instance of the general equation (5.1) that we consider (an example of such an instance being (5.3)), but are independent of the internal structure of the group (be it abelian, compact, semisimple or whatever), so they hold on any group. Often the equation cannot be solved by a single procedure on all types of groups, even when you take the relations into account, so to continue you may here have to use internal properties of the group, for example that it is abelian (Exercise 5.5) or compact (Exercise 5.11), or maybe you must do some ad hoc computations (Example 5.11). The remarks above are illustrated by our treatment of the sine addition formula in Exercise 5.12. Other illustrations are presented in Exercises 5.5 and 5.11.

5.3 Regularity of the solutions

Theorem 5.2 was formulated and proved algebraically without any regularity assumptions. That is remedied in Corollary 5.5 where we discuss continuity and differentiability of the solutions. The results can be applied to Lie groups like $G = \mathbb{R}$. Differentiability is useful, because it allows us to transfer problems about functional equations to differential equations. And mathematicians possess a rich stock of methods with which they can handle differential equations.

Corollary 5.5. *Let G be a locally compact group. Let $f : G \to \mathbb{C}$ solve Levi-Civita's functional equation* (5.1) *for some $g_1, \dots, g_N, h_1, \dots, h_N : G \to \mathbb{C}$.*

(a) *If f is measurable with respect to a left Haar measure on G, then f is a matrix-coefficient of a continuous finite-dimensional representation of G. In particular $f \in C(G)$.*

(b) *Let G be a Lie group. If f is measurable with respect to a left Haar measure on G, then f is a matrix-coefficient of a C^∞ finite-dimensional representation of G. In particular $f \in C^\infty(G)$.*

We assume in the rest of this corollary that both $g_1, \dots, g_N$ and $h_1, \dots, h_N$ are linearly independent, and let ρ be the representation in Theorem 5.2(b).

(c) *f is continuous $\Leftrightarrow$ g is continuous $\Leftrightarrow$ h is continuous $\Leftrightarrow$ ρ is continuous.*

(d) *Let G is a Lie group. Then $f \in C^\infty \Leftrightarrow g \in C^\infty \Leftrightarrow h \in C^\infty \Leftrightarrow \rho \in C^\infty$.*

Proof. (a) and (b) We are done if $f = 0$, so we may assume that $g_1, \dots, g_N$ and $h_1, \dots, h_N : S \to \mathbb{C}$ are linearly independent. Let ρ be the representation in Theorem 5.2(b). It follows from Theorem 5.2(d) that ρ is measurable and hence continuous (Theorem D.4(a)). And then the formula (5.2) shows f is continuous.

If G is a Lie group we refer to Proposition D.6.

(c) Using Theorem 5.2(d) we see that f continuous implies ρ continuous. And ρ continuous implies by (5.2) that g and h are continuous. Conversely, if g is continuous, then so is f (take $y = e$ in (5.1)). Similarly if h is continuous.

(d) is proved in the same manner as (c). □

On a topological, abelian group G the continuous solutions of (5.1) (assuming that both $g_1, \dots, g_N$ and $h_1, \dots, h_N$ are linearly independent) are particularly nice: They are *normal exponential polynomials*, i.e., linear combinations of functions of the form χP, where χ is a continuous character of G and P is a polynomial in continuous, additive functions on G. Indeed, we get from Theorem 5.2 that

$$\operatorname{span}\{R(y)f \mid y \in G\} = \operatorname{span}\{g_1, g_2, \dots, g_N\} = \operatorname{span}\{h_1, h_2, \dots, h_N\},$$

to which we apply a result from spectral synthesis ([197, Theorem 10.1]) that says:

Theorem 5.6. *Any finite dimensional, translation invariant linear space of continuous, complex valued functions on a topological abelian group consists of normal exponential polynomials.*

The formulas in Theorem 5.2(c) combined with Theorem 5.7 give a direct proof that continuous solutions of the Levi-Civita functional equation on $\mathbb{R}^n$ are normal exponential polynomials and as such very smooth.

Theorem 5.7. *Matrix-coefficients of continuous, finite-dimensional representations of $\mathbb{R}^n$ are normal exponential polynomials, i.e., linear combinations of functions on $\mathbb{R}^n$ of the form $x \mapsto p(x)e^{\langle \lambda, x \rangle}$, where $\lambda \in \mathbb{C}^n$ and p is a polynomial.*

Proof. Let ρ be a continuous representation of $\mathbb{R}^n$ on $\mathbb{C}^m$. Since

$$\rho(x_1, x_2, \dots, x_n) = \rho(x_1, 0, \dots, 0)\, \rho(0, x_2, \dots, 0) \cdots \rho(0, 0, \dots, x_n)$$

we may assume that $n = 1$. Being C^∞ (by Proposition D.6) ρ has the form

$$\rho(x) = e^{xA}, \qquad \text{where } A \in M(m \times m, \mathbb{C}).$$

The Jordan normal form reveals that $A = S + N$, where S is diagonalizable, N nilpotent and $SN = NS$ (see for instance [104, Exercise 4 of §57]). We may assume S is a diagonal matrix with, say, $\alpha_1, \alpha_2, \ldots, \alpha_m$ in the diagonal. $N^l = 0$ for some $l \in \mathbb{N}$. Now the computation

$$\rho(x) = e^{xA} = e^{xS+xN} = e^{xS}e^{xN} = \begin{pmatrix} e^{\alpha_1 x} & \ldots & 0 \\ \vdots & \ddots & \vdots \\ 0 & \ldots & e^{\alpha_m x} \end{pmatrix} \sum_{k=0}^{l-1} x^k \frac{N^k}{k!}$$

proves the theorem. □

5.4 Two special cases

The two innocently looking Levi-Civita functional equations

$$f(xy) = f(x)h(y) + f(y), \qquad x, y \in S, \tag{5.3}$$

$$f(xy) = f(x) + g(x)f(y), \qquad x, y \in S, \tag{5.4}$$

where $f, h : S \to \mathbb{C}$, resp. $f, g : S \to \mathbb{C}$ are the unknowns on the semigroup S, occur often (see the exercises). Therefore we shall study them in this section. They are closely related on groups: If (f, h) is a solution of (5.3), then $(\check{f}, g = \check{h})$ is a solution of (5.4) and vice versa. So on groups it suffices to study (5.3). The special case of $h = 1$ is the additive Cauchy equation the solutions of which vary very much from group to group: You might compare Corollary 2.4 with Example 2.12. So its solutions must be discussed for each type of group individually. The functional equation (5.3) is even more complicated than the additive Cauchy equation: The solutions of the additive Cauchy equation are central on any group, but this is not generally so for (5.3): In Example 5.11 the function f is not central in (5.3). That is the reason why Proposition 5.8 only gives the complete solution of (5.3) for f central which is of course the case if the underlying semigroup S is abelian. The equations also illustrate the fact that it is usually much harder to solve functional equations on non-abelian groups than on abelian, also when the number of terms is small.

Proposition 5.8. *Let S be a semigroup, and let $f, h : S \to \mathbb{C}$.*

(a) *If the pair (f, h) satisfies (5.3) and $f \neq 0$, then h is multiplicative.*

(b) *If $h = 1$ then the solutions f of (5.3) are the additive functions.*

(c) *If h is multiplicative, then $f = c(h-1)$ is a solution of (5.3) for any $c \in \mathbb{C}$.*

(d) *For fixed multiplicative h, the functional equation (5.3) has only the following central solutions f: If $h = 1$ the additive functions, and if $h \neq 1$ the functions of the form $f = c(h-1)$ where $c \in \mathbb{C}$ is an arbitrary constant.*

Proof. We do not use Theorem 5.2 for the proof, because it is easier to proceed directly.

(a) To see that h is multiplicative exploit that $f(x(yz)) = f((xy)z)$ for all $x, y, z \in S$. We leave the simple computations to the reader.

(b) and (c) are left to the reader.

(d) That f is central implies that $f(x)h(y) + f(y) = f(y)h(x) + f(x)$, from which we get the identity $f(x)[h(y) - 1] = f(y))[h(x) - 1]$. If $h \neq 1$ we choose $y_0 \in S$ such that $h(y_0) \neq 1$ and get from the identity that $f = c(h-1)$, where $c = f(y_0)/(h(y_0) - 1)$. If $h = 1$ we get from (5.3) that f is additive. □

Example 5.9. By help of Proposition 5.8 we find that the continuous solutions $f, h : \mathbb{R} \to \mathbb{C}$ of the functional equation

$$f(x+y) = f(x)h(y) + f(y), \qquad x, y \in \mathbb{R},$$

are

(a) $h \in C(\mathbb{R})$ arbitrary and $f = 0$.

(b) $h = 0$ and f a non-zero constant.

(c) $h = 1$ and $f(x) = \alpha x$ for some $\alpha \in \mathbb{C} \setminus \{0\}$.

(d) $h(x) = e^{\lambda x}$ where $\lambda \in \mathbb{C} \setminus \{0\}$, and $f(x) = c(e^{\lambda x} - 1)$ where $c \in \mathbb{C} \setminus \{0\}$.

It is a heuristic principle that nilpotent groups behave very much like abelian groups, so that properties of abelian groups often extend to nilpotent groups. This principle holds for solutions of the functional equation (5.3) as shown by Proposition 5.10.

Proposition 5.10. *If $f, h : G \to \mathbb{C}$ is a solution of* (5.3) *on a nilpotent group G, then f is central.*

Proof. Let $\mathcal{C}^0G \supseteq \mathcal{C}^1G \supseteq \cdots \supseteq \mathcal{C}^nG = \{e\}$ be the descending central series for G with $n \in \mathbb{N}$ is chosen so that $\mathcal{C}^{n-1}G \neq \{e\}$. The proof goes by induction on the length n of the descending central series. The proposition is true if $n = 0$ or $n = 1$, because in those cases G is abelian. So assume that the lemma holds for some $n \geq 1$. We shall then prove it when the length of the descending central series is $n+1$. In this case $\mathcal{C}^nG \subseteq Z(G)$. We may assume that $f \neq 0$, $h \neq 0$ and $h \neq 1$.

If there exists a $z_0 \in Z(G)$ such that $h(z_0) \neq 1$ then we find for any $x \in G$ from $f(xz_0) = f(z_0x)$ that $f(x)h(z_0) + f(z_0) = f(z_0)h(x) + f(x)$, which implies that

$$f(x) = \frac{f(z_0)}{h(z_0) - 1}\,(h(x) - 1).$$

Since h is central, so is f.

We may thus assume that $h = 1$ on $Z(G)$. Since $h \neq 1$ there exists an $x_0 \in G$ such that $h(x_0) \neq 1$. For any $z \in Z(G)$ we get from $f(x_0z) = f(xz_0)$ that $f(z) = 0$. Now, for any $x \in G$ and $z \in Z(G)$ we find $f(xz) = f(x)h(z) + f(z) = f(x)$, which shows that f is a function on $G/Z(G)$. Clearly so is h, because $h = 1$ on $Z(G)$. It follows that (f, h) is a solution of (5.3) on $G/\mathcal{C}^nG$. The central descending series of this quotient group is

$$\mathcal{C}^0G/\mathcal{C}^nG \supseteq \mathcal{C}^1G/\mathcal{C}^nG \supseteq \cdots \supseteq \mathcal{C}^nG/\mathcal{C}^nG = \{e\mathcal{C}^nG\}$$

so it has length n, and so the induction hypothesis applies. □

Example 5.11. We can solve the functional equation (5.3) on abelian and more generally on nilpotent groups because f is central on such groups. But f is not central in general: Consider the $(ax+b)$-group G from Examples A.17(i), where we for brevity write

$$(a, b) := \begin{pmatrix} a & b \\ 0 & 1 \end{pmatrix} \qquad \text{for } a > 0 \text{ and } b \in \mathbb{R}.$$

The pair of functions $f(a, b) := a^{-1}b$, $h(a, b) := a^{-1}$, $(a, b) \in G$, satisfies (5.3). The corresponding representation ρ of G on $\mathbb{C}^2$ is

$$\rho\begin{pmatrix} a & b \\ 0 & 1 \end{pmatrix} = \begin{pmatrix} 1/a & 0 \\ b/a & 1 \end{pmatrix}.$$

But f is not central. (Exercise 5.14 lists all continuous solutions of (5.3) on the $(ax+b)$-group.)

For results about (5.3) on semidirect products see Exercise 5.14 and the comment to it.

5.5 Exercises

Exercise 5.1.

(a) Show that the R-invariant, 1-dimensional subspaces of the complex functions on $\mathbb{R}$ are the spaces $\mathbb{C}\chi$, where χ ranges over the characters of $\mathbb{R}$.

(b) Show that the R-invariant, 1-dimensional subspaces of $C(\mathbb{R})$ are the spaces $\mathbb{C}\chi$, where χ ranges over the continuous characters of $\mathbb{R}$.

(c) Let $f \in C(\mathbb{R})$ be a joint eigenvector for all $R(y)$, $y \in \mathbb{R}$, i.e., $f \neq 0$ and there is a function $\chi : \mathbb{R} \to \mathbb{C}$ such that $R(y)f = \chi(y)f$ for all $y \in \mathbb{R}$. Show that f has the form $f(x) = ce^{\lambda x}$, $x \in \mathbb{R}$, where $c, \lambda \in \mathbb{C}$ are constants. What is the χ corresponding to this f?

Exercise 5.2. Let us for any $\lambda \in \mathbb{C}$ define $e_\lambda(x) := e^{\lambda x}$, $x \in \mathbb{R}$. Let $\lambda_1, \ldots, \lambda_N \in \mathbb{C}$. Show that the subspace $\operatorname{span}\{e_{\lambda_1}, \ldots, e_{\lambda_N}\}$ of $C(\mathbb{R})$ is invariant under the right regular representation. What is its dimension?

Exercise 5.3. Let M be a monoid. Let $\{f, g, h\}$ be a solution of (5.1) such that both $g_1, \ldots, g_N$ and $h_1, \ldots, h_N$ are linearly independent. Show that f is bounded $\Leftrightarrow$ g is bounded $\Leftrightarrow$ h is bounded $\Leftrightarrow$ ρ is bounded.

Exercise 5.4. Let M be an abelian monoid. Let $\{f, g, h\}$ be a solution of (5.1) such that both $g_1, \ldots, g_N$ and $h_1, \ldots, h_N$ are linearly independent. Show that $\operatorname{span}\{g_1, g_2, \ldots, g_N\} = \operatorname{span}\{h_1, h_2, \ldots, h_N\}$.

Exercise 5.5. Find the continuous complex-valued solutions of the functional equation

$$f(x + y) = f(x)h_1(y) + h_2(y), \qquad x, y \in \mathbb{R}. \tag{5.5}$$

Proceed as follows: Consider until further notice the more general functional equation

$$f(xy) = f(x)h_1(y) + h_2(y), \qquad x, y \in G, \tag{5.6}$$

where G is any group. It contains the additive and the multiplicative Cauchy equations and (5.3) as special instances.

(a) Discuss first the case of f constant.

Answer: $(f, h_1, h_2) = (\alpha, h, \alpha(1-h))$, where $\alpha \in \mathbb{C}$ is an arbitrary constant and h is any function on G.

(b) Solve next the case where f is not constant, and where h_1 and h_2 are proportional.

Answer: There exist an $\alpha \in \mathbb{C} \setminus \{0\}$ and a character $\chi \neq 1$ on G such that $f = \alpha\chi$, $h_1 = \chi$ and $h_2 = 0$.

From now on we assume that f is not constant and that h_1 and h_2 are linearly independent.

(c) Show that the matrix ρ from Theorem 5.2(b) is

$$\rho(x) = \begin{pmatrix} h_1(x) & 0 \\ h_2(x) & 1 \end{pmatrix}, \qquad x \in G.$$

(d) Conclude that h_1 is a character, and that

$$h_2(xy) = h_2(x)h_1(y) + h_2(y) \qquad \text{for all } x, y \in G. \tag{5.7}$$

(e) Assuming furthermore that G is abelian, solve the equation (5.6).

Answer:

(i) $(f, h_1, h_2) = (\beta\chi - \alpha, \chi, \alpha(\chi - 1))$, where $\alpha, \beta \in \mathbb{C} \setminus \{0\}$ are arbitrary constants and $\chi \neq 1$ is a character on G, or

(ii) $(f, h_1, h_2) = (a + \alpha, 1, a)$, where $\alpha \in \mathbb{C}$ is an arbitrary constant and $a : G \to \mathbb{C}$ is an additive function, $a \neq 0$.

(f) Find the continuous solutions of (5.5).

Exercise 5.6. Let G be a group, let $N \in \mathbb{N}$ and consider the Levi-Civita equation

$$f(xy) = \sum_{l=1}^{N} h_l^\dagger(x)h_l(y), \qquad x, y \in G,$$

where $\phi^\dagger(x) := \overline{\phi(x^{-1})}$, $x \in G$, for any function $\phi : G \to \mathbb{C}$. We assume that the functions $h_1, h_2, \ldots, h_N$ are linearly independent.

Note that $g = h^\dagger$ in the notation of (5.1). Let ρ be the representation of G on $\mathbb{C}^N$ from Theorem 5.2(b).

(a) Show that ρ is unitary, i.e., that $\rho(x^{-1})^* = \rho(x)$ for all $x \in G$.

Hint: Use the solution formulas (5.2) to infer that $\rho(x)h(y) = \rho(x^{-1})^*h(y)$ for all $x, y \in G$, and use that $\operatorname{span}\{h(y) \mid y \in G\} = \mathbb{C}^N$ by Lemma C.2(c).

(b) Assume furthermore that G is abelian. Show that there exist a unitary $N \times N$ matrix U, a vector $\zeta \in \mathbb{C}^N \setminus \{0\}$ and N different unitary characters $\chi_1, \chi_2, \dots, \chi_N$ on G such that

$$f(x) = \left\langle \begin{pmatrix} \chi_1(x) & \dots & 0 \\ \vdots & \ddots & \vdots \\ 0 & \dots & \chi_N(x) \end{pmatrix} \zeta, \zeta \right\rangle, \qquad x \in G,$$

and

$$h(x) = U \begin{pmatrix} \chi_1(x) & \dots & 0 \\ \vdots & \ddots & \vdots \\ 0 & \dots & \chi_N(x) \end{pmatrix} \zeta, \qquad x \in G.$$

Hint: Since $\{\rho(x) \mid x \in G\}$ is a commuting set of normal matrices, its elements can be simultaneously diagonalized in the sense that there exists a unitary $N \times N$ matrix U and maps $\chi_1, \chi_2, \dots, \chi_N : G \to \mathbb{C}$ such that

$$\rho(x) = U \begin{pmatrix} \chi_1(x) & \dots & 0 \\ \vdots & \ddots & \vdots \\ 0 & \dots & \chi_N(x) \end{pmatrix} U^{-1}, \qquad x \in G.$$

Exercise 5.7. Let G be an abelian group. Consider the Levi-Civita equation

$$f(xy) = f(x) + f(y) + g(x)h(y), \qquad x, y \in G, \tag{5.8}$$

where $f, g, h : G \to \mathbb{C}$ are functions to be determined. We disregard the cases of linear dependence, so we assume that f, 1 and g are linearly independent and that so are 1, f and h.

(a) Show that the matrix-valued function ρ from Theorem 5.2(b) is

$$\rho(y) = \begin{pmatrix} 1 & 0 & \rho_{13}(y) \\ f(y) & 1 & \rho_{23}(y) \\ h(y) & 0 & \chi(y) \end{pmatrix}, \qquad y \in G,$$

for some functions $\rho_{13}, \rho_{23}, \chi : G \to \mathbb{C}$.

(b) Infer from (5.8) and $\rho(xy) = \rho(x)\rho(y)$ that $\rho_{13} = 0$ and $\rho_{23} = g$. Conclude that we in addition to the original equation (5.8) get that

$$g(xy) = g(x)\chi(y) + g(y),$$
$$h(xy) = h(x) + \chi(x)h(y)$$

and

$$\chi(xy) = \chi(x)\chi(y) \qquad \text{for all } x, y \in G.$$

(c) Note that $\chi \neq 0$, because $\chi = 0$ implies that h is constant. Conclude that χ is a character of G.

(d) Complete the exercise using that G is abelian. (The result of Exercise 2.12 might be useful.)

Answer: There are two types of solutions:

(i) There exist constants $\alpha, \beta \in \mathbb{C} \setminus \{0\}$, an additive function $a : G \to \mathbb{C}$ and a character χ of G, $\chi \neq 1$, such that for all $x \in G$

$$f(x) = \alpha\beta(\chi(x) - 1) + a(x), \qquad g(x) = \alpha(\chi(x) - 1),$$
$$h(x) = \beta(\chi(x) - 1).$$

(ii) There exist two additive functions $a_0, a : G \to \mathbb{C}$ with $a_0 \neq 0$ and a constant $c \in \mathbb{C} \setminus \{0\}$ such that

$$f = \tfrac{c}{2}a_0^2 + a, \quad g = a_0, \quad h = ca_0.$$

Exercise 5.8. Recall that we in Exercise 3.24 made $P := \mathbb{R} \setminus \{-1\}$ into an abelian group by the composition rule $x \circ y := x + y + xy$ for $x, y \in \mathbb{R}$. Consider the functional equation

$$f(x \circ y) = p(x) + q(y) + g(x)h(y), \qquad x, y \in P,$$

where $f, p, q, g, h : P \to \mathbb{C}$ are the unknown functions.

(a) Show that

$$F(x \circ y) = F(x) + F(y) + G(x)H(y) \qquad \text{for all } x, y \in P,$$

where $F = f - f(0)$, $G = g - g(0)$ and $H = h - h(0)$.

(b) Interchange x and y to relate G and H.

(c) Finish the exercise by using the result of Exercise 5.7.

Exercise 5.9. Find the continuous solutions of (5.4) on $G = \mathbb{R}$.

Exercise 5.10.

(a) Find the continuous solutions of the functional equation

$$f(xy) = f(x)h(y) + f(y), \qquad x, y \in \mathbb{R}^*.$$

Hint: The equation is (5.3).

(b) Find the continuous solutions of the functional equation

$$f(xy) = f(x) + g(x)f(y), \qquad x, y \in \mathbb{R}^*.$$

Exercise 5.11. Let G be a compact group. Find the continuous solutions $f, h_1, h_2 \in C(G)$ of

$$f(xy) = f(x)h_1(y) + h_2(y), \qquad x, y \in G.$$

Hint: Integrate the functional equation over G with respect to the normalized Haar measure dx to get that $h_2 = \alpha(1 - h_1)$ for some $\alpha \in \mathbb{C}$.

Answer:

(i) $(f, h_1, h_2) = (\alpha + \beta h, h, \alpha(1 - h))$, where $h \in C(G)$ is multiplicative, and $\alpha, \beta \in \mathbb{C}$ are arbitrary constants, and

(ii) $(f, h_1, h_2) = (\alpha, h, \alpha(1 - h))$ where $\alpha \in \mathbb{C}$ and $h \in C(G)$ are arbitrary.

Exercise 5.12. Let G be a group. We shall in this exercise treat the sine addition formula (4.3), i.e.,

$$f(xy) = f(x)g(y) + g(x)f(y) \qquad \text{for all } x, y \in G.$$

So we shall rederive Theorem 4.1 by applying the theory of the Levi-Civita functional equation to the special case of (4.3).

(a) Solve (4.3) when f and g are linearly dependent.

Assume from now on that f and g are linearly independent.

(b) Show that the matrix ρ from Theorem 5.2(b) has the form

$$\rho(y) = \begin{pmatrix} g(y) & \rho_{12}(y) \\ f(y) & \rho_{22}(y) \end{pmatrix} \qquad \text{for } y \in G,$$

where $\rho_{12}, \rho_{22} : G \to \mathbb{C}$.

(c) Using the functional equation and that ρ is a homomorphism to show that $\rho_{22} = g$ and that

$$g(xy) = g(x)g(y) + \rho_{12}(x)f(y) = f(x)\rho_{12}(y) + g(x)g(y).$$

Infer that there is a constant $\alpha \in \mathbb{C}$ such that $\rho_{12} = \alpha^2 f$ (cf. the formula (4.4)).

(d) Conclude that

$$\rho = \begin{pmatrix} g & \alpha^2 f \\ f & g \end{pmatrix}.$$

(e) Assume first that $\alpha \neq 0$. Show that ρ can be diagonalized. More precisely that

$$\begin{pmatrix} \alpha & -\alpha \\ 1 & 1 \end{pmatrix}^{-1} \rho \begin{pmatrix} \alpha & -\alpha \\ 1 & 1 \end{pmatrix} = \begin{pmatrix} g+\alpha f & 0 \\ 0 & g-\alpha f \end{pmatrix}.$$

(f) Finish the proof of Theorem 4.1 for $\alpha \neq 0$.

(g) Finish the proof of Theorem 4.1 by treating the case of $\alpha = 0$: Note by help of the formula in (d) that g is a character in this case.

Exercise 5.13. Find the solutions $f, g_1, h_1, h_2 \in C(\mathbb{R})$ of the functional equation

$$f(x+y) = g_1(x)h_1(y) + h_2(y), \qquad x, y \in \mathbb{R}.$$

Hint: Try to reduce it to (5.3). There will be a number of special cases.

Exercise 5.14. In the present exercise we derive formulas for the solutions of (5.3) on a semidirect product $G = N \times_s Q$ like in Definition A.18 and apply these formulas to the $(ax+b)$-group. Recall that the $(ax+b)$-group is a semidirect product (Example A.22(b)). Let $f, h : G \to \mathbb{C}$ be a solution of (5.3).

(a) Prove that $f(nq) = f(n)h(q) + f(q)$ for all $n \in N$ and $q \in Q$.

(b) Derive that $f(qnq^{-1})h(q) - f(n) = f(q)[h(n) - 1]$ for all $n \in N$ and $q \in Q$.

(a) and (b) are valid on any semidirect product.

We next consider a particular case by asking the reader to find all continuous solutions (f, h) of the functional equation (5.3) on the $(ax+b)$-group.

Answer: There are the following five cases:

1. $f = 0$. In this case h is an arbitrary continuous function.
2. f is a constant $\neq 0$. In this case $h = 0$.
3. $f(a,b) = c(a^s - 1)$ and $h(a,b) = a^s$ where $c, s \in \mathbb{C}$ are arbitrary constants.
4. $f(a,b) = c \log a$ where $c \in \mathbb{C}$ is an arbitrary constant, and $h = 1$.
5. $f(a,b) = Ca^{-1}b + c(a^{-1} - 1)$ where $C, c \in \mathbb{C}$ are arbitrary constants, and $h(a,b) = a^{-1}$.

Hints: Assuming $f \neq 0$ and $h \neq 0$ then h is a character by Proposition 5.8(a). The continuous characters on the $(ax+b)$-group are according to Example 3.13 the functions of the form $h(a,b) = a^\lambda$, where $\lambda \in \mathbb{C}$. Start by showing that the relations (a) and (b) for solutions on the $(ax+b)$-group reduce to the following for all $a > 0$ and $x \in \mathbb{R}$ when $h \neq 0$:

$$f\begin{pmatrix} a & x \\ 0 & 1 \end{pmatrix} = f\begin{pmatrix} 1 & x \\ 0 & 1 \end{pmatrix} h\begin{pmatrix} a & 0 \\ 0 & 1 \end{pmatrix} + f\begin{pmatrix} a & 0 \\ 0 & 1 \end{pmatrix},$$

and

$$f\begin{pmatrix} 1 & x \\ 0 & 1 \end{pmatrix} = f\begin{pmatrix} 1 & ax \\ 0 & 1 \end{pmatrix} h\begin{pmatrix} a & 0 \\ 0 & 1 \end{pmatrix}.$$

Proposition 5.8(d) will be handy here, because the factors N and Q are abelian in the case of the $(ax+b)$-group.

5.6 Notes and remarks

Let $N \in \mathbb{N}$, and let X and Y be non-empty sets. A decomposition of a function $F : X \times Y \to \mathbb{C}$ into the form

$$F(x,y) = \sum_{l=1}^{N} g_l(x)h_l(y), \qquad (x,y) \in X \times Y,$$

in which both $g_1, \dots, g_N$ and $h_1, \dots, h_N$ are linearly independent is called a *minimal decomposition.* It is essentially unique by Rassias and Šimša [163, Section 2.3]. In particular the linear independence guarantees that N cannot be replaced by another $N \in \mathbb{N}$.

An and Yang [17] call the functional equation

$$f(xy) + f(yx) = \sum_{l=1}^{N} g_l(x)h_l(y), \qquad x, y \in G,$$

where G is a compact group, Levi-Civita's functional equation.

Chung, Kannappan and Ng [45] solve the Levi-Civita equation

$$f(xy) = f(x)g(y) + f(y)g(x) + h(x)h(y), \qquad x, y \in G,$$

where G is a group. The results of Friis and Stetkær [89] extend those of [45] for G abelian, replacing the left hand side by $(f(x+y) + f(x-y))/2$.

The functional equation (5.1) was for each $N \leq 4$ thoroughly worked out on abelian groups with explicit formulas for the solutions by Székelyhidi [197, Chapter 10]. See in particular [197, Theorem 10.4].

Theorem 5.2 and Corollary 5.5 are expansions of Shulman [171, Lemma 4].

Theorem 5.7: Shulman [173] studied regularity of the solutions of the following generalization of Levi-Civita's equation

$$\sum_{i=1}^{m} a_i(g)b_i(hg) = \sum_{j=1}^{n} u_j(g)v_j(h), \qquad g, h \in G,$$

where G is a connected topological group, and $a_i, b_i, u_j, v_j \in C(G)$ for $i = 1, \ldots, m$ and $j = 1, \ldots, n$. A typical result is [173, Corollary 3.1]: If the functions $a_1, a_2, \ldots, a_m$ are linearly independent then $b_1, b_2, \ldots, b_m$ are matrix coefficients of a continuous finite-dimensional representation of G; for $G = \mathbb{R}^n$ this means that they are normal exponential polynomials (her terminology is quasipolynomials) on $\mathbb{R}^n$.

Equation (5.3): This functional equation was in the abelian case studied by Aczél and Dhombres [6, Ch. 15] in connection with the generalization $f(x+y) = g(x)h_1(y) + h_2(y)$ of Pexider's functional equation $f(x+y) = g(x) + h(y)$. Solutions on semidirect products of groups, that need not be abelian, have been studied by Bruce Ebanks (private communication). See Exercise 5.14 for a couple of his general results.

Exercise 5.6: See O'Connor [152, Theorem] and Gajda [94, Theorem 2]. They allow $h_1, h_2, \ldots, h_N$ to be linearly dependent.

Exercise 5.7: Exercise 2.16 is a special case of Exercise 5.7.

Exercise 5.13: The equation in the exercise is known as the first Vincze functional equation. The second Vincze functional equation is

$$f(xy) = g_1(x)h_1(y) + h_2(y), \qquad x, y \in \mathbb{R}.$$

Exercise 5.14: See the above comment to the functional equation (5.3).

Chapter 6

The Symmetrized Sine Addition Formula

We will in this chapter study the symmetrized sine addition formula (6.1). It is interesting, because it has connections to many other functional equations, for instance to the pre-d'Alembert functional equation (see Lemma 8.8).

The most important result of the present chapter is Lemma 6.1, due to its many useful algebraic identities for solutions of the symmetrized sine addition formula.

6.1 Introduction

The *symmetrized sine addition formula* is

$$w(xy) + w(yx) = 2w(x)g(y) + 2w(y)g(x), \qquad x, y \in S, \qquad (6.1)$$

where S is a topological semigroup and the unknown functions $w, g \in C(S)$.

When w is central the symmetrized sine addition formula (6.1) reduces to the sine addition formula (4.3) that we solved in Section 4.2. Hence its name. (6.1) is much harder to handle than the sine addition formula. Even on compact groups we encounter non-abelian phenomena like irreducible 2-dimensional representations (Theorem 6.9), which was not the case for the sine addition formula the solutions of which are abelian functions.

The case of $g = 1$ in the symmetrized sine addition formula (6.1) is the symmetrized additive Cauchy equation (2.3), which we discussed in Section 2.4. On groups its solutions are according to Proposition 2.17 the same as those solutions w of Jensen's functional equation $w(xy)+w(xy^{-1}) = 2w(x)$ for which $w(e) = 0$. In our studies of Jensen's functional equation in Chapter 12 we will present examples of non-abelian solutions w of (6.1), so (6.1) does not in general reduce to the sine addition formula (4.3).

Since an entire chapter is devoted to Jensen's functional equation and since the symmetrized sine addition formula contains Jensen's functional equation we must expect it has a rather long and complicated theory.

The symmetrized sine addition formula was solved for by Davison in [64] under the assumption of g being a non-abelian pre-d'Alembert function on a monoid. We describe his results in Chapter 8 (Theorem 8.24). On compact groups the symmetrized sine addition formula was solved by An and Yang [17, Theorem 5.9]. We cite their result below as Theorem 6.9.

We must impose conditions on w, g or S. For example, we obtain the form of the g part of the solutions (w, g) of (6.1) on a group assuming that g is abelian, but a complete description of the corresponding w's still eludes us. So the theory of (6.1) is not satisfactory except on compact and abelian groups, it has not yet been fully worked out, and so it is still in the messy stage. Thus the present chapter is a description of work in progress.

Some results in later chapters will be derived from properties of solutions of the symmetrized sine addition formula. For instance Lemma 8.22 in the proof of which we use Corollary 6.2. And Theorem 10.1(d) and (e) that follow from Proposition 6.5.

6.2 Key formulas and results

This section collects formulas for the solutions of the symmetrized sine addition formula on general semigroups and groups. The functions d, $\widetilde{\Delta}$ and Δ from Definition B.6 enter.

Lemma 6.1. *Let S be a semigroup. For any solution (w, g) of (6.1) we have for all $x, y, z \in S$ the formulas:*

$$w(x^2) = 2w(x)g(x), \tag{6.2}$$

$$\begin{aligned} w(xyz) = g(xy)w(z) + w(xy)g(z) + w(x)g(yz) \\ + w(yz)g(x) - w(zx)g(y) - w(y)g(zx), \end{aligned} \tag{6.3}$$

$$\begin{aligned} w(yzx) + w(xzy) = w(x)[g(yz) + g(zy)] + w(y)[g(xz) + g(zx)] \\ - w(z)[g(xy) + g(yx) - 4g(x)g(y)], \end{aligned} \tag{6.4}$$

$$w(xzx) = d(x)w(z) + w(x)[g(xz) + g(zx)], \tag{6.5}$$

$$w(x^2y) + w(yx^2) = 2w(xyx) - 2w(x)[g_x(y) + g_y(x)] + 4w(y)g_x(x), \tag{6.6}$$

and

$$\begin{aligned}
w(yx)[g(xyz) + g(zxy)] + w(xy)[g(yxz) + g(zyx)] \\
= w(x)[g(xyzy) + g(yzyx) + (g(xz) + g(zx))d(y)] \\
+ w(y)[g(yxzx) + g(xzxy) + (g(yz) + g(zy))d(x)] \\
+ w(z)\widetilde{\Delta}(x, y).
\end{aligned} \tag{6.7}$$

Furthermore, if $w \neq 0$, *then*

$$d(xy) - d(x)d(y) = \widetilde{\Delta}(x, y)/2 = 0, \qquad \text{when } x \text{ and } y \text{ commute,} \tag{6.8}$$

and in particular

$$d(x^n) = d(x)^n \qquad \text{for all } x \in S \text{ and } n \in \mathbb{N}. \tag{6.9}$$

Proof. Equation (6.2): Put $y = x$ in (6.1).

Equation (6.3): From (6.1) we get the two identities

$$\begin{aligned}
w(xyz) + w(yzx) &= 2w(x)g(yz) + 2w(yz)g(x), \\
w(yzx) + w(zxy) &= 2w(y)g(zx) + 2w(zx)g(y).
\end{aligned}$$

The first one comes about when we replace y by yz in (6.1). The second one is the first one except for a cyclic permutation of x, y, z. When we add the two identities and use (6.1) we obtain

$$\begin{aligned}
w(yzx) + w(xy)g(z) + w(z)g(xy) \\
= w(x)g(yz) + w(yz)g(x) + w(y)g(zx) + w(zx)g(y),
\end{aligned}$$

which is (6.3) except for a permutation of x, y and z.

Equation (6.4): When we add the just derived identity to the one we get by interchanging x and y we get that

$$\begin{aligned}
w(yzx) + w(xzy) + [w(xy) + w(yx)]g(z) + w(z)[g(xy) + g(yx)] \\
= w(x)g(yz) + w(y)g(xz) + g(x)[w(yz) + w(zy)] + w(y)g(zx) \\
+ w(x)g(zy) + g(y)[w(xz) + w(zx)],
\end{aligned}$$

that we by help of (6.1) reduce to (6.4).

Equation (6.5): When we in (6.4) put $y = x$ we get by the definition of d (Definition B.6) the identity (6.5).

Equation (6.6): Putting $z = x$ in (6.4) we obtain

$$\begin{aligned} w(yx^2) + w(x^2y) &= w(x)[g(yx) + g(xy)] + w(y)[g(x^2) + g(x^2)] \\ &\quad - w(x)[g(xy) + g(yx) - 4g(x)g(y)] \\ &= 4w(x)g(x)g(y) + 2w(y)g(x^2). \end{aligned}$$

Subtracting 2 times (6.5) with $z = y$ from this we get (6.6) as follows:

$$\begin{aligned} &w(x^2y) + w(yx^2) - 2w(xyx) \\ &\quad = 4w(x)g(x)g(y) + 2w(y)g(x^2) - 2d(x)w(y) - 2w(x)[g(xy) + g(yx)] \\ &\quad = -2w(x)[g(xy) + g(yx) - 2g(x)g(y)] + 2w(y)[g(x^2) - d(x)] \\ &\quad = -2w(x)[g_x(y) + g_y(x)] + 2w(y)[g(x^2) - 2g(x)^2 + g(x^2)] \\ &\quad = -2w(x)[g_x(y) + g_y(x)] + 4w(y)g_x(x). \end{aligned}$$

Equation (6.7): In (6.5) we replace z by yzy and get

$$w(xyzyx) = d(x)w(yzy) + w(x)[g(xyzy) + g(yzyx)],$$

in which we interchange x and y and get

$$w(yxzxy) = d(y)w(xzx) + w(y)[g(yxzx) + g(xzxy)].$$

When we add the last two identities and apply (6.5) we find that

$$\begin{aligned} &w(xyzyx) + w(yxzxy) \\ &\quad = w(x)[g(xyzy) + g(yzyx) + (g(xz) + g(zx))d(y)] \\ &\qquad + w(y)[g(yxzx) + g(xzxy) + (g(yz) + g(zy))d(x)] \\ &\qquad + 2w(z)d(x)d(y). \end{aligned}$$

When we compute the left hand side by help of (6.4), more precisely when we replace x by yx and y by xy in (6.4) we get

$$\begin{aligned} &w(yx)[g(xyz) + g(zxy)] + w(xy)[g(yxz) + g(zyx)] \\ &\qquad - w(z)[g(yx^2y) + g(xy^2x) - 4g(yx)g(xy)] \\ &\quad = w(x)[g(xyzy) + g(yzyx) + (g(xz) + g(zx))d(y)] \\ &\qquad + w(y)[g(yxzx) + g(xzxy) + (g(yz) + g(zy))d(x)] \\ &\qquad + 2w(z)d(x)d(y), \end{aligned}$$

which is a reformulation of (6.7).

Equation (6.8): Lemma B.7(d) gives us the identity $d(x)d(y) - d(xy) = \widetilde{\Delta}(x, y)/2$, so it is left to show either that $d(x)d(y) = d(xy)$ or that $\widetilde{\Delta}(x, y)$ vanishes. To do so we consider the subsemigroup $\langle x, y\rangle$ of S generated by x and y. It is abelian, so on it (6.1) reduces to (4.3) with $f = w$. If $w|_{\langle x,y\rangle} \neq 0$ we read from Theorem 4.1(f) that $d|_{\langle x,y\rangle} : \langle x, y\rangle \to \mathbb{C}$ is multiplicative, so that $d(x)d(y) = d(xy)$. If $w|_{\langle x,y\rangle} = 0$ then (6.7) reduces to $w(z)\widetilde{\Delta}(x, y) = 0$ for all $z \in S$. We get $\widetilde{\Delta}(x, y) = 0$, since $w \neq 0$. □

The following corollary, which admittedly is unmotivated here, will be needed at a crucial point later (to prove $\dim W(g) = 3$ in Lemma 8.22).

Corollary 6.2. *Let* (w, g) *be a solution of* (6.1) *on a semigroup* S*, and let* $a, b \in S$ *be so that* $\widetilde{\Delta}(a, b) \neq 0$*. Then*

$$w(a) = w(b) = w(ab) = 0 \quad \Rightarrow \quad w = 0.$$

Proof. From $w(ab) + w(ba) = 2w(a)g(b) + w(b)g(a)$ we see that also $w(ba) = 0$. Putting $x = a$ and $y = b$ in (6.7) we find that $0 = w(z)\widetilde{\Delta}(a, b)$ for all $z \in S$, and so, since $\widetilde{\Delta}(a, b) \neq 0$, that $w = 0$. □

On groups the function $d(x) = 2g(x)^2 - g(x^2)$ (from Definition B.6) occurs naturally in connection with the symmetrized sine addition formula as shown by Lemma 6.3. The lemma says in particular that d vanishes nowhere, and that w is odd and g even with respect to d in a certain sense. The notions of odd and even with respect to d reduce to the notions of odd and even when $d = 1$. These properties of w and g will be handy in later computations.

Lemma 6.3. *Let* G *be a group, and let* $w, g : G \to \mathbb{C}$ *be linearly independent functions satisfying* (6.1).

(a) $w(e) = 0$, $g(e) = 1$ *and* $d(e) = 1$.

(b) $d : G \to \mathbb{C}^*$*. Furthermore* $d(x^n) = d(x)^n$ *for all* $x \in G$ *and* $n \in \mathbb{Z}$*, so in particular* $d(x^{-1}) = d(x)^{-1}$ *for all* $x \in G$.

(c) $w(x^{-1}) = -d(x)^{-1}w(x)$ *for all* $x \in G$.

(d) $g(x^{-1}) = d(x)^{-1}g(x)$ *for all* $x \in G$.

Proof. (a) Putting $y = e$ in (6.1) we get that $(1 - g(e))w - w(e)g = 0$. The coefficients are 0 due to the linear independence, i.e., $g(e) = 1$ and $w(e) = 0$. By the definition of d we get $d(e) = 2g(e)^2 - g(e) = 2 - 1 = 1$.

(b) d is multiplicative on $\langle x\rangle$ by (6.8) and $\neq 0$ by (a), so it is a character on $\langle x\rangle$ (Lemma 3.4(a)). Hence (b).

(c) is a result of the following computation:

$$\begin{aligned}w(x) &= \tfrac{1}{2}(w(x^{-1}x^2) + w(x^2x^{-1})) = w(x^{-1})g(x^2) + w(x^2)g(x^{-1})\\ &= w(x^{-1})g(x^2) + 2w(x)g(x)g(x^{-1})\\ &= w(x^{-1})g(x^2) + g(x)[2w(x)g(x^{-1}) + 2w(x^{-1})g(x)]\\ &\quad - 2w(x^{-1})g(x)^2\\ &= w(x^{-1})[g(x^2) - 2g(x)^2] + g(x)[w(xx^{-1}) + w(x^{-1}x)]\\ &= -d(x)w(x^{-1}) + 2g(x)w(e) = -d(x)w(x^{-1}).\end{aligned}$$

(d) Putting $y = x^{-1}$ in (6.1) we get from (c) that

$$0 = w(x)g(x^{-1}) + w(x^{-1})g(x) = w(x)[g(x^{-1}) - g(x)d(x)^{-1}],$$

which shows that the desired formula holds for all $x \in S$ such that $w(x) \neq 0$. We are left with the points $x = x_0 \in G$ for which $w(x_0) = 0$. For them (6.5) reduces to $w(x_0zx_0) = d(x_0)w(z)$. When we here put $z = x_0^{-1}y$, resp. $z = yx_0^{-1}$, and add the resulting two identities we find that

$$2w(y)g(x_0) + 2w(x_0)g(y) = d(x_0)[2w(x_0^{-1})g(y) + 2w(y)g(x_0^{-1})].$$

However, $w(x_0) = 0$ implies $w(x_0^{-1}) = -d(x_0^{-1})w(x_0) = 0$, so the last identity reduces to $2w(y)g(x_0) = d(x_0)2w(y)g(x_0^{-1})$ for all $y \in G$. The result follows, since $w \neq 0$. □

The functions w and g were linearly independent in Lemma 6.3. The next lemma covers the degenerate case in which they are linearly dependent.

Lemma 6.4. *Let M be a monoid with neutral element e, and assume that the pair $w, g : M \to \mathbb{C}$, where $w \neq 0$, satisfies* (6.1). *Assume furthermore that w and g are linearly dependent.*

(a) *$w = cg$ for some constant $c \in \mathbb{C} \setminus \{0\}$.*

(b) *$2g$ is multiplicative function on M such that $g(e) = \frac{1}{2}$.*

(c) *$d(x) = 2g(x)^2 - g(x^2) = 0$ for all $x \in M$.*

Proof. Taking $y = e$ in (6.1) we see that $g = 0$ implies $w = 0$. This contradicts our hypothesis $w \neq 0$, so $g \neq 0$. We may thus write $w = cg$ for some constant $c \in \mathbb{C} \setminus \{0\}$. Substituting this into (6.1) and dividing c

away we get $g(xy) + g(yx) = 4g(x)g(y)$ for all $x, y \in M$. This says that $2g$ is a solution of the symmetrized multiplicative Cauchy equation (3.10). Theorem 3.21 tells us that $2g$ is a multiplicative function of M into $\mathbb{C}$. Finally $d(x) = 2g(x)^2 - g(x^2) = 2g(x)^2 - \frac{1}{2}2g(x^2) = 2g(x)^2 - \frac{1}{2}(2g(x)^2) = 0$. □

For any function $g : S \to \mathbb{C}$ on a semigroup S we use the notation $g_x(y) := g(xy) - g(x)g(y)$, $x, y \in S$. From the definition of d we get that $g(x)^2 - d(x) = g_x(x)$ for all $x \in S$. Let $z_0 \in S$. Then $g(z_0)^2 = d(z_0)$ if and only if $g_{z_0}(z_0) = 0$. The condition $g(z_0)^2 \neq d(z_0)$ in the following proposition is also present in Propositions 8.14 and 9.23. The other condition $z_0 \in Z(w)$ holds if z_0 lies in the center $Z(S)$ of S, so the result is interesting when $Z(S)$ is not trivial, for instance if S is a step 2 nilpotent group. The proposition is used in the proof of Proposition 10.5.

Proposition 6.5. *Let (w, g) be a solution of the symmetrized sine addition formula (6.1) on a monoid M such that $w \neq 0$.*

(a) *$Z(w) \subseteq Z(g)$.*

(b) *Let $z_0 \in Z(w)$. If $g(z_0)^2 \neq d(z_0)$, then there exist two different multiplicative functions $\chi, \rho : M \to \mathbb{C}$ and a constant $\alpha \in \mathbb{C} \setminus \{0\}$ such that $w = \alpha(\chi - \rho)$ and $g = (\chi + \rho)/2$. The pair $\{\chi, \rho\}$ of multiplicative functions is unique.*

(c) *Let $z_0 \in Z(w)$. If $g(z_0)^2 = d(z_0)$, then $g(xz_0) = g(x)g(z_0)$ for all $x \in M$.*

Proof. We use Lemma B.4(c) without explicit mentioning.

(a) Let $z_0 \in Z(w)$. We shall show $g(xuz_0) = g(xz_0u)$ for all $x, u \in M$, i.e., that u and z_0 may be interchanged in $g(xuz_0)$. Replacing x by xuz_0 in (6.1) we find

$$2w(y)g(xuz_0) = w(yxuz_0) + w(xuz_0y) - 2g(y)w(xuz_0).$$

Since $z_0 \in Z(w)$ we may interchange u and z_0 on the right and hence also on the left hand side. We are done, because $w \neq 0$.

(b) and (c) Since $z_0 \in Z(w)$ we have for the left hand side of the symmetrized sine addition formula (6.1) that

$$w((xz_0)y) + w(y(xz_0)) = w(x(yz_0)) + w((yz_0)x).$$

Thus the right hand sides agree, i.e.,

$$2w(xz_0)g(y) + 2w(y)g(xz_0) = 2w(x)g(yz_0) + 2w(yz_0)g(x).$$

Substituting

$$2w(xz_0) = w(xz_0) + w(z_0x) = 2w(x)g(z_0) + 2w(z_0)g(x)$$

and similarly for $2w(yz_0)$ into this we find that $w(x)g_{z_0}(y) = w(y)g_{z_0}(x)$. Since $w \neq 0$ we infer that g_{z_0} and w are proportional: $g_{z_0} = cw$ for some $c \in \mathbb{C}$. Now the functional equation implies that

$$g_{z_0}(xy) + g_{z_0}(yx) = 2g_{z_0}(x)g(y) + 2g_{z_0}(y)g(x) \quad \text{for all } x, y \in M. \quad (6.10)$$

As a motivation for our way of action we think of g as a kind of cosine. From the addition formula for cosine the quantity $g_{z_0}(x)g_{z_0}(y)$ should be an analogue of $\sin x \sin y$. So should $g(xy) + g(yx) - 2g(x)g(y)$. Therefore we expect the expressions are proportional. Our expectation is correct. More precisely we shall derive the identity

$$2g_{z_0}(x)g_{z_0}(y) = g_{z_0}(z_0)[g(xy) + g(yx) - 2g(x)g(y)] \qquad \text{for all } x, y \in M. \quad (6.11)$$

To do so we use (6.10) and its special case of $y = z_0$:

$$g_{z_0}(z_0)g(x) = g_{z_0}(z_0x) - g(z_0)g_{z_0}(x).$$

The exercise consists in separating x and y in the terms $g(xy) + g(yx)$ of the right hand side of (6.11). We proceed as follows:

$$\begin{aligned}
g_{z_0}(z_0)[g(xy) + g(yx)] &= g_{z_0}(z_0xy) + g_{z_0}(z_0yx) - g(z_0)[g_{z_0}(xy) + g_{z_0}(yx)] \\
&= g_{z_0}(xz_0y) + g_{z_0}(yxz_0) - 2g(z_0)[g_{z_0}(x)g(y) + g_{z_0}(y)g(x)] \\
&= 2g_{z_0}(xz_0)g(y) + 2g_{z_0}(y)g(xz_0) - 2g_{z_0}(x)g(z_0)g(y) - 2g_{z_0}(y)g(x)g(z_0) \\
&= 2g(y)g(z_0)g_{z_0}(x) + 2g(y)g_{z_0}(z_0)g(x) + 2g_{z_0}(y)g(xz_0) \\
&\quad - 2g_{z_0}(x)g(z_0)g(y) - 2g_{z_0}(y)g(x)g(z_0) \\
&= 2g_{z_0}(z_0)g(x)g(y) + 2g_{z_0}(y)[g(xz_0) - g(x)g(z_0)] \\
&= 2g_{z_0}(z_0)g(x)g(y) + 2g_{z_0}(y)g_{z_0}(x),
\end{aligned}$$

from which (6.11) follows easily.

(b) When $g_{z_0}(z_0) \neq 0$ then (6.11) means that

$$g(xy)+g(yx)-2g(x)g(y) = \frac{2}{g_{z_0}(z_0)}\, g_{z_0}(x)g_{z_0}(y) = 2\lambda^2 g_{z_0}(x)g_{z_0}(y), \quad (6.12)$$

where $\lambda \in \mathbb{C}$ is chosen such that $\lambda^2 = 1/g_{z_0}(z_0)$. In particular $\lambda \neq 0$. Since g is a kind of cosine and g_{z_0} is a kind of sine, we can hope from the formula

$e^{ix} = \cos x + i \sin x$ that $\chi := g + \lambda g_{z_0}$ is a kind of exponential function and hence multiplicative. Indeed it is, because using (6.10) and (6.12) we compute that

$$\begin{aligned}\chi(xy) + \chi(yx) &= g(xy) + g(yx) + \lambda(g_{z_0}(xy) + g_{z_0}(yx)) \\ &= 2g(x)g(y) + 2\lambda^2 g_{z_0}(x)g_{z_0}(y) + 2\lambda(g_{z_0}(x)g(y) + g_{z_0}(y)g(x)) \\ &= 2[g(x) + \lambda g_{z_0}(x)][g(y) + \lambda g_{z_0}(y)] = 2\chi(x)\chi(y).\end{aligned}$$

It follows from Theorem 3.21 that χ is multiplicative. Similarly we find that $\rho := g - \lambda g_{z_0} : M \to \mathbb{C}$ is multiplicative. We get (b), the first statement because w is proportional to $g_{z_0} \neq 0$. The uniqueness can be read off Corollary 3.19.

(c) Here $g_{z_0}(z_0) = 0$, so (6.11) reduces to $2g_{z_0}(x)g_{z_0}(y) = 0$. This implies $g_{z_0} = 0$, which is (c). □

6.3 The case of w being central

If w is central then the symmetrized sine addition formula (6.1) reduces to $w(xy) = w(x)g(y) + w(y)g(x)$ which we recognize as the sine addition formula (4.3) from Section 4.2. Theorem 4.1 solved that equation. It might be mentioned that, except for the trivial case of $w = 0$, we get from Theorem 4.1(e) that both w and g are abelian functions.

6.4 The case of g being abelian

Consider a solution (w, g) of the symmetrized sine addition formula (6.1) such that $w \neq 0$ and g is abelian. Proposition 6.7 says that g has the form $g = (\chi_1 + \chi_2)/2$ where χ_1 and χ_2 are multiplicative functions, but regrettably little is known about w. It is for instance not true generally that is w central: The Heisenberg group provides a counter-example with $g = 1$ (Example 12.4).

In our study of the pre-d'Alembert's functional equation (Chapter 8) we shall in Lemma 8.23(b) use the following corollary of Theorem 4.1. It involves the right regular representation R (see Definition E.15).

Lemma 6.6. *Let g be a central function on a monoid M. Let*

$$W(g) := \{w \in \mathbb{C}^M \mid w(xy) + w(yx) = 2w(x)g(y) + 2w(y)g(x) \text{ for all } x, y \in M\}.$$

If there exists a 1-dimensional, R-invariant subspace of $\mathbb{C}g + W(g)$ then g is abelian.

Proof. We leave it to the reader to show that any 1-dimensional, R-invariant subspace W of $\mathbb{C}^M$ has the form $W = \mathbb{C}\chi$, where $\chi : M \to \mathbb{C}$ is multiplicative and $\chi \neq 0$. By hypothesis we may write χ in the form $\chi = \alpha g + w$, where $\alpha \in \mathbb{C}$ and $w \in W(g)$.

Since both χ and g are central so is $w = \chi - \alpha g$. This implies that $w(xy) = w(x)g(y) + w(y)g(x)$ for all $x, y \in M$. If $w \neq 0$ we read from Theorem 4.1(e) that g is abelian. If $w = 0$ we have $\chi = \alpha g$. Here $\alpha \neq 0$, because $\chi \neq 0$. Since χ is abelian, so is $g = \chi/\alpha$. □

We proceed to derive the indicated information about an abelian g:

Theorem 6.7. *Let G be a group, and let $w, g : G \to \mathbb{C}$ be a solution of (6.1) such that $w \neq 0$ and g is abelian.*

There exist multiplicative functions $\chi_1, \chi_2 : G \to \mathbb{C}$, at least one of which is non-zero, such that $g = (\chi_1 + \chi_2)/2$ and $d = \chi_1\chi_2$. In particular $d : G \to \mathbb{C}$ is multiplicative, and $\Delta = 0$ and $\widetilde{\Delta} = 0$.

Proof. The conclusions are easily seen to hold if w and g are linearly dependent (Lemma 6.4). Thus we may and will during the rest of the proof assume that w and g are linearly independent, so that we can use the results of Lemma 6.3.

Since g is abelian and hence central, (6.5) becomes

$$d(x)w(z) = w(xzx) - 2w(x)g(xz) \qquad \text{for all } x, z \in G,$$

from which we find that

$$\begin{aligned}
[d(xz) - d(x)d(z)]w(z^{-1}) &= w(xzz^{-1}xz) - 2w(xz)g(xzz^{-1}) + d(x)w(z) \\
&= w(x^2z) - 2w(xz)g(x) + w(xzx) - 2w(x)g(xz) \\
&= w(x^2z) + w(xzx) - (2w(xz)g(x) + 2w(x)g(xz)) \\
&= w(x^2z) + w(xzx) - (w(xzx) + w(xxz)) = 0.
\end{aligned}$$

By Lemma 6.3(c) this implies that

$$[d(xz) - d(x)d(z)]w(z) = 0 \qquad \text{for all } x, z \in G.$$

Interchanging x and z we get that $[d(xz) - d(x)d(z)]w(x) = 0$ for all $x, z \in G$ because d is central. To get that d is a homomorphism it is now left to show that $d(ab) = d(a)d(b)$ when $w(a) = w(b) = 0$. We may actually also assume that $w(ab) = 0$. Indeed, from the computation

$$\begin{aligned}
&[d(xz - d(x)d(z)]w(z^{-1}x^{-1}) \\
&\quad = -w(xz) - d(x)[w(zz^{-1}x^{-1}z) - 2w(z)g(zz^{-1}x^{-1})] \\
&\quad = -w(xz) - d(x)[w(x^{-1}z) - 2w(z)g(x^{-1})] \\
&\quad = -w(xz) - w(xx^{-1}zx) + 2w(x)g(xx^{-1}z) + 2w(z)g(x) \\
&\quad = -w(xz) - w(zx) + 2w(x)g(z) + 2w(z)g(x) = 0
\end{aligned}$$

we get, again by Lemma 6.3, that

$$[d(xz - d(x)d(z)]w(xz) = -[d(xz - d(x)d(z)]w(z^{-1}x^{-1})d(xz)^{-1} = 0.$$

Now Corollary 6.2 tells us that $\widetilde{\Delta}(a, b) = 0$. We get from Lemma B.7(b) that $d(ab) = d(a)d(b)$. We have thus proved that d is multiplicative.

Using the multiplicativity of d and that g is abelian in the definition (B.2) of $\widetilde{\Delta}$ we find that $\widetilde{\Delta} = 0$. And then $\Delta = 0$ according to Lemma B.7(c).

It suffices now to prove that there exist multiplicative functions $\chi_1, \chi_2 : G \to \mathbb{C}$, such that $g = (\chi_1 + \chi_2)/2$. If $\chi_1 = \chi_2 = 0$ we get $g = 0$ which implies $w = 0$, contrary to our assumption. So either χ_1 or χ_2 must be different from 0. And then the statement about d follows from the formula $d = \chi_1\chi_2$ that we derive from $g = (\chi_1 + \chi_2)/2$.

We disengage g from its association with w by the following claim

$$g(xy) + d(y)g(xy^{-1}) = 2g(x)g(y) \qquad \text{for all } x, y \in G.$$

We proceed to prove the claim: Since $\widetilde{\Delta} = 0$ and g is abelian the formula (6.7) simplifies to

$$\begin{aligned}
&2[w(xy) + w(yx)]g(xyz) \\
&\quad = 2w(x)[g(xy^2z) + g(xz)d(y)] + 2w(y)[g(x^2yz) + g(yz)d(x)],
\end{aligned}$$

which according to the functional equation reduces to

$$\begin{aligned}
&w(x)[g(xy^2z) + g(xz)d(y) - 2g(xyz)g(y)] \\
&\qquad + w(y)[g(x^2yz) + g(yz)d(x) - 2g(x)g(xyz)] = 0.
\end{aligned}$$

When we put $z = y^{-1}$ we obtain by help of the definition of d that

$$w(x)[g(xy) + d(y)g(xy^{-1}) - 2g(x)g(y)] = 0.$$

Here we switch x and y. Using the formula $g(x^{-1}) = d(x)^{-1}g(x)$ (Lemma 6.3(d)) we get

$$w(y)[g(xy) + d(y)g(xy^{-1}) - 2g(x)g(y)] = 0.$$

When we take the two identities together we see that it is left to verify the claim for those $x, y \in G$ for which $w(x) = w(y) = 0$. So let $x_0, y_0 \in G$ be such that $w(x_0) = w(y_0) = 0$. For all $z \in G$ we get from (6.4) that

$$w(y_0zx_0) + w(x_0zy_0) = 2w(z)[2g(x_0)g(y_0) - g(x_0y_0)]. \tag{6.13}$$

Since $w(y_0^{-1}) = -d(y_0)^{-1}w(y_0) = 0$ (Lemma 6.3(c)) we may replace y_0 by y_0^{-1} in (6.13) which gives us

$$\begin{aligned} w(y_0^{-1}zx_0) + w(x_0zy_0^{-1}) &= 2w(z)[2g(x_0)g(y_0^{-1}) - g(x_0y_0^{-1})] \\ &= 2w(z)[2g(x_0)d(y_0^{-1})g(y_0) - g(x_0y_0^{-1})], \end{aligned}$$

so that

$$d(y_0)[w(y_0^{-1}zx_0) + w(x_0zy_0^{-1})] = 2w(z)[2g(x_0)g(y_0) - d(y_0)g(x_0y_0^{-1})]. \tag{6.14}$$

Adding (6.13) and (6.14) we get a formula containing the expression in the claim:

$$\begin{aligned} &2w(z)[4g(x_0)g(y_0) - g(x_0y_0) - d(y_0)g(x_0y_0^{-1})] \\ &\quad = w(y_0zx_0) + w(x_0zy_0) + d(y_0)[w(y_0^{-1}zx_0) + w(x_0zy_0^{-1})]. \end{aligned}$$

From the formula (6.5) we get since $w(x_0) = w(y_0) = 0$ that $w(x_0zx_0) = d(x_0)w(z)$ and $w(y_0zy_0) = d(y_0)w(z)$ for all $z \in G$. Substituting this into the identity above we get

$$\begin{aligned} &2w(z)[4g(x_0)g(y_0) - g(x_0y_0) - d(y_0)g(x_0y_0^{-1})] \\ &\quad = w(y_0zx_0) + w(x_0zy_0) + w(zx_0y_0) + w(y_0x_0z) \\ &\quad = w(y_0zx_0) + w(zx_0y_0) + w(x_0zy_0) + w(y_0x_0z) \\ &\quad = 2w(y_0)g(zx_0) + 2w(zx_0)g(y_0) + 2w(y_0)g(x_0z) + 2w(x_0z)g(y_0) \\ &\quad = 2[w(zx_0) + w(x_0z)]g(y_0) = 4[w(z)g(x_0) + w(x_0)g(z)]g(y_0) \\ &\quad = 4w(z)g(x_0)g(y_0). \end{aligned}$$

Subtracting the right hand side from the left hand side we are left with

$$2w(z)[2g(x_0)g(y_0) - g(x_0y_0) - d(y_0)g(x_0y_0^{-1})] = 0 \qquad \text{for all } z \in G,$$

which gives the claim, because $w \neq 0$.

$d \neq 0$ by Lemma 6.3(b). Being multiplicative d is thus a character on G (Lemma 3.4(a)). Copying the proof of Proposition 9.17(c) we find that g is a solution of the pre-d'Alembert functional equation (8.1). Being also abelian, g has the desired form by Theorem 8.13. □

6.5 The functional equation on a semigroup with an involution

The next theorem reveals some connections between the symmetrized sine addition formula and Wilson's long functional equation.

Theorem 6.8. *Let S be a semigroup with an involution $\tau : S \to S$. Let $\{w, g\}$ be a solution of (6.1) such that $w \neq 0$. If w is odd with respect to τ, then*

(a) *g is even with respect to τ, i.e., $g \circ \tau = g$, and*

(b) *w is a solution of Wilson's long functional equation corresponding to g, i.e.,*

$$w(xy) + w(yx) + w(x\tau(y)) + w(\tau(y)x) = 4w(x)g(y), \qquad x, y \in S.$$

Proof. (a) Using that w is odd we compute with (6.1) as follows:

$$\begin{aligned} 2w(x)g(\tau(y)) - 2w(y)g(x) &= 2w(x)g(\tau(y)) + 2w(\tau(y))g(x) \\ &= w(x\tau(y)) + w(\tau(y)x) = -[w(y\tau(x)) + w(\tau(x)y)] \\ &= -2w(y)g(\tau(x)) - 2w(\tau(x))g(y) = -2w(y)g(\tau(x)) + 2w(x)g(y). \end{aligned}$$

Comparing the left and right hand sides we see that

$$w(x)[g(y) - g(\tau(y))] = -w(y)[g(x) - g(\tau(x))] \qquad \text{for all } x, y \in S.$$

By Exercise 1.1 the only solutions $a, b : S \to \mathbb{C}$ of the functional equation $a(x)b(y) = -a(y)b(x)$ are $a = 0$ and b arbitrary, and $b = 0$ and a arbitrary. Combining this with $w \neq 0$, we conclude that $g \circ \tau - g = 0$.

(b) The proof of (b) consists of the following simple computation:

$$\begin{aligned}
&w(xy) + w(yx) + w(x\tau(y)) + w(\tau(y)x) \\
&\quad = 2w(x)g(y) + 2w(y)g(x) + 2w(x)g(\tau(y)) + 2w(\tau(y))g(x) \\
&\quad = 2w(x)g(y) + 2w(y)g(x) + 2w(x)g(y) - 2w(y)g(x) = 4w(x)g(y). \quad \square
\end{aligned}$$

6.6 The equation on compact groups

On compact groups the symmetrized sine addition formula (6.1) was solved by An and Yang. We are content with citing their result, since we do not need it later.

Theorem 6.9. *Let G be a compact group. Let $\{w, g\}$ be a continuous solution of the functional equation* (6.1) *such that $w \neq 0$. One of the following three possibilities occur:*

(i) *There exist a continuous character χ of G and a constant $c \in \mathbb{C} \setminus \{0\}$ such that $g = \chi/2$ and $w = c\chi$.*

(ii) *There exist a continuous irreducible representation $\pi : G \to U(2)$ and a non-zero matrix $X \in M(2 \times 2, \mathbb{C})$ with $\operatorname{tr} X = 0$ such that $g = \frac{1}{2}\operatorname{tr}(\pi(x))$ and $w(x) = \operatorname{tr}(X\pi(x))$ for all $x \in G$.*

(iii) *There exist two different continuous characters χ and χ' of G and a constant $c \in \mathbb{C} \setminus \{0\}$ such that $g = (\chi + \chi')/2$ and $w = c(\chi - \chi')$.*

6.7 Notes and remarks

Most of the material of the present chapter has been adapted from Dilian Yang [216].

The symmetrized sine addition formula was considered by Davison [64], but in his set up g was a given non-abelian pre-d'Alembert function. We do not assume g is a pre-d'Alembert function.

Proposition 6.5: See the comment to Proposition 9.30.

Theorem 6.9 is due to An and Yang [17, Theorem 5.8].

Chapter 7

Equations with Symmetric Right Hand Side

7.1 Discussion and results

We shall in later chapters study d'Alembert's functional equation

$$g(xy) + g(xy^{-1}) = 2g(x)g(y) \qquad \text{for all } x, y \in G,$$

and the quadratic functional equation

$$f(xy) + f(xy^{-1}) = 2f(x) + 2f(y) \qquad \text{for all } x, y \in G,$$

on possibly non-abelian groups G with unknowns $g : G \to \mathbb{C}$ and $f : G \to \mathbb{C}$ respectively. They are natural extensions of the functional equations (1.1) and (1.3) that we met in the Introduction (on the group $G = \mathbb{R}$). However there are other natural extensions of (1.1) and (1.3) from $\mathbb{R}$ to non-abelian groups. For example the following new functional equations

$$f(xy) + f(y^{-1}x) = 2f(x)f(y) \qquad \text{for all } x, y \in G,$$

and

$$f(xy) + f(y^{-1}x) = 2f(x) + 2f(y) \qquad \text{for all } x, y \in G,$$

in which the term $f(xy^{-1})$ on the left hand side of the original equations has been replaced by $f(y^{-1}x)$. Of course, the old and the corresponding new equations coincide if G is abelian, but more is true due to a common feature of the functional equations above: Their right hand sides are symmetric, i.e., do not change when x and y are interchanged.

(1) They have the same solutions on any group (Theorem 7.1), so there is no reason to make separate studies of the new functional equations.
(2) The solutions are central and even (= inversion invariant).

It is not true in general that solutions f of trigonometric functional equations of the form $f(xy) + f(xy^{-1}) = \cdots$ are central. Jensen's equation provides a counter-example (see the discussion after Definition 12.3), but, of course, its right hand side is not symmetric.

You might wonder whether it would suffice to consider functional equations of the form $f(xy) + f(xy^{-1}) = \cdots$ for varying right hand sides. But no, functional equations of the form $f(xy) + f(y^{-1}x) = \cdots$ like Fréchet's and Drygas' equations do play a role. So both forms must be taken into account.

Theorem 7.1. *Let S be a semigroup, $\tau : S \to S$ an involution of S, H an abelian group and $F : S \times S \to H$ a symmetric map.*

(a) *Any τ-invariant solution $f : S \to H$ of*

$$f(xy) + f(x\tau(y)) = F(x, y) \qquad \text{for all } x, y \in S, \tag{7.1}$$

or of

$$f(xy) + f(\tau(y)x) = F(x, y) \qquad \text{for all } x, y \in S \tag{7.2}$$

is central.

(b) *If S is a monoid, then every solution $f : S \to H$ of (7.1) or of (7.2) is τ-invariant and central, so (7.1) and (7.2) have the same set of solutions $f : S \to H$.*

Proof. (a) Let us assume that $f : S \to H$ is a solution of (7.1). We recall that f is said to be τ-invariant if $f \circ \tau = f$. By the symmetry of F we get for all $x, y \in S$ that

$$f(xy) + f(x\tau(y)) = f(yx) + f(y\tau(x)) = f(yx) + (f \circ \tau)(x\tau(y)), \tag{7.3}$$

from which we see that $f = f \circ \tau$ implies f is central. A similar argument works if f satisfies (7.2) instead of (7.1).

(b) Let $f : S \to H$ be a solution of (7.1). Putting $x = e$ (the neutral element) in the identity (7.3) and noting that $\tau(e) = e$ (Lemma A.31) we get that $f(\tau(y)) = f(y)$, i.e., that f is τ-invariant. And then f is central by (a). It is trivial that a central solution of (7.1) is a solution of (7.2). A similar argument works if f satisfies (7.2) instead of (7.1). $\square$

7.2 Exercises

Exercise 7.1. Let G be a group. Assume that $g, f, h : G \to \mathbb{C}$ satisfy

$$g(xy) + g(xy^{-1}) = 2f(x)g(y) + 2f(y)g(x) + h(x)h(y), \qquad x, y \in G.$$

Show that g is even and central.

Exercise 7.2. Let $\mu : G \to \mathbb{C}^*$ be a character on a group G. Let $g : G \to \mathbb{C}$ satisfy

$$g(xy) + \mu(y)g(xy^{-1}) = F(x, y) \qquad \text{for all } x, y \in G,$$

where $F : G \times G \to \mathbb{C}$ is symmetric.

(a) Show that $\check{g} = g/\mu$.

(b) Show that g is central.

Hint: Adapt the proof of Theorem 7.1.

7.3 Notes and remarks

Theorem 7.1: For more general range spaces than the ones in Theorem 7.1 consult Friis and Stetkær [91, Section 9].

Equation (7.1): The μ-d'Alembert functional equation (9.11) extends the left hand side of the functional equation $f(xy) + f(x\tau(y)) = F(x, y)$ by incorporating a character μ like we do in Exercise 7.2.

Exercise 7.2: A special instance of the equation in Exercise 7.2 is the μ-d'Alembert functional equation

$$g(xy) + \mu(y)g(xy^{-1}) = 2g(x)g(y) \qquad \text{for all } x, y \in G,$$

some properties of which are discussed in Proposition 9.17.

Chapter 8

The Pre-d'Alembert Functional Equation

8.1 Introduction

In this chapter we solve the pre-d'Alembert functional equation (8.1) on monoids. On groups it is closely related to d'Alembert's functional equation (Theorem 9.19). In particular, any solution of d'Alembert's functional equation is a solution of the pre-d'Alembert functional equation, so the results of the present chapter apply to d'Alembert's functional equation.

Although the pre-d'Alembert functional equation looks rather forbidding at first glance, it can nevertheless be solved, and its solutions are surprisingly simple to describe. Even so, our study of the pre-d'Alembert equation is longer and more complicated than our studies of the functional equations that we have encountered so far, like the sine addition formula. Not only are some of the proofs tricky, but genuinely non-abelian phenomena like 2-dimensional, irreducible representations turn up and are inherent in the theory.

In Section 8.3 we show how pre-d'Alembert functions relate to the symmetrized sine addition formula from Chapter 6. In Section 9.3 we describe how they relate to d'Alembert functions and more generally to solutions of the extension (9.11) of d'Alembert's functional equation.

In Section 8.4 we find the abelian pre-d'Alembert functions, while the harder problem of characterizing the non-abelian ones is solved by Davison's structure theorem (Theorem 8.26) which is the most important result of the present chapter.

8.2 Definitions and examples

Definition 8.1. Let S be a semigroup. A function $g : S \to \mathbb{C}$ is said to be a solution of or to satisfy *the pre-d'Alembert functional equation* if

$$g(xyz) + g(xzy) = 2g(x)g(yz) + 2g(y)g(xz) + 2g(z)g(xy) - 4g(x)g(y)g(z) \qquad \text{for all } x, y, z \in S. \tag{8.1}$$

A solution g of the pre-d'Alembert functional equation on a monoid with neutral element e is said to be a *pre-d'Alembert function* if $g(e) = 1$. Note that we define pre-d'Alembert functions on monoids, not on general semigroups.

The following non-trivial examples of pre-d'Alembert functions are typical for our later results.

Example 8.2. If $\chi_1, \chi_2 : S \to \mathbb{C}$ are multiplicative functions on a semigroup S, then $g := (\chi_1 + \chi_2)/2$ is a solution of the pre-d'Alembert functional equation by an elementary computation that we leave out. This is actually the general form of an abelian solution g of the pre-d'Alembert functional equation (Theorem 8.13). In particular $g = 0$, $g = \frac{1}{2}$ and $g = 1$ are solutions of the pre-d'Alembert functional equation. Of these three only $g = 1$ is a pre-d'Alembert function (when S is a monoid).

Example 8.3. On the multiplicative monoid $M(2 \times 2, \mathbb{C})$ of 2×2 complex matrices the function $g(x) := \frac{1}{2} \operatorname{tr} x$, $x \in M(2 \times 2, \mathbb{C})$, is a pre-d'Alembert function, i.e., $g(I) = 1$ and

$$\operatorname{tr}(xyz) + \operatorname{tr}(xzy) = \operatorname{tr} x \operatorname{tr}(yz) + \operatorname{tr} y \operatorname{tr}(xz) + \operatorname{tr} z \operatorname{tr}(xy) - \operatorname{tr} x \operatorname{tr} y \operatorname{tr} z \tag{8.2}$$

for all $x, y, z \in M(2 \times 2, \mathbb{C})$.

Proof of (8.2). It is easy to see that if (8.2) holds for some x, y, z then it also holds for $x, y, z + \alpha I$, where $\alpha \in \mathbb{C}$ is arbitrary. Thus we may assume that $\operatorname{tr} z = 0$. Similarly we may assume that $\operatorname{tr} x = \operatorname{tr} y = 0$, so that it is left to show that $\operatorname{tr}(x[yz + zy]) = 0$ when $\operatorname{tr} x = \operatorname{tr} y = \operatorname{tr} z = 0$. But if y and z are 2×2 matrices of trace 0, then a small computation shows that $yz + zy$ is a constant times I, so the result follows, because $\operatorname{tr} x = 0$. □

It is immediate to extend the above from $M(2 \times 2, \mathbb{C})$ to the multiplicative monoid $\mathcal{L}(V)$ where V is any 2-dimensional vector space: $g(x) := \frac{1}{2} \operatorname{tr} x$, $x \in \mathcal{L}(V)$ is a pre-d'Alembert function on $\mathcal{L}(V)$.

Example 8.4. Example 8.3 may be extended: If ρ is a representation of a monoid M on $\mathbb{C}^2$, then $g(x) := \frac{1}{2} \operatorname{tr} \rho(x)$, $x \in M$, is a pre-d'Alembert function on M (follows from Example 8.3).

Surprisingly, all pre-d'Alembert functions on M have the form presented in Example 8.4. This is the content of Davison's structure theorem (Theorem 8.26). In the process of proving it we gain more information about the interplay between g and ρ.

Example 8.5. Any d'Alembert function is a solution of the pre-d'Alembert functional equation (Theorem 9.17(c) or Corollary 9.18(c)). So $g(x) = \cos x$ is an example of a pre-d'Alembert function on the group $\mathbb{R}$. Section 9.2 contains further examples of d'Alembert functions.

Example 8.6. Any μ-d'Alembert function is a solution of the pre-d'Alembert functional equation (Theorem 9.17(c)). A converse is true on groups: On a group any pre-d'Alembert function is a μ-d'Alembert function for some μ (Theorem 9.19).

8.3 Key properties of solutions

In this section we collect a few indispensable formulas and results for solutions of the pre-d'Alembert functional equation on a monoid. Some of the derivations are technically intricate, but: No pain, no gain!

For definitions and general properties of the functions g_x, $\widetilde{\Delta}$, Δ and d occurring below see Section B.2.

Lemma 8.7. *Let g be a solution of the pre-d'Alembert functional equation* (8.1) *on a semigroup.*

(a) *g is central.*

(b) *Let the semigroup be a monoid M with neutral element e. There are only three possibilities for $g(e)$, viz., $g(e) = 0$, $g(e) = \frac{1}{2}$ and $g(e) = 1$. If $g(e) = 0$, then $g = 0$. If $g(e) = \frac{1}{2}$ then $2g : M \to \mathbb{C}$ is a multiplicative function.*

Proof. (a) Interchange of y and z does not change the left hand side of (8.1), so from the right hand side we get

$$\begin{aligned}&2g(x)g(yz) + 2g(y)g(xz) + 2g(z)g(xy) - 4g(x)g(y)g(z)\\&\quad = 2g(x)g(zy) + 2g(z)g(xy) + 2g(y)g(xz) - 4g(x)g(y)g(z),\end{aligned}$$

that reduces to $g(x)[g(yz) - g(zy)] = 0$, which implies (a).

(b) Putting $x = y = z = e$ in (8.1) we find that $g(e)$ is root in a third order equation with the roots 0, $\frac{1}{2}$ and 1.

If $g(e) = 0$ we see that $g = 0$ by putting $y = z = e$ in (8.1).

If $g(e) = \frac{1}{2}$ we put $x = e$ in (8.1) and find that $g(zy) = 2g(y)g(z)$. This identity shows that $2g$ is multiplicative. □

That pre-d'Alembert functions are central (Lemma 8.7(a)) is crucial for our development of the theory of pre-d'Alembert functions on non-abelian groups, because it is used time and time again in computations.

Lemma 8.7(b) shows that $g(e) = 1$ is the only interesting possibility for a solution of the pre-d'Alembert functional equation (8.1): The two other possibilities give rather trivial solutions. That is why we impose the condition $g(e) = 1$ in the definition of pre-d'Alembert functions.

If $g = \cos$ then g_x is a constant times sin, so the pair $\{g_x, g\}$ satisfies the sine addition formula. This is a special instance of a more general phenomenon: Lemma 8.8 points out the corresponding general relation between the pre-d'Alembert functional equation (8.1) and the symmetrized sine addition formula (6.1). The lemma enables us to draw upon the results of Chapter 6.

Lemma 8.8. *The solutions of the pre-d'Alembert functional equation* (8.1) *on a semigroup S are the central functions $g : S \to \mathbb{C}$ that for each $x \in WS$ satisfy the symmetrized sine addition formula* (6.1) *with $w = g_x$, i.e.,*

$$g_x(yz) + g_x(zy) = 2g_x(y)g(z) + 2g_x(z)g(y) \qquad \text{for all } y, z \in S. \tag{8.3}$$

Proof. By a simple calculation we find for any function $g : S \to \mathbb{C}$ the identity

$$\begin{aligned} g_x(yz) &+ g_x(zy) - 2g_x(y)g(z) - 2g_x(z)g(y) \\ &= g(xyz) + g(xzy) - g(x)[g(yz) + g(zy)] \\ &\quad - 2g(xy)g(z) - 2g(xz)g(y) + 4g(x)g(y)g(z), \qquad x, y, z \in S. \end{aligned} \tag{8.4}$$

The proof of the lemma is now straightforward from Lemma 8.7(a): For any central function g the right hand side of (8.4) is the difference between the left and right hand sides of (8.1). □

About the following lemma we note that (a) is used many times in computations. If g is a non-zero solution of d'Alembert's functional equation

$g(xy) + g(xy^{-1}) = 2g(x)g(y)$ on a group, then $d = 1$. (b) is a non-obvious extension of this. Theorem 8.26(b) reveals in a transparent way why d is multiplicative, but it requires a knowledge of the structure of g that we have not yet obtained. In (c) we connect the functions $\widetilde{\Delta}$, Δ and g_x for pre-d'Alembert functions.

Lemma 8.9. *Let $g : S \to \mathbb{C}$ is a solution of the pre-d'Alembert functional equation* (8.1) *on a semigroup S.*

(a) $g_x(y^2) = 2g_x(y)g(y)$ *for all $x, y \in S$.*

(b) $d : S \to \mathbb{C}$ *is multiplicative.*

(c) $\widetilde{\Delta}(x, y)/4 = \Delta(x, y) = g_x(x)g_y(y) - g_x(y)^2$ *for all $x, y \in S$.*

Proof. (a) Put $z = y$ in (8.3).

(b) We put $z = xy$ in (8.1) and get

$$\begin{aligned} &g((xy)^2) + g(x^2y^2) \\ &\qquad = 2g(x)g(yxy) + 2g(y)g(xyx) + 2g(xy)^2 - 4g(x)g(y)g(xy), \end{aligned}$$

from which we find, using the definition of $d(xy)$ and that g is central, that

$$d(xy) = g(x^2y^2) - 2[g(x)g(xy^2) + g(y)g(x^2y)] + 4g(x)g(y)g(xy),$$

where we compute the 3 terms on the right hand side with a view to separate x and y as they are in $d(x)d(y)$. We use (a) and that $g_x(y) = g_y(x)$:

$$\begin{aligned} &g(x^2y^2) \\ &\qquad = g_{x^2}(y^2) + g(x^2)g(y^2) = 2g_{x^2}(y)g(y) + g(x^2)g(y^2) \\ &\qquad = 2g_y(x^2)g(y) + g(x^2)g(y^2) = 4g_y(x)g(x)g(y) + g(x^2)g(y^2) \\ &\qquad = 4g_x(y)g(x)g(y) + g(x^2)g(y^2), \\ &g(x)g(xy^2) + g(y)g(x^2y) \\ &\qquad = g(x)[g_x(y^2) + g(x)g(y^2)] + g(y)[g_y(x^2) + g(x^2)g(y)] \\ &\qquad = 2g(x)g_x(y)g(y) + g(x)^2g(y^2) + 2g(y)g_y(x)g(x) + g(x^2)g(y)^2 \\ &\qquad = 4g_x(y)g(x)g(y) + g(x)^2g(y^2) + g(x^2)g(y)^2, \end{aligned}$$

and $g(x)g(y)g(xy) = g_x(y)g(x)g(y) + g(x)^2g(y)^2$.

Combining these expressions for the three terms we get (b) as follows:

$$\begin{aligned} d(xy) &= 4g_x(y)g(x)g(y) + g(x^2)g(y^2) \\ &\quad - 8g_x(y)g(x)g(y) - 2g(x)^2g(y^2) - 2g(x^2)g(y)^2 \\ &\quad + 4g_x(y)g(x)g(y) + 4g(x)^2g(y)^2 \\ &= g(x^2)g(y^2) - 2g(x)^2g(y^2) - 2g(x^2)g(y)^2 + 4g(x)^2g(y)^2 \\ &= [2g(x)^2 - g(x^2)][2g(y)^2 - g(y^2)] = d(x)d(y). \end{aligned}$$

(c) Since g is central and d is multiplicative (by (b)) the first equality is Lemma B.7(c).

According to Lemma 8.8 we may take $w = g_y$ in (6.6). Small computations give us the second equality, using again that g is central. □

Although the criterion of the following lemma is formulated as a condition ensuring g is abelian, it is only used in the non-abelian case.

Lemma 8.10. *Let g be a solution of the pre-d'Alembert functional equation* (8.1) *on a semigroup. Then $\Delta = 0$ if and only if g is abelian.*

Proof. Let us denote the semigroup S. It is trivial from the definition of Δ that g abelian implies $\Delta = 0$. It is left to show that $\Delta = 0$ implies that g is abelian, so assume that $\Delta = 0$.

First an observation for use below: If $g_c(c) = 0$ for a $c \in S$ then $c \in Z(g)$. Proof of the observation: For any $x \in S$ we have

$$0 = \Delta(x,c) = g_x(x)g_c(c) - g_x(c)^2 = -g_x(c)^2 = -[g(xc) - g(x)g(c)]^2,$$

so that $g(xc) = g(x)g(c)$. Finally for any $x, y \in S$ we get $g(xyc) = g(xy)g(c)$ $= g(yx)g(c) = g(yxc) = g(xcy)$.

In the expression $g_x(x)g_y(y) - g_x(y)^2 = \Delta(x,y) = 0$ in Lemma 8.9 we replace y by yz and zy respectively and get that $g_x(x)g_{yz}(yz) - g_x(yz)^2 = 0$ and $g_x(x)g_{zy}(zy) - g_x(zy)^2 = 0$. From g being central we find that $g_{yz}(yz) = g_{zy}(zy)$, so that $g_x(yz)^2 = g_x(zy)^2$ or $[g_x(yz) - g_x(zy)][g_x(yz) + g_x(zy)] = 0$. Using the definition of g_x on the first factor and Lemma 8.8 on the second factor we obtain that

$$[g(xyz) - g(xzy)][g_x(y)g(z) + g_x(z)g(y)] = 0 \quad \text{for all } x, y, z \in S. \tag{8.5}$$

Interchange of x and y, respectively x and z, only changes the first factor by a sign, so we find respectively that

$$[g(xyz) - g(xzy)][g_x(y)g(z) + g_y(z)g(x)] = 0, \tag{8.6}$$

and

$$[g(xyz) - g(xzy)][g_y(z)g(x) + g_x(z)g(y)] = 0. \tag{8.7}$$

Subtracting (8.5) from the sum of (8.6) and (8.7) we obtain that

$$[g(xyz) - g(xzy)]g(x)g_y(z) = 0 \qquad \text{for all } x, y, z \in S. \tag{8.8}$$

If $g_y(z) = 0$ then we read from $0 = \Delta(y, z) = g_y(y)g_z(z) - g_y(z)^2 = g_y(y)g_z(z)$ that either $g_y(y) = 0$ or $g_z(z) = 0$. In either case $g(xyz) - g(xzy) = 0$ by the observation at the beginning of the proof. Thus (8.8) reduces to

$$[g(xyz) - g(xzy)]g(x) = 0 \qquad \text{for all } x, y, z \in MS. \tag{8.9}$$

The factor $g(x)$ in (8.9) may be replaced by $g(y)$ and $g(z)$ (symmetry!).

We shall show that the set $D := \{(x, y, z) \in S \times S \times S \mid g(xyz) \neq g(xzy)\}$ is empty. Let us note that for any $(x, y, z) \in D$ we have that

$$g(x) = g(y) = g(z) = 0, \tag{8.10}$$
$$g(xy) \neq 0, \quad \text{and} \quad g(xyz) \neq 0, \tag{8.11}$$

where the first statement comes from (8.9) and the second one from the observation above: $0 \neq g_x(y) = g(xy) - g(x)g(y) = g(xy) - 0 = g(xy)$. About the third one we see from the pre-d'Alembert functional equation (8.1) that $g(xyz) + g(xzy) = 0$. Thus if $g(xyz) = 0$ then so is $g(xzy)$, and hence $g(xyz) = g(xzy)$, contradicting that $g(xyz) \neq g(xzy)$.

Assuming there exists an element $(x_0, y_0, z_0) \in D$ we shall arrive at a contradiction. If $(x_0, y_0, x_0y_0z_0) \in D$, then $g(x_0y_0z_0) = 0$ by (8.10), contradicting (8.11). Hence $(x_0, y_0, x_0y_0z_0) \notin D$, so that $g(x_0y_0(x_0y_0z_0)) = g(x_0(x_0y_0z_0)y_0)$. The left hand side is (use (6.2) with $w = g_{z_0}$ in the computations)

$$\begin{aligned} g((x_0y_0)^2z_0) &= g_{z_0}((x_0y_0)^2) + g(z_0)g((x_0y_0)^2) \\ &= 2g_{z_0}(x_0y_0)g(x_0y_0) + 0 \cdot g((x_0y_0)^2) \\ &= 2[g(z_0x_0y_0) - g(z_0)g(x_0y_0)]g(x_0y_0) \\ &= 2g(z_0x_0y_0)g(x_0y_0) = 2g(x_0y_0z_0)g(x_0y_0) \neq 0 \end{aligned}$$

with the $\neq 0$ due to (8.11). We now get the desired contradiction, because the right hand side vanishes:

$$\begin{aligned} g(x_0^2 y_0 z_0 y_0) &= g_{y_0 z_0 y_0}(x_0^2) + g(y_0 z_0 y_0) g(x_0^2) \\ &= 2 g_{y_0 z_0 y_0}(x_0) g(x_0) + g(z_0 y_0^2) g(x_0^2) \\ &= 0 + [g_{z_0}(y_0^2) + g(z_0) g(y_0^2)] g(x_0^2) \\ &= g_{z_0}(y_0^2) g(x_0^2) = 2 g_{z_0}(y_0) g(y_0) g(x_0^2) = 0. \end{aligned}$$ □

Proposition 8.11 is an inverse to Proposition 9.17 on groups with the involution being the group inversion.

Proposition 8.11. *Any pre-d'Alembert function g on a group G satisfies*

$$g(xy) + d(y) g(xy^{-1}) = 2g(x)g(y) \qquad \text{for all } x, y \in G, \tag{8.12}$$

where $d(x) := 2g(x)^2 - g(x^2)$, $x \in G$, is a character on G. So g is a d-d'Alembert function on G.

Proof. Replacing x by xy^{-1} and putting $z = y$ in (8.1) we get (8.12). That d is multiplicative is contained in Lemma 8.9. That d is a character is due to $g(e) = 1$ because that implies $d(e) = 1$. □

We postpone the proof of the following general, quantitative result to the next section, because Theorem 8.13 is used in the proof.

Proposition 8.12. *If $g : S \to \mathbb{C}$ is a bounded solution of the pre-d'Alembert functional equation on a semigroup S, then $|g(x)| \leq 1$ for all $x \in S$.*

8.4 Abelian pre-d'Alembert functions

The next theorem describes all abelian solutions of the pre-d'Alembert functional equation. Theorem 6.7 is the corresponding result for solutions of the symmetrized sine addition formula (6.1) on groups.

Theorem 8.13. *If g is an abelian solution of the pre-d'Alembert functional equation on a semigroup S, then there exist multiplicative functions $\chi_1, \chi_2 : S \to \mathbb{C}$ such that $g = (\chi_1 + \chi_2)/2$. They are unique, except that they can be interchanged.*

Conversely, if $\chi_1, \chi_2 : S \to \mathbb{C}$ are multiplicative functions, then $g := (\chi_1 + \chi_2)/2$ is an abelian solution of the pre-d'Alembert functional equation on S.

Furthermore g is bounded if and only if χ_1 and χ_2 are bounded.

If S is a topological semigroup, then g is continuous if and only if χ_1 and χ_2 are continuous.

Proof. The pair $g_x, g : S \to \mathbb{C}$ is a solution of the symmetrized sine addition formula (6.1) for all $x \in S$. g being abelian, (8.3) reduces to the sine addition formula $g_x(yz) = g_x(y)g(z) + g_x(z)g(y)$ that we solved in Theorem 4.1. If $g_x = 0$ for all $x \in S$ then g is multiplicative, so we may take $\chi_1 = \chi_2 = g$. If $g_x \neq 0$ for some $x \in S$ then the result is contained in Theorem 4.1(b).

The conversely statement is left to the reader.

For the remaining statements we refer to Theorem 3.18. □

Proof of Proposition 8.12. Let $x \in S$ be given. The restriction of g to the abelian subsemigroup $S_x := \{x^n \mid n \in \mathbb{N}\}$ of S has by Theorem 8.13 the form $g|_{S_x} = (\chi_1 + \chi_2)/2$, where $\chi_1, \chi_2 : S_x \to \mathbb{C}$ are multiplicative. χ_1 and χ_2 are bounded, g being bounded (Theorem 3.18(c)). By Lemma A.27 they are bounded by 1. Hence $|g(x)| = |(\chi_1(x) + \chi_2(x))/2| \leq 1$. □

Proposition 8.14(a) contains a criterion for a solution of the pre-d'Alembert function to be abelian on a possibly non-commutative monoid. A related result is Proposition 6.5.

Proposition 8.14. *Let g be a solution of the pre-d'Alembert functional equation on a semigroup S.*

(a) *If there is a $z_0 \in Z(g)$ such that $g(z_0)^2 \neq d(z_0)$, then there exist two different multiplicative functions $\chi_1, \chi_2 : S \to \mathbb{C}$ such that $g = (\chi_1 + \chi_2)/2$.*

(b) *If $z_0 \in Z(g)$ and $g(z_0)^2 = d(z_0)$, then $g(xz_0) = g(x)g(z_0)$ for all $x \in S$.*

Proof. (a) Since $z_0 \in Z(g)$ and g is central (Lemma 8.7(a)) we get that g_{z_0} is central. Being furthermore a solution of (8.3) (Lemma 8.8) we see that g_{z_0} is a solution of the sine addition formula (4.3), and by assumption $g_{z_0}(z_0) = g(z_0^2) - g(z_0)^2 = -d(z_0) + g(z_0)^2 \neq 0$. It follows from Theorem 4.1(e) that g has the form $g = (\chi_1 + \chi_2)/2$, where $\chi_1, \chi_2 : S \to \mathbb{C}$ are multiplicative. $\chi_1 = \chi_2$ implies that g is multiplicative. In that case we have $g(z_0)^2 = d(z_0)$, a contradiction. So $\chi_1 \neq \chi_2$.

(b) The requirement $g(z_0)^2 = d(z_0)$ means that $g_{z_0}(z_0) = 0$. If $g_{z_0} = 0$ we are done, so we shall arrive at a contradiction from the assumption $g_{z_0} \neq 0$.

As in part (a) we find that g_{z_0} is a non-zero solution of the sine addition formula. By Theorem 4.1(b) $g = (\chi_1 + \chi_2)/2$, where $\chi_1, \chi_2 : S \to \mathbb{C}$ are multiplicative. For g of this form a small computation shows that

$$g_{z_0} = \frac{\chi_1(z_0) - \chi_2(z_0)}{2} \frac{\chi_1 - \chi_2}{2}.$$

Now $g_{z_0}(z_0) = 0$ implies that

$$g_{z_0}(z_0) = \Big(\frac{\chi_1(z_0) - \chi_2(z_0)}{2}\Big)^2 = 0, \quad \text{so} \quad \frac{\chi_1(z_0) - \chi_2(z_0)}{2} = 0.$$

But then $g_{z_0} = 0$, contradicting our assumption $g_{z_0} \neq 0$. □

8.5 When is a pre-d'Alembert function on a group abelian?

Let g be a pre-d'Alembert function on a group G. It is of interest to have criteria in terms of g and/or G for when g is abelian, because then we know how g looks like (Theorem 8.13). As we shall see, on some non-abelian groups like the Heisenberg group and the $(ax+b)$-group all pre-d'Alembert functions are abelian, so this phenomenon is not restricted to abelian groups. These groups are important in other connections and naturally occurring; they are not artificial constructions made with the purpose that all d'Alembert functions on them shall be abelian. This motivates us to seek sufficient, general conditions for g to be abelian, conditions that cover at least the examples above. This is done in the present section.

Proposition 8.15. *A pre-d'Alembert function g on a group G is abelian if and only if $g([x, y]) = 1$ for all $x, y \in G$.*

Proof. Using (8.12) in the following computation

$$\begin{aligned} d(xy)g([x,y]) &= d(yx)g(xyx^{-1}y^{-1}) = d(yx)g(xy(yx)^{-1}) \\ &= 2g(xy)g(yx) - g(xyyx) = 2g(xy)^2 - g(x^2y^2) \\ &= g((xy)^2) + d(xy) - g(x^2y^2) = d(xy) - 2\Delta(x,y) \end{aligned}$$

we get the formula $2\Delta(x, y) = d(xy)(1 - g([x, y]))$. Combining this with Lemma 8.10 we get the proposition. □

Proposition 10.3 presents a version of Proposition 8.15 for d'Alembert's long functional equation.

The condition $g([x, y]) = 1$ for all $x, y \in G$ of Proposition 8.15 is both necessary and sufficient. It is equivalent to $g \equiv 1$ on all of $[G, G]$. Theorem 8.16 weakens the condition to $g|_{[G,G]}$ being abelian. Like Proposition 8.15, but in contrast to Theorem 8.29, it requires no topology on the group and hence no continuity of the d'Alembert function in question. It applies in particular, if the commutator subgroup is abelian, so all pre-d'Alembert functions on the Heisenberg group and the $(ax + b)$-group are abelian, because for both groups $[G, G]$ is isomorphic to the abelian group $\mathbb{R}$. Thus it may happen for a non-abelian group that all its pre-d'Alembert functions are abelian.

Theorem 8.16. *Let G be a group which is generated by its squares. Let g be a solution of the pre-d'Alembert functional equation on G such that $g|_{[G,G]}$ is abelian. Then g is abelian.*

Proof. According to Lemma 8.7(b) we may assume that g is a pre-d'Alembert function.

Consider the set $\mathcal{N}$ of subgroups N of G such that $N \supseteq [G, G]$ and $g|_N$ is abelian. Order $\mathcal{N}$ by inclusion. Applying Zorn's lemma we find that $\mathcal{N}$ has a maximal element $N_{\max} \in \mathcal{N}$. The following Lemma 8.17 implies that $N_{\max} = G$. Thus $g = g|_G$ is abelian. □

Lemma 8.17. *Let G be a group which is generated by its squares. Let N be a subgroup of G such that $[G, G] \subseteq N$. Let g be a pre-d'Alembert function on G such that $g|_N$ is abelian. Let $x_0 \in G$ and put $N_0 := \langle N, x_0 \rangle$. Then $g|_{N_0}$ is abelian.*

Proof. Note that N is normal. If γ is a character on N and $x \in G$, we let $x \cdot \gamma$ denote the character on N defined by $(x \cdot \gamma)(n) := \gamma(x^{-1}nx)$, $n \in N$. Note that $(xy) \cdot \gamma = x \cdot (y \cdot \gamma)$ for all $x, y \in G$.

Since $g|_N$ is an abelian pre-d'Alembert function we may write it as $g|_N = (\chi_1 + \chi_2)/2$, where $\chi_1, \chi_2 : N \to \mathbb{C}^*$ are characters on N (Theorem 8.13). The multiplicative functions from Theorem 8.13 are characters, because $g(e) = 1$ for any pre-d'Alembert function.

Let $x \in G$. Since g is central (Lemma 8.7(a)) we find for any $n \in N$ that

$$\begin{aligned}\frac{\chi_1(n) + \chi_2(n)}{2} &= g(n) = g(x^{-1}nx) \\ &= \frac{\chi_1(x^{-1}nx) + \chi_2(x^{-1}nx)}{2} = \frac{(x \cdot \chi_1)(n) + (x \cdot \chi_2(n)}{2},\end{aligned}$$

so that $\chi_1 + \chi_2 = x \cdot \chi_1 + x \cdot \chi_2$. According to Corollary 3.19 there are only two possibilities: (i) $x \cdot \chi_1 = \chi_1$ and $x \cdot \chi_2 = \chi_2$ or (ii) $x \cdot \chi_1 = \chi_2$ and $x \cdot \chi_2 = \chi_1$.

In either case $x^2 \cdot \chi_1 = \chi_1$ and $x^2 \cdot \chi_2 = \chi_2$. It follows that $x \cdot \chi_1 = \chi_1$ and $x \cdot \chi_2 = \chi_2$ for all $x \in G$, because G is generated by its squares.

We will use Proposition 8.15, so we compute the commutator of two arbitrary elements in N_0. Since N is normal they have the form $n_1x_1, n_2x_2 \in N_0$, where $n_1, n_2 \in N$ and $x_1, x_2 \in \langle x_0 \rangle$. Now,

$$[n_1x_1, n_2x_2] = n_1(x_1n_2x_1^{-1})(x_1x_2x_1^{-1})n_1^{-1}(x_1x_2x_1^{-1})^{-1}[x_1, x_2]n_2^{-1}.$$

We may apply the character χ_1 to this, because χ_1 is defined on N and $[x_1, x_2] \in [G, G] \subseteq N$. Using $x \cdot \chi_1 = \chi_1$ for all $x \in G$ we find

$$\begin{aligned}
&\chi_1([n_1x_1, n_2x_2]) \\
&\quad = \chi_1(n_1)\chi_1(x_1n_2x_1^{-1})\chi_1((x_1x_2x_1^{-1})n_1^{-1}(x_1x_2x_1^{-1})^{-1}) \\
&\qquad \times \chi_1([x_1, x_2])\chi_1(n_2^{-1}) \\
&\quad = \chi_1(n_1)(x_1^{-1} \cdot \chi_1)(n_2)((x_1x_2x_1^{-1})^{-1} \cdot \chi_1)(n_1^{-1})\chi_1([x_1, x_2])\chi_1(n_2^{-1}) \\
&\quad = \chi_1(n_1)\chi_1(n_2)\chi_1(n_1^{-1})\chi_1([x_1, x_2])\chi_1(n_2^{-1}) = \chi_1([x_1, x_2]).
\end{aligned}$$

Similarly $\chi_2([n_1x_1, n_2x_2]) = \chi_2([x_1, x_2])$, so that

$$g([n_1x_1, n_2x_2]) = \frac{\chi_1 + \chi_2}{2}([n_1x_1, n_2x_2]) = \frac{\chi_1 + \chi_2}{2}([x_1, x_2]) = 1,$$

the last equality because x_1 and x_2 commute, being in the cyclic group $\langle x_0 \rangle$. Finally, $g|_{N_0}$ is abelian according to the last statement of Proposition 8.15. □

Corollary 8.18. *Pre-d'Alembert functions are abelian on solvable, connected Lie groups.*

Proof. Let g be a pre-d'Alembert function on a connected solvable Lie group G. We recall the notation $\mathcal{D}(H) := [H, H]$ for the derived group of a group H. That G is solvable means that the descending series $G \supseteq \mathcal{D}(G) \supseteq \mathcal{D}^2(G) \supseteq \cdots$ terminates after a finite number of steps, so that $\mathcal{D}^n(G) = \{e\}$ for some $n \in \mathbb{N}$. We prove the corollary by induction on n. It is true for $n = 1$, because here G is abelian. Assume it is true for some $n \geq 1$. Let $\mathcal{D}^{n+1}(G) = \{e\}$. Noting that $[G, G]$ is a connected, solvable Lie group, we get by the induction hypothesis that g is abelian on $[G, G]$. We now refer to Theorem 8.16. □

It is immediate from Corollary 8.18 that pre-d'Alembert functions on the Heisenberg group and the $(ax + b)$-group are abelian. Theorem 8.29 is a variant of the corollary involving topological groups that need not be Lie groups. See Theorem 10.6 for a related result about solutions of d'Alembert's long functional equation.

Remark 8.19. The assumption that G is generated by its squares cannot be deleted from Theorems 8.16 and Corollary 8.18: The classical d'Alembert function g_0 on Q_8 (Example 9.10) is a counter-example.

8.6 Translates of pre-d'Alembert functions

We now start a discussion of pre-d'Alembert functions that need not be abelian, so let g be a continuous pre-d'Alembert function on a topological monoid M. Our road to the description of g goes via the space of translates of g.

In the abelian case the solutions of d'Alembert's functional equation were built from multiplicative functions (Theorem 8.13), i.e., by 1-dimensional representations of the monoid. In analogy with non-abelian harmonic analysis it is to be expected that the multiplicative functions shall be replaced by characters of irreducible representations in the non-abelian case. This is indeed so. To set the stage let us remark that given a function F on a semigroup or even just on $\mathbb{R}$ it is often expedient to involve the semigroup action by embedding F in the space of functions spanned by the translates of F. We did this when we discussed Levi-Civita's functional equation in Chapter 5. We shall do the same here for the function g, where the pertinent function space is $T(g) := \text{span}\{R(x)g \mid x \in M\} \subseteq C(M)$, R denoting the right regular representation. The vector spaces on which the representations work are subspaces of $T(g)$ and the corresponding representations are restrictions of R to these subspaces. $T(g)$ is closely related to the symmetrized sine addition formula (6.1) through the vector space

$$\begin{aligned} W(g) := \{w \in C(M) \mid & \, w(xy) + w(yx) \\ & = 2w(x)g(y) + 2w(y)g(x) \text{ for all } x, y \in M\}. \end{aligned}$$

How the relation goes will be revealed below and in the next section.

Lemma 8.20. *Let g be a continuous pre-d'Alembert function on a topological monoid M.*

(a) $\dim T(g) = 1$, 2 *or* 4.

(b) *$T(g)$ is invariant under both the right regular representation R and under the left regular antirepresentation L.*

(c) *$R|_{T(g)}$ is a continuous representation of M on $T(g)$.*

(d) *$T(g) \subseteq \mathbb{C}g + W(g)$, and the sum on the right is direct.*

Proof. (a) If g is abelian we get in the notation of Theorem 8.13 that $T(g) \subseteq \text{span}\{\chi_1, \chi_2\}$, so that $\dim T(g) \leq 2$. If g is non-abelian then we refer to Lemma 8.22(c), the proof of which does not depend on the present proof.

(b) By definition $T(g)$ is invariant under R. It is also invariant under left translations, because g is central (Lemma 8.7(a)).

(c) is Proposition E.16(a).

(d) We have seen in Lemma 8.8 that $g_x \in W(g)$ for all $x \in M$. Since g is central this says that $R(x)g - g(x)g \in W(g)$, from which we get that $R(x)g \in \mathbb{C}g + W(g)$. It is left to show that the sum is direct. But $g(e) = 1$ by definition so that $w(e) = 0$ for all $w \in W(g)$. (Put $x = y = e$ in the definition of $W(g)$.) □

8.7 Non-abelian pre-d'Alembert functions

Throughout Section 8.7 we let $g \in C(M)$ be a continuous, non-abelian pre-d'Alembert function on a topological monoid M. The corresponding function $\Delta = \Delta_g : M \times M \to \mathbb{C}$ is not identically 0 (Lemma 8.10), and we fix a pair of points $a, b \in M$ such that $\Delta(a, b) \neq 0$.

We shall in the Examples 9.10, 9.11 and 9.13 see that there exist non-abelian d'Alembert functions on certain groups. Naturally we want to characterize all such d'Alembert functions. Actually we are more ambitious: We want to characterize all pre-d'Alembert functions on monoids. We have already found all abelian solutions of the pre-d'Alembert functional equation (Theorem 8.13), so only the non-abelian ones are left.

The purpose of this section is to characterize the non-abelian pre-d'Alembert functions on M. A benefit of our treatment is that it also gives us the solutions w of the symmetrized sine addition formula (4.3) for the given g.

Lemma 8.21. *The three functions*

$$w_1 := \frac{g_b(b)}{\Delta(a,b)}\, g_a - \frac{g_a(b)}{\Delta(a,b)}\, g_b + \frac{g(b)}{\Delta(a,b)}\, \frac{g_{ab} - g_{ba}}{2},$$

$$w_2 := -\frac{g_a(b)}{\Delta(a,b)}\, g_a + \frac{g_a(a)}{\Delta(a,b)}\, g_b + \frac{g(a)}{\Delta(a,b)}\, \frac{g_{ab} - g_{ba}}{2},$$

$$w_3 := -\frac{1}{\Delta(a,b)}\, \frac{g_{ab} - g_{ba}}{2} = -\frac{1}{\Delta(a,b)}\, \frac{(R_{ab} - R_{ba})g}{2}$$

in span$\{g_x \mid x \in M\}$ *have the properties*

$$\begin{aligned} w_1(a) &= 1, & w_1(b) &= 0, & w_1(ab) &= 0, \\ w_2(a) &= 0, & w_2(b) &= 1, & w_2(ab) &= 0, \\ w_3(a) &= 0, & w_3(b) &= 0, & w_3(ab) &= 1. \end{aligned}$$

Proof. The proof relies on the following formulas, most of which depend on the fact that g is central:

$$\begin{aligned} g_a(b) &= g_b(a), \\ g_{ab}(a) - g_{ba}(a) &= g_{ab}(b) - g_{ba}(b) = 0, \\ g_{ab}(ab) - g_{ba}(ab) &= -2\Delta(a,b) && \text{by Definition (B.3)}, \\ g_a(a)g_b(b) - g_a(b)^2 &= \Delta(a,b) && \text{from Lemma 8.9}, \\ g_x(xy) = g_x(yx) &= g(x)g_x(y) + g_x(x)g(y) && \text{for all } x, y \in M. \end{aligned}$$

For the last identity note first that $g_x(xy) = g_x(yx)$ by the centrality of g, and then replace y by x and z by y in Lemma 8.8.

We leave the rest of Lemma 8.21 to the reader. □

We continue by deriving some properties first of $T(g)$ and then of the restriction of the right regular representation R to $T(g)$. We have already noted some properties in Lemma 8.20.

Lemma 8.22.

(a) $W(g) = \text{span}\{g_x \mid x \in M\} = \text{span}\{w_1, w_2, w_3\}$.

(b) $T(g) = \mathbb{C}g + W(g)$, *where the sum is direct.*

(c) $\dim T(g) = 4$ *and* $\dim W(g) = 3$.

(d) $\{f \in T(g) \mid f \text{ is central}\} = \mathbb{C}g$.

Proof. (a) Lemma 8.8 gives the inclusion span$\{g_x \mid x \in M\} \subseteq W(g)$. To prove the converse inclusion we let $w \in W(g)$ be arbitrary. We use the notation of Lemma 8.21. w and $w(a)w_1 + w(b)w_2 + w(ab)w_3$ assume the same values at a, b and ab, so they coincide everywhere on M according to Corollary 6.2. (Note by help of Lemma 8.9 that $\widetilde{\Delta} = 4\Delta \neq 0$.) Thus $w = w(a)w_1 + w(b)w_2 + w(ab)w_3 \in \text{span}\{g_x \mid x \in M\}$.

(b) By Lemma 8.20(d) it suffices to prove that $g \in T(g)$ and $W(g) \subseteq T(g)$. The last is equivalent to $g_x \in T(g)$ by (a). And $g = R(e)g \in T(g)$ and $g_x = R(x)g - g(x)g \in T(g) + T(g) \subseteq T(g)$.

(c) A basis of $T(g)$ is $\{g, w_1, w_2, w_3\}$.

(d) Since g is central it suffices by (b) to show that any central $w \in W(g)$ is 0. For such w we get from the definition of $W(g)$ that $w(xy) = w(x)g(y) + w(y)g(x)$, which means that (w, g) is a solution of the sine addition formula. If $w \neq 0$ then g is abelian according to Theorem 4.1(e), contradicting our assumption about g being non-abelian. Hence $w = 0$. □

We go on to properties of the representation $R|_{T(g)}$ of M on $T(g)$.

Lemma 8.23.

(a) *If $W \neq \{0\}$ is an R-invariant subspace of $T(g)$, then the character $\chi_{R|W}$ of the representation $R|W$ is $\chi_{R|W} = \dim(W)\, g$.*

(b) *There are no 1-dimensional, R-invariant subspaces of $T(g)$.*

Proof. (a) Let W be an R-invariant subspace of $T(g)$. The character $\chi_{R|W}$ is in $C(R|_W)$, hence in $C(R|_{T(g)})$ which equals $T(g)$ by Proposition E.16(b). Thus $\chi_{R|_W} \in T(g)$. Characters being central, Lemma 8.22(d) tells us that $\chi_{R|_W} = \alpha g$ for some constant $\alpha \in \mathbb{C}$. Finally, $\alpha = \alpha g(e) = \chi_{R|_W}(e) = \operatorname{tr}((R|_W)(e)) = \operatorname{tr} I_W = \dim W$.

(b) We shall arrive at a contradiction assuming the existence of a 1-dimensional, R-invariant subspace W of $T(g)$. From (a) we read that $g = \chi_{R|W}$, so $\chi_{R|W}$ is non-abelian. On the other hand a character of a 1-dimensional representation is multiplicative and hence abelian. □

A conditioned reflex for mathematicians doing representation theory is to study irreducible subrepresentations of any given representation. So that is what we do for the representation $R|_{T(g)}$ of M on $T(g)$. Note that $\dim T(g) = 4$ (Lemma 8.22(c)), so that we are deal with finite-dimensional vector spaces, not infinite-dimensional.

Choose an R-invariant subspace $W \neq \{0\}$ of $T(g)$ such that $\rho := R|_W$ is irreducible. Such a subspace exists (take W as an R-invariant subspace of minimal dimension ≥ 1). Let χ_ρ denote the character of ρ (i.e., $\chi_\rho(x) = \operatorname{tr} \rho(x)$ for all $x \in M$), and $C(\rho)$ the space of matrix-coefficients of ρ. For any $A \in \mathcal{L}(W)$ we define $c_A \in C(\rho)$ by $c_A(x) := \frac{1}{2} \operatorname{tr}(A\rho(x))$, $x \in M$, so $c : \mathcal{L}(W) \to C(\rho)$. We put $\operatorname{sl}(W) := \{A \in \mathcal{L}(W) \mid \operatorname{tr} A = 0\}$. Finally we recall that $d(x) := 2g(x)^2 - g(x^2)$, $x \in M$. Theorem 8.24 and Proposition 8.25 describe in terms of ρ how the non-abelian pre-d'Alembert function g looks like and what its associated space $W(g)$ is.

Theorem 8.24.

(a) *ρ is an irreducible, continuous representation of M on W.*

(b) $\dim W = 2$.

(c) $g = \frac{1}{2}\chi_\rho$. *This formula determines the irreducible representation ρ up to equivalence from g.*

(d) $d = \det \rho$ *and* $T(g) = C(\rho)$.

(e) *The map $A \mapsto c_A$ is an isomorphism of $\mathcal{L}(W) = \mathbb{C}I \oplus \mathrm{sl}(W)$ onto $T(g) = \mathbb{C}g \oplus W(g)$, mapping $\mathbb{C}I$ onto $\mathbb{C}g$ and $\mathrm{sl}(W)$ onto $W(g)$. In particular $W(g) = \{c_A \mid A \in \mathrm{sl}(W)\}$.*

(f) *g is bounded $\Leftrightarrow$ ρ is bounded. If M is a group then g is bounded $\Leftrightarrow$ ρ is equivalent to a unitary representation.*

Proof. (a) ρ is chosen irreducible. Lemma 8.20(c) tells us that it is continuous.

(b) We get $C(\rho) = C(R|_W) \subseteq C(R|_{T(g)}) = T(g)$ by Proposition E.16(b). Using Corollary E.12, Lemma 8.23(b) and that $\dim T(g) = 4$ we infer

$$\dim T(g) \geq \dim C(\rho) = (\dim \rho)^2 = (\dim W)^2 \geq 2^2 = 4 = \dim T(g),$$

which proves not just (b), but also the statement $T(g) = C(\rho)$ in (d).

(c) That $\chi_\rho = 2g$ is immediate from Lemma 8.23(a). The essential uniqueness follows from Theorem E.13(b).

(d) Using (c) and the easily verified fact that $(\mathrm{tr}\, A)^2 - \mathrm{tr}(A^2) = 2 \det A$ for all 2×2 matrices A we find that

$$d(x) = 2g(x)^2 - g(x^2) = \tfrac{1}{2}[(\mathrm{tr}\, \rho(x))^2 - \mathrm{tr}(\rho(x)^2)] = \det \rho(x).$$

(e) Noting that $T(g) = C(\rho)$ we get from Corollary E.12 that the map is an isomorphism.

That the map sends $\mathbb{C}I$ onto $\mathbb{C}g$ is obvious from $\chi_\rho = 2g$. For the final statement of (e) it suffices by dimension arguments to prove that $\mathrm{tr}\, A = 0 \Rightarrow f := \mathrm{tr}(A\rho(-)) \in W(g)$, so assume that $\mathrm{tr}\, A = 0$. Writing $f = \alpha g + w$, where $\alpha \in \mathbb{C}$ and $w \in W(g)$ (Lemma 8.22(a)) it now suffices to prove that $\alpha = 0$. Evaluating at $x = e$ we find that $\alpha = \alpha g(e) + w(e) = f(e) = \frac{1}{2}\mathrm{tr}(A\rho(e)) = \frac{1}{2}\mathrm{tr}(A) = 0$.

(f) If ρ is bounded, then so is $g = \frac{1}{2}\chi_\rho$. Assuming conversely that g is bounded we shall prove that the matrix-coefficients of ρ are bounded. If g is bounded, then so are all translates of it, and so $T(g) = \mathrm{span}\{R(x)g \mid x \in M\}$ consists of bounded functions. And $C(\rho) = T(g)$.

A result by Weil (see [111, 22.23(c)]) says that there exists an invertible matrix A such that $A^{-1}\rho(x)A$ is unitary for all $x \in M$, when ρ is bounded and M a group. □

We finally note a partial converse of Theorem 8.24(c):

Proposition 8.25. *Let ρ is a 2-dimensional irreducible representation of M. Then $g := \frac{1}{2}\operatorname{tr}\rho = \frac{1}{2}\chi_\rho$ is pre-d'Alembert function. Furthermore g is non-abelian $\Leftrightarrow$ ρ is irreducible.*

Proof. It was established in Example 8.4 that $g := \frac{1}{2}\operatorname{tr}\rho$ is a pre-d'Alembert function.

Let g be non-abelian. Then there exists an irreducible 2-dimensional representation ρ' of M such that $g = \frac{1}{2}\chi_{\rho'}$ (Theorem 8.24(c)), so $\chi_{\rho'} = \chi_\rho$. If ρ is not irreducible χ_ρ can be split into a sum of characters of 1-dimensional representations. But that contradicts Theorem E.13(b). So ρ is irreducible.

Assume conversely that g is abelian. Then (Theorem 8.13) $g = \frac{1}{2}\chi_{\rho'}$, where

$$\rho'(x) := \begin{pmatrix} \chi_1(x) & 0 \\ 0 & \chi_2(x) \end{pmatrix}, \qquad x \in M,$$

is clearly not irreducible. From $\chi_\rho = \chi_{\rho'}$ we conclude from Theorem E.13(b) that ρ is not irreducible either. □

8.8 Davison's structure theorem

We extract the essence of the present chapter in this section. In contrast to Section 8.7 we discuss general pre-d'Alembert functions, not just non-abelian, so we also take the results of Section 8.4 into account.

Let M denote a topological monoid with a neutral element e. Example 8.3 gave the first glimpse of Davison's structure theorem (Theorem 8.26) that describes the pre-d'Alembert functions on M in terms of characters of 2-dimensional representations of M.

The function $d : M \to \mathbb{C}$, given by $d(x) := 2g(x)^2 - g(x^2)$, $x \in M$, is multiplicative (Lemma 8.9), but the reason for this algebraic property has been a mystery until now. Theorem 8.26(b) reveals why $d : M \to \mathbb{C}$ is multiplicative.

Replacing the 2-dimensional vector space W from Section 8.7 by the isomorphic vector space $\mathbb{C}^2$ we arrive at our version of Davison's structure theorem:

Theorem 8.26 (Davison).

(a) *The continuous pre-d'Alembert functions on M are the functions of the form $g = \frac{1}{2}\chi_\rho = \frac{1}{2}\operatorname{tr}\rho$, where ρ ranges over the continuous representations of M on $\mathbb{C}^2$.*

Below we let $g = \frac{1}{2}\chi_\rho$, where ρ is a continuous representation of M on $\mathbb{C}^2$.

(b) $d(x) = \det\rho(x)$ *for all $x \in M$.*

(c) *g is non-abelian if and only if ρ is irreducible. If g is non-abelian, then ρ is unique up to equivalence of representations.*

(d) *If g is abelian, then $g = (\chi_1 + \chi_2)/2$ where $\chi_1, \chi_2 \in C(M)$ are multiplicative functions such that $\chi_1(e) = \chi_2(e) = 1$. The multiplicative functions χ_1 and χ_2 are unique, except that they may be interchanged.*

Proof. Contained in the previous sections of the present chapter. □

Theorem 8.26 is called a structure theorem because it tells us what the structure of a pre-d'Alembert function is: It is the composition of a 2-dimensional representation with the normalized trace. The structure theorem does not tell us how to find a 2-dimensional representation ρ for a prescribed pre-d'Alembert function g. It might be added that the representation ρ is not uniquely determined: It can be replaced by the algebraically equivalent representation $x \mapsto A\rho(x)A^{-1}$ for any $A \in GL(\mathbb{C}^2)$, because that leaves the trace unchanged. For non-abelian solutions this is the only possibility according to Theorem 8.26(c), but this is not so in general: If $a : G \to \mathbb{C}$ is an additive function, $a \neq 0$, then the representations

$$\rho_1(x) := \begin{pmatrix} 1 & 0 \\ 0 & 1 \end{pmatrix} \quad \text{and} \quad \rho_2(x) := \begin{pmatrix} 1 & a(x) \\ 0 & 1 \end{pmatrix}$$

have the same trace, but they are inequivalent.

Mathematicians have studied group representations intensively, so a great deal about representations of any specific group G can be found in the literature. A knowledge of the 2-dimensional representations of G will tell us what the possible pre-d'Alembert functions on G are. It might be added that only few groups have irreducible representations of dimension 2. See Examples 9.13 and 9.14 for illustrations.

Corollary 8.27. *Let G be a topological group. If $g \in C(G)$ is a bounded pre-d'Alembert function, then there exists a continuous unitary representation ρ of G on $\mathbb{C}^2$ such that $g = \frac{1}{2}\chi_\rho$.*

Proof. It suffices to prove that there is a bounded, continuous, representation ρ of G on $\mathbb{C}^2$ such that $g = \frac{1}{2}\chi_\rho$. Indeed, according to [111, 22.23(c)] in that case there exists an invertible matrix A such that $A^{-1}\rho(x)A$ is unitary for all $x \in G$. We may thus replace ρ by $x \mapsto A^{-1}\rho(x)A \in U(2)$.

If g is non-abelian, we refer to Theorem 8.24(f).

Let us next assume that g is abelian. Here ρ may according to the proof of Theorem 8.26 be chosen in the form

$$\rho(x) = \begin{pmatrix} \chi_1(x) & 0 \\ 0 & \chi_2(x) \end{pmatrix}, \qquad x \in G.$$

Since $g = (\chi_1 + \chi_2)/2$ is bounded, so are χ_1 and χ_2 by Proposition 3.18(c). □

If G in Corollary 8.27 is compact, in particular if it is finite, then any continuous complex-valued function on it is bounded. So on a compact group any continuous pre-d'Alembert function g has the form $g = \frac{1}{2}\operatorname{tr}\rho$, where ρ is a continuous, unitary representation of G on $\mathbb{C}^2$.

In the following theorem the definition of solvable (Definition A.8) has to be modified so that we can apply Lie's theorem. The modification is not necessary for Lie groups as we shall see in Corollary 8.18.

Definition 8.28. Let G be a Hausdorff topological group. Let $G^{(1)}$ be the smallest closed subgroup of G containing $[G, G]$ and define successively $G^{(k+1)} := (G^{(k)})'$. If there is an $n \in \mathbb{N}$ such that $G^{(n)} = \{e\}$ we say that G is a *topologically solvable* group.

Theorem 8.29. *Any continuous pre-d'Alembert function on a connected, topologically solvable group is abelian.*

Proof. By Lie's theorem (see [112, Theorem 29.42]) any strongly continuous, finite-dimensional, irreducible representation of such a group is 1-dimensional. In particular it has no 2-dimensional, continuous, irreducible representations. We are done by Davison's structure theorem (Theorem 8.26). □

Remark 8.30. We have only worked with continuous solutions g of the pre-d'Alembert function, but giving M the discrete topology our work covers all pre-d'Alembert functions. The fact that the solutions of the pre-d'Alembert functional equation have a very special form enables us to prove that any measurable solution g on a locally compact group G is continuous:

If g is abelian then g has the form $g = (\chi_1 + \chi_2)/2$, where χ_1 and χ_2 are characters. They are measurable by Theorem 3.18(f) and consequently continuous (Proposition D.4(b)).

If g is non-abelian we write $g = \frac{1}{2}\operatorname{tr}\rho$ as in Theorem 8.24, where ρ is a 2-dimensional, irreducible representation of G. Since $C(\rho) = T(g)$ (Theorem 8.24(c)) we find that the matrix-coefficients of ρ are in the span $T(g)$ of the translates of the measurable function g, so they are measurable. By Proposition D.4(d) ρ is continuous. Hence so is $g = (\operatorname{tr}\rho)/2$.

8.9 Exercises

Exercise 8.1. Show that the continuous, real-valued solutions of the pre-d'Alembert functional equation on the real line are

1. $g = 0$,
2. $g(x) = e^{\alpha x}\cos(\beta x)$, $x \in \mathbb{R}$, where $\alpha, \beta \in \mathbb{R}$, and
3. $g(x) = e^{\alpha x}/2$, $x \in \mathbb{R}$, where $\alpha \in \mathbb{R}$.

Exercise 8.2. What is the function $d : S \to \mathbb{C}$ in Theorem 8.13?

Exercise 8.3. In this exercise we shall fill in the missing arguments in the proof of the following

Proposition 8.31. *If g is a solution of the pre-d'Alembert functional equation on a semigroup S such that $g(x^2) = g(x)^2$ for all $x \in S$, then g is multiplicative.*

Proof. The conclusion of the proposition may be expressed as $g_x(y) = 0$ for all $x, y \in S$, and that is what we shall prove.

(a) Prove that $d(x) = g(x)^2$ for all $x \in S$.

(b) Show that $g(xy)^2 = g(x)^2 g(y)^2$ for all $x, y \in S$.
Hint: Lemma 8.9(b).

(c) Deduce that $[g(xy) - g(x)g(y)][g(xy) + g(x)g(y)] = 0$ or equivalently that

$$g_x(y)[g_x(y) + 2g(x)g(y)] = 0 \qquad \text{for all } x, y \in S. \tag{8.13}$$

(d) Show that

$$g_x(y)[g_x(y) + g(x)g(y)] = 0 \qquad \text{for all } x, y \in S. \tag{8.14}$$

□

Hint: Replace y by y^2 in (8.13) and use Lemma 8.9(a). Then distinguish between the two cases $g(y) \neq 0$ and $g(y) = 0$. In the last case (8.14) is the same as (8.13).

(e) Show that $g_x(y) = 0$.

Hint: Combine (8.13) and (8.14).

Exercise 8.4. In Theorem 8.24 we deduced that there exists a 2-dimensional irreducible R-invariant subspace of $T(g)$ is 2-dimensional, but we did not give any explicit construction of such a space. Here is one. We use the notation of Lemma 8.21. Furthermore that $(L'(x)F)(y) := F(xy)$ for all $x, y \in M$.

Fix $\sigma \in \mathbb{C}$ such that $\sigma^2 = -\Delta$, and define

$$T^{\pm}(g) := \{f \in T(g) \mid \Lambda f = \pm\sigma f\}, \qquad \text{where } \Lambda := \frac{L'(ab) - L'(ba)}{2}.$$

(a) Show that $T^+(g)$ and $T^-(g)$ are R-invariant subspaces of $T(g)$ that form direct sum.

(b) Show that

$$\left[\frac{L'(xy) - L'(yx)}{2}\right]^2 g = -\Delta(x,y)g \qquad \text{for all } x, y \in M.$$

(c) Show that $\Lambda^2 g = \sigma^2 g$ and as a consequence that $\Lambda^2 f = \sigma^2 f$ for all $f \in T(g)$.

(d) Show that $\Lambda(g + \sigma w_3) = \sigma(g + \sigma w_3)$ and $\Lambda(g - \sigma w_3) = -\sigma(g - \sigma w_3)$, so that $T^+(g) \neq \{0\}$ and $T^-(g) \neq \{0\}$.

(e) Deduce that $T^+(g)$ is a 2-dimensional, irreducible subspace of $T(g)$.

8.10 Notes and remarks

Most of the results of this and the next chapter are due to Davison [63, 64]. In particular the crucial Lemmas 8.9, 8.10 and 8.22, where we have followed the treatment of [64]. Having these lemmas we rely for the rest partly on the theory of matrix-elements of representations. By doing so we shortcut and simplify the exposition, making it less computational and more transparent.

Definition 8.1:

(1) The pre-d'Alembert functional equation got its name in Davison's paper [64], but it occurred earlier in another paper of his [63, Proposition 2.1]. In his discussion of pre-d'Alembert functions he was inspired by analogous, known properties of the trace function as found in, e.g., Thurston [200, Problem 2.6.9].

(2) Davison's definition of the pre-d'Alembert functional equation is formally different from ours: Instead of our term $2g(y)g(xz)$ in (8.1) he has $2g(y)g(zx)$. However, on monoids g is central in either case, so the two definitions agree on monoids. We have chosen the form (8.1), because it has the pleasant property that any solution g of it is central, also on semigroups.

Lemma 8.9: If $x \in S$ and $y \in S$ commute, or if $y \in Z(g)$, then $\Delta = 0$, which according to Lemma 8.9 has the consequence that $g_x(y)^2 = g_x(x)g_y(y)$ or equivalently that

$$(g(xy) - g(x)g(y))^2 = [g(x)^2 - d(x)]\,[g(y)^2 - d(y)]. \tag{8.15}$$

This identity was for d'Alembert functions on groups derived in Corovei [50, Lemma 1]. (8.15) implies Davison [63, Proposition 2.2].

Lemma 8.10: The criterion of Lemma 8.10 is due to Davison [64, Proposition 2.7]. Stetkær [195, Theorem 3.7(c)] has it for d'Alembert functions.

Proposition 8.11 is due to Davison [64, Proposition 2.11].

Proposition 8.14: We encounter the conditions $g(z_0)^2 \neq d(z_0)$ (equivalently $g_{z_0}(z_0) \neq 0$) and $g(z_0)^2 = d(z_0)$ of Proposition 8.14 in other surroundings: For the symmetrized sine addition formula in Proposition 6.5. For d'Alembert's functional equation in Proposition 9.23 and Proposition 9.30. For d'Alembert's long functional equations in Theorem 10.1(d) and (e).

Proposition 8.15: In both [129, Lemma 4.1] and [132, Result 3.17a] Kannappan derives the identity (9.17), i.e., $g = (\chi + \check{\chi})/2$, under the assumption that g is 1 on the entire commutator group $[G, G]$ ([132, Result 3.17a] imposes in addition the condition that $[G, G]$ is a torsion group). Our Proposition 8.15 requires only that g is 1 on each commutator $[x, y]$.

Corollary 8.18: Peter Friis [90, Corollary 2.8] has Corollary 8.18 for d'Alembert functions on nilpotent, connected Lie groups. Actually for the more general case of non-zero solutions of d'Alembert's long functional equation [90, Theorem 2.6].

Remark 8.30: Járai [116, Section 22.5] derives the result for d'Alembert functions under an additional assumption about the underlying locally compact group G.

Proposition 8.31 is adapted from Davison [64, Proposition 3.3].

Chapter 9

D'Alembert's Functional Equation

9.1 Introduction

Although d'Alembert's functional equation

$$g(x+y)+g(x-y)=2g(x)g(y) \qquad \text{for all } x,y\in\mathbb{R}, \tag{9.1}$$

for functions $g:\mathbb{R}\to\mathbb{C}$ on the real line has it roots back in d'Alembert's investigations of vibrating strings [13] from 1750, it is more pertinent for us to note that the equation is closely connected to the trigonometrical functions and hence to the group structure of the real line. Indeed, one solution of (9.1) is $g(x)=\cos x$, another is $g(x)=\cosh x$. This accounts for the fact that the functional equation is sometimes called *the cosine equation*. As we shall see, the general continuous solution $g\neq 0$ of it is

$$g(x)=\frac{e^{\alpha x}+e^{-\alpha x}}{2}, \qquad x\in\mathbb{R}, \text{ where } \alpha\in\mathbb{C}.$$

With $\alpha=i$ we get the solution $\cos x$, and with $\alpha=1$ the solution $\cosh x$.

The obvious extension of (9.1) from $\mathbb{R}$ to an abelian group $(G,+)$ is the functional equation

$$g(x+y)+g(x-y)=2g(x)g(y) \qquad \text{for all } x,y\in G, \tag{9.2}$$

where $g:G\to\mathbb{C}$ is the unknown. At this point we cannot resist to mention Kannappan's amazing extension of the result on $\mathbb{R}$ to a general abelian group: The non-zero solutions of (9.2) are the functions of the form $g(x)=(\chi(x)+\chi(-x))/2$, $x\in G$, where χ is a character in G (Corollary 9.22). $\chi(x)=e^{\alpha x}$ gives the result cited above about $G=\mathbb{R}$. The generality of Kannappan's result is remarkable: It holds for any abelian group and no assumptions about continuity or measurability are imposed.

If the group G is not assumed abelian we consider the solutions $g : G \to \mathbb{C}$ of

$$g(xy) + g(xy^{-1}) = 2g(x)g(y) \qquad \text{for all } x, y \in G. \tag{9.3}$$

The aim of this chapter is to find the solutions $g : G \to \mathbb{C}$ of (9.3) and some of their properties. Much of the discussion still works, when we replace the group inversion $x \mapsto x^{-1}$ by a general involution τ (the notion of an involution is defined in Definition A.30) as follows:

Definition 9.1. *D'Alembert's functional equation* on a semigroup S with involution $\tau : S \to S$ (abbreviated *d'Alembert's functional equation* or often just *d'Alembert's equation*) is the functional equation

$$g(xy) + g(x\tau(y)) = 2g(x)g(y) \qquad \text{for all } x, y \in S, \tag{9.4}$$

where we want to find $g : S \to \mathbb{C}$.

A *d'Alembert function* is a non-zero solution of (9.4) (where S and τ should be clear from the context or stated).

We call the special case (9.3) of (9.4) the *classical d'Alembert functional equation* and its non-zero solutions the *classical d'Alembert functions.*

If S is a monoid with neutral element e then a solution g of (9.4) is non-zero, i.e., is a d'Alembert function, if and only if $g(e) = 1$ (Put $y = e$ in (9.4)). So on a monoid a d'Alembert function g can alternatively be defined as a solution of (9.4) for which $g(e) = 1$.

D'Alembert's functional equation has always at least one non-zero solution, to wit the constant function $g = 1$.

Although our principal interest lies in the case where the underlying semigroup is a topological group, we will not impose that as a blanket assumption. Actually, most of our results will be derived for monoids, and our way of proceeding is mainly algebraic.

The manner in which the involution τ enters the solution formulas reveals the basic structure of these formulas in a more transparent way than the classical group inversion can (for an illustration compare Corollary 9.22 with Kannappan's original formula $g(x) = (\chi(x) + \chi(-x))/2$).

Our treatment of d'Alembert's functional equation relies heavily on our study of pre-d'Alembert functions in Chapter 8, because we exploit that d'Alembert functions are pre-d'Alembert functions (Proposition 9.17(c)). So the d'Alembert functions have all the properties of pre-d'Alembert functions. This explains why the results of the present chapter are for a large part corollaries of the corresponding results of Chapter 8. The most important results are Kannappan's theorem (Corollary 9.22) and Davison's structure theorem (Theorem 9.26). Kannappan's result covers the abelian case of d'Alembert's functional equation, while Davison's result takes care of the non-abelian.

9.2 Examples of d'Alembert functions

In this section we present selected examples of continuous d'Alembert functions on specific groups. The examples throw light on later general results about d'Alembert functions, and some of them serve as counter-examples that reveal how far these results remain true and where they break down. Examples 9.10 and 9.11 show that non-abelian, classical d'Alembert functions exist. Example 9.3, in which $S = \mathbb{R}$, discloses that continuous, classical d'Alembert functions may be real-valued, bounded or unbounded.

Lemma 9.2 below produces abelian d'Alembert functions of a special form that exist on any semigroup with involution. These functions will be put into perspective in our discussion of Kannappan's result (Section 9.4), because we show there that all abelian d'Alembert functions have this form.

Lemma 9.2. *Let $\tau : S \to S$ be an involution of a semigroup S. If $\chi : S \to \mathbb{C}$ is multiplicative, then*

$$g := \frac{\chi + \chi \circ \tau}{2} \tag{9.5}$$

is an abelian solution of d'Alembert's functional equation (9.4).

Proof. Simple computations that we leave to the reader. □

Except for Example 9.11 we let throughout the rest of this section τ be the group inversion $\tau(x) = x^{-1}$ so that we deal with the classical functional equation (9.3).

Example 9.3. We will in this example and the next one (Example 9.5) compute all continuous solutions g of d'Alembert's classical functional equation on $\mathbb{R}^n$, more precisely of the functional equation

$$g(x+y) + g(x-y) = 2g(x)g(y) \qquad \text{for all } x, y \in \mathbb{R}^n. \tag{9.6}$$

In other words, we will find all continuous, classical d'Alembert functions on $\mathbb{R}^n$. In the present example we discuss the case of $n = 1$. The result

(Proposition 9.4) is a special case of the one of Example 9.5 and the method of Example 9.5 works also for $n = 1$. Nevertheless we present here for $n = 1$ a direct derivation. The reason is that we want to give a concrete illustration of how differential equations can assist in solving functional equations. This is not the first time we do so, because we used differential equations to find the continuous characters of $\mathbb{R}$ (Example 3.7(a)).

Let $g \in C(\mathbb{R})$ be a non-zero solution of (9.6). Then $g \in C^\infty(\mathbb{R})$ by Example D.8, so it is allowed to differentiate (9.6). We do so with respect to y at $y = 0$ and arrive at the differential equation $g'' = g''(0)g$.

Assume first that $g''(0) = 0$, so that $g'' = 0$. This means that $g(x) = Ax + B$ where $A, B \in \mathbb{C}$ are constants. Substituting this into the functional equation (9.6) we find that $A = 0$ and $B = 1$, i.e., $g = 1$.

Assume next that $g''(0) \neq 0$. Choosing $\alpha \in \mathbb{C} \setminus \{0\}$ such that $\alpha^2 = g''(0)$ we get from the differential equation that g has the form

$$g(x) = Ae^{\alpha x} + Be^{-\alpha x} \qquad \text{for all } x \in \mathbb{R},$$

where $A, B \in \mathbb{C}$ are complex constants. Due to the fact that the left hand side of (9.6) does not change, if y is replaced by $-y$ we see that $g(x) = g(-x)$, which gives us that

$$Ae^{\alpha x} + Be^{-\alpha x} = Ae^{-\alpha x} + Be^{\alpha x} \qquad \text{for all } x \in \mathbb{R},$$

that we reformulate to

$$(A - B)(e^{\alpha x} - e^{-\alpha x}) = 0 \qquad \text{for all } x \in \mathbb{R}.$$

This implies that $A = B$. Finally, from $g(0) = 1$ (put $y = 0$ in (9.6)) we get that $A = \frac{1}{2}$, so that

$$g(x) = \frac{e^{\alpha x} + e^{-\alpha x}}{2}, \qquad x \in \mathbb{R}, \tag{9.7}$$

is the only possible solution in this case.

Conversely the g from (9.7) is a continuous non-zero solution of (9.6) for any $\alpha \in \mathbb{C}$. This is immediate from Lemma 9.2, but is also easily checked directly.

Writing $\alpha = i\lambda$ we sum up the discussion above as follows:

Proposition 9.4. *The continuous, non-zero solutions of d'Alembert's functional equation* (9.1) *on the real line are the functions* $g_\lambda(x) := \cos(\lambda x)$, $x \in \mathbb{R}$, *where* λ *ranges over* $\mathbb{C}$.

The only real-valued solutions among the ones listed in Proposition 9.4 are $g_r(x) = \cos(rx)$ and $g_{ri}(x) = \cosh(rx)$, where $r \in \mathbb{R}$.

Example 9.5. The non-zero, continuous solutions of d'Alembert's classical functional equation (9.6) on $\mathbb{R}^n$ are the functions of the form

$$g_\lambda(x) := \frac{e^{i\langle\lambda,x\rangle} + e^{-i\langle\lambda,x\rangle}}{2} = \cos\langle\lambda, x\rangle, \qquad x \in \mathbb{R}^n,$$

where λ ranges over $\mathbb{C}^n$.

To see this combine Corollary 9.22 with Example 3.7(b).

Example 9.6. The continuous, non-zero solutions of d'Alembert's classical functional equation (9.3) on the multiplicative group $(\mathbb{R}^+, \cdot)$ are the functions of the form

$$g_\alpha(t) := \frac{t^\alpha + t^{-\alpha}}{2}, \qquad t \in \mathbb{R}^+, \tag{9.8}$$

where α ranges over $\mathbb{C}$.

This follows from Proposition 9.4 because $x \mapsto \exp x$ is a topological isomorphism of $(\mathbb{R}, +)$ onto $(\mathbb{R}^+, \cdot)$.

Example 9.7. Davison [62] noted that any classical d'Alembert function g on the group $(\mathbb{Z}, +)$ is given by the values of the Chebyshev polynomials $\{T_n\}_{n=0}^\infty$ at a point of $\mathbb{C}$. More precisely that the formula $g(n) = T_n(g(1))$ holds for $n \geq 0$. In this example we derive and generalize Davison's formula.

The *Chebyshev polynomials* (of the first kind) T_n, $n = 0, 1, \ldots$, are defined by $T_0(s) = 1$, $T_1(s) = s$ and the recurrence relation $T_{n+1}(s) = 2sT_n(s) - T_{n-1}(s)$ for $s \in \mathbb{C}$ and $n \geq 1$. T_n is a polynomial of degree n with leading coefficient 2^{n-1} for $n \geq 1$. An easy induction on n gives the formula

$$T_n\Big(\frac{z + z^{-1}}{2}\Big) = \frac{z^n + z^{-n}}{2} \qquad \text{for } z \in \mathbb{C} \setminus \{0\}, \tag{9.9}$$

of which a particular case is

$$T_n(\cos x) := \cos(nx), \qquad x \in \mathbb{R}, n = 0, 1, \ldots \tag{9.10}$$

(9.10) is often taken as the definition of the Chebyshev polynomials.

Our theory tells us (combine Corollary 9.22 with Example 3.11) that the classical d'Alembert functions on $\mathbb{Z}$ are the functions of the form $g(n) := (z^n + z^{-n})/2$, $n \in \mathbb{Z}$, where z ranges over $\mathbb{C}^*$. From this we obtain Davison's formula $g(n) = T_n(g(1))$ by (9.9). We conclude that the classical d'Alembert functions g on $\mathbb{Z}$ are the functions $g(n) = T_n((z + z^{-1})/2)$, $n = 0, 1, 2, \ldots$, where z ranges over $\mathbb{C}^*$. An example is $g(n) = (-1)^n$ corresponding to $z = -1$.

Cosine is a d'Alembert function on the group $\mathbb{R}$, so it is perhaps not surprising that the relation (9.10) carries over to any group as shown by Proposition 9.8.

Proposition 9.8. *If g is a classical d'Alembert function on a group G then $T_n(g(x)) = g(x^n)$ for any $x \in G$ and any $n = 0, 1, \ldots$*

Proof. Let $x \in G$. We know from Proposition 9.17(e) that $g(x^0) = g(e) = 1$, so $T_0(g(x)) = g(x^0)$. Furthermore $g(x^1) = g(x) = T_1(g(x))$, because $T_1(s) = s$. To prove the result for the remaining indices we proceed by induction: Assume that $T_n(g(x)) = g(x^n)$ and $T_{n-1}(g(x)) = g(x^{n-1})$ for some $n \geq 1$. Then we get from

$$g(x^{n+1}) + g(x^{n-1}) = g(x^n x) + g(x^n x^{-1}) = 2g(x^n)g(x)$$

by the recurrence relations for the Chebyshev polynomials that

$$\begin{aligned} g(x^{n+1}) &= 2g(x^n)g(x) - g(x^{n-1}) \\ &= 2g(x)T_n(g(x)) - T_{n-1}(g(x)) = T_{n+1}(g(x)), \end{aligned}$$

finishing the proof. □

Taking $G = (\mathbb{Z}, +)$ in Proposition 9.8 we rederive Davison's formula. Thus Proposition 9.8 extends Davison's result.

Example 9.9. In this example we consider the $(ax+b)$-group G. We exhibit all continuous, classical d'Alembert functions g on it.

By Theorem 8.16 (or Corollary 8.18) g is abelian, so by Kannappan's theorem (Corollary 9.22) there exists a continuous character χ of G such that $g = (\chi + \check{\chi})/2$. From Example 3.13 we conclude that the continuous, classical d'Alembert functions g on G are the functions of the form

$$g_\alpha \begin{pmatrix} a & b \\ 0 & 1 \end{pmatrix} := \frac{a^\alpha + a^{-\alpha}}{2}, \qquad \begin{pmatrix} a & b \\ 0 & 1 \end{pmatrix} \in G,$$

where α ranges over $\mathbb{C}$.

Example 9.10. In this example we find all classical d'Alembert functions on the quaternion group $Q_8 = \{\pm 1, \pm i, \pm j, \pm k\}$. Although the group is small (it has order 8), it is nevertheless interesting, because it is not abelian and not generated by its squares. Thus neither Kannappan's result (Corollary 9.22) nor Theorem 8.16 nor Theorem 9.25 apply. Actually there is a non-abelian classical d'Alembert function g_0 on Q_8 as we shall see.

In Example A.17(h) we introduced a representation Π of the multiplicative monoid $\mathbb{H}$ on $\mathbb{C}^2$ by

$$\Pi(a + bi + cj + dk) = \begin{pmatrix} a + ib & c + id \\ -(c - id) & a - ib \end{pmatrix}.$$

From Lemma 9.16 we read that

$$g := \tfrac{1}{2} \operatorname{tr} \Pi, \qquad \text{i.e., } g(a + bi + cj + dk) = a,$$

is a non-zero solution of the μ-d'Alembert functional equation

$$g(xy) + \|y\|^2 g(xy^{-1}) = 2g(x)g(y), \qquad x, y \in \mathbb{H} \setminus \{0\}.$$

In particular that g is a μ-d'Alembert function on $\mathbb{H} \setminus \{0\}$.

The restriction g_0 of g to Q_8 is a d'Alembert function on Q_8. It is non-abelian: $g_0(ijk) = g_0(-1) = -1$, while $g_0(ikj) = g_0(1) = 1$.

To find the other d'Alembert functions on Q_8 we analyze the situation. Let g be a classical d'Alembert function on Q_8. Note that $g(1) = 1$ (Proposition 9.17(e)). Taking $x = y = -1$ in (9.3) we get that $g(-1)^2 = 1$, implying that $g(-1) \in \{\pm 1\}$, so there are two possibilities

(1) If $g(-1) = 1$. Here we see that g is identically 1 on the commutator group $[Q_8, Q_8] = \{\pm 1\}$. Then g is abelian (Theorem 8.15) and so according to Kannappan's result (Corollary 9.22) given in terms of the characters of Q_8 by the formula (9.17). And we computed the characters in Example 3.15.

(2) If $g(-1) = -1$. Here we get from $-1 = g(-1) = g((\pm i)^2) = 2g(\pm i)^2 - 1$ that $g(\pm i) = 0$, and similarly that $g(\pm j) = g(\pm k) = 0$. Thus $g = g_0$.

Conclusion. There are five classical d'Alembert functions on Q_8, four abelian and one non-abelian, viz.,

(i) $g = 1$, which is abelian.

(ii) $g(\pm 1) = 1$ with one of the following 3 choices for the remaining values

(a) $g(\pm i) = 1$, $g(\pm j) = g(\pm k) = -1$.
(b) $g(\pm j) = 1$, $g(\pm i) = g(\pm k) = -1$.
(c) $g(\pm k) = 1$, $g(\pm i) = g(\pm j) = -1$.

These three functions are abelian.

(iii) g_0, which is the only non-abelian d'Alembert function on Q_8.

The Wilson functions corresponding to g_0 are computed in [192, Example 11.3].

Example 9.11 puts g_0 into a wider perspective by describing how it extends to d'Alembert functions on certain groups containing Q_8.

Example 9.11. Consider the matrix group $\mathrm{GL}(2,\mathbb{C})$. By Lemma 9.16 (with $\pi(x) = x$) the normalized trace $x \mapsto \frac{1}{2}\operatorname{tr}(x)$, $x \in M(2\times 2,\mathbb{C})$, restricts to a continuous, non-zero solution g' of the μ-d'Alembert functional equation

$$g'(xy) + (\det y)g'(xy^{-1}) = 2g'(x)g'(y), \qquad x, y \in \mathrm{GL}(2,\mathbb{C}).$$

So $g := g'|_{\mathrm{SL}(2,\mathbb{C})}$ is a classical d'Alembert function on $\mathrm{SL}(2,\mathbb{C})$.

Consequently, the restriction of g to the subgroup $\mathrm{SU}(2)$ of $\mathrm{SL}(2,\mathbb{C})$ is a non-zero, continuous solution of d'Alembert's classical functional equation (9.3) on $\mathrm{SU}(2)$. The topological isomorphism Π between the group of unit quaternions and $\mathrm{SU}(2)$ described in Example A.17(h) transfers g from $\mathrm{SU}(2)$ to a d'Alembert function $g_1 = g \circ \Pi$ on the group of unit quaternions. A computation reveals that $g_1(a + bi + cj + dk) = a$, which shows that g_1 is the g of Example 9.10 restricted to the group of unit quaternions.

Since g_0 is non-abelian, so is any extension of g_0 to a group larger than Q_8. In particular g_1 is non-abelian on the group of unit quaternions, and consequently so is g on $\mathrm{SU}(2)$ and $\mathrm{SL}(2,\mathbb{C})$.

It may be checked directly that g is non-abelian on $\mathrm{SU}(2)$ and $\mathrm{SL}(2,\mathbb{C})$: With

$$a := \begin{pmatrix} i & 0 \\ 0 & -i \end{pmatrix}, \quad b := \begin{pmatrix} 0 & 1 \\ -1 & 0 \end{pmatrix}, \quad c := \begin{pmatrix} 0 & i \\ i & 0 \end{pmatrix},$$

we find that $g(abc) = -1$ is different from $g(acb) = 1$.

The normalized trace is also an non-abelian d'Alembert function on the group $\mathrm{SL}(2,\mathbb{R})$. The same arguments as for $\mathrm{SL}(2,\mathbb{C})$ give that it is a d'Alembert function. To see that it is non-abelian we consider the matrices

$$x := \begin{pmatrix} 1 & 1 \\ 0 & 1 \end{pmatrix}, \quad y := \begin{pmatrix} 2 & 0 \\ 0 & \frac{1}{2} \end{pmatrix}, \quad z := \begin{pmatrix} 0 & 1 \\ -1 & 0 \end{pmatrix},$$

and note that $g(xyz) = -\frac{1}{4}$ while $g(xzy) = -1$.

Example 9.12. Consider the matrix group $\mathrm{GL}(2,\mathbb{C})$ with the adjugate map adj from Example A.33(j) as the involution τ. We leave it to the reader to show that the normalized trace

$$g(x) := \tfrac{1}{2}\,\mathrm{tr}(x), \qquad x \in \mathrm{GL}(2,\mathbb{C}),$$

is a continuous d'Alembert function. (Exercise 9.6.)

Example 9.13. The continuous irreducible representations of $\mathrm{SU}(2)$ are well known (see [112, §29]). Up to equivalence there is only one 2-dimensional such representation π. As a representation on $\mathbb{C}^2$ it is quite simple:

$$\pi\begin{pmatrix} \alpha & \beta \\ -\overline{\beta} & \overline{\alpha} \end{pmatrix}$$
$$= \text{the operator on } \mathbb{C}^2 \text{ corresponding to the matrix } \begin{pmatrix} \alpha & \beta \\ -\overline{\beta} & \overline{\alpha} \end{pmatrix}.$$

By Davison's structure theorem (Theorem 8.26) there is one and only one non-abelian, classical d'Alembert function on $\mathrm{SU}(2)$, viz.,

$$g_\pi(U) = \tfrac{1}{2}\,\mathrm{tr}(\pi(U)) = \tfrac{1}{2}\,\mathrm{tr}\,U, \qquad U \in \mathrm{SU}(2).$$

The formula shows that g_π is the function g from Example 9.11.

The abelian d'Alembert functions can according to Kannappan's result (Corollary 9.22) be expressed in terms of characters of $\mathrm{SU}(2)$. But $[\mathrm{SU}(2), \mathrm{SU}(2)] = \mathrm{SU}(2)$ (see Example A.17(g)), so the only character is 1.

Conclusion. $\mathrm{SU}(2)$ has only two continuous, classical d'Alembert functions, viz., $g = \frac{1}{2}\,\mathrm{tr}$ and 1.

Example 9.14. Consider the group $\mathrm{SU}(n+1)$, where $n \geq 2$. The smallest dimension > 1 that the continuous irreducible representations of $\mathrm{SU}(n+1)$ have, is $n+1$ according to Weyl's dimension formula (see [33, p. 266]), so $\mathrm{SU}(n+1)$ has no continuous 2-dimensional representations. It follows from Davison's structure theorem (Theorem 8.26) that all continuous, classical d'Alembert functions on $\mathrm{SU}(n+1)$ are abelian. Arguing like for $\mathrm{SU}(2)$ in Example 9.13 we conclude that there is only one continuous, classical d'Alembert function on $\mathrm{SU}(n+1)$, viz., the constant function 1.

9.3 μ-d'Alembert functions

In this section we derive some general properties of d'Alembert functions. Actually we work with a cosine-exponemntial functional equation, the μ-d'Alembert functional equation, which is an extension of d'Alembert's functional equation. It occurs in the literature (see the comments to Definition 9.15 in the section *Notes and remarks*) and is closely related to pre-d'Alembert functions (see Theorem 9.19).

Definition 9.15. Let S be a semigroup with an involution $\tau : S \to S$. Let $\mu : S \to \mathbb{C}^*$ be a multiplicative function such that $\mu(x\tau(x)) = 1$ for all $x \in S$. A *μ-d'Alembert function* on G is a non-zero solution $g : S \to \mathbb{C}$ of the *μ-d'Alembert functional equation*

$$g(xy) + \mu(y)g(x\tau(y)) = 2g(x)g(y), \qquad x, y \in S. \tag{9.11}$$

D'Alembert's functional equation is the case of $\mu = 1$.

Let us first note some special solutions of (9.11) on groups:

Lemma 9.16. *If π is a 2-dimensional representation of a group G, then $g := \frac{1}{2}\operatorname{tr}\pi$ is a non-zero solution of*

$$g(xy) + \det(\pi(y))g(xy^{-1}) = 2g(x)g(y), \qquad x, y \in G. \tag{9.12}$$

Proof. By the Cayley-Hamilton theorem any $Y \in \mathrm{GL}(2,\mathbb{C}) \subseteq M(2\times 2,\mathbb{C})$ is annulled by its characteristic polynomial, so that $Y^2-(\operatorname{tr}Y)Y+(\det Y)I = 0$. Multiply this by XY^{-1} and take the trace. □

Proposition 9.17 lists some properties of the solutions of (9.11). In particular we see that any d'Alembert function on a monoid is a pre-d'Alembert function, so that we can apply the results in Chapter 8. That solutions are central (Proposition 9.17(b)) is used time and time again, showing that centrality is an indispensable property of d'Alembert functions. The condition in the proposition that $\mu(x\tau(x)) = 1$ for all $x \in S$ is satisfied for any character μ if S is a group and τ the group inversion, because in that case the condition reduces to $\mu(e) = 1$.

Let us take the classical d'Alembert functional equation on the real line with the solution $g = \cos$ as an illustration. Point (b) below (that g is central) is trivially true, because the underlying group $\mathbb{R}$ is abelian. Point (a) below says that cos is even and (e) that $\cos 0 = 1$. The function d from (d) is identically 1.

Proposition 9.17. *Let S be a semigroup with an involution $\tau : S \to S$. Let $\mu : S \to \mathbb{C}^*$ be a multiplicative function such that $\mu(x\tau(x)) = 1$ for all $x \in S$. Let finally $g : S \to \mathbb{C}$ be a solution of (9.11), i.e.,*

$$g(xy) + \mu(y)g(x\tau(y)) = 2g(x)g(y), \qquad x, y \in S.$$

(a) $g \circ \tau = g/\mu$.

(b) *g is central.*

(c) *g is a solution of the pre-d'Alembert functional equation (8.1).*

(d) *The function $d : S \to \mathbb{C}$, given by $d(x) := 2g(x)^2 - g(x^2)$ for $x \in S$, is multiplicative.*

(e) *If S is a monoid and $g \neq 0$, then $g(e) = d(e) = 1$ where e denotes the neutral element of S.*

Proof. (a) and (b) Replacing y by $\tau(y)$ in the functional equation and multiplying the result by $\mu(y)$ we get by a computation that

$$\begin{aligned}\mu(y)2g(x)g(\tau(y)) &= \mu(y)g(x\tau(y)) + \mu(y)\mu(\tau(y))g(xy)\\ &= \mu(y)g(x\tau(y)) + g(xy) = 2g(x)g(y),\end{aligned}$$

which implies the invariance formula (a), while the computation

$$\begin{aligned}g(xy) + \mu(y)g(x\tau(y)) &= 2g(x)g(y) = 2g(y)g(x) = g(yx) + \mu(x)g(y\tau(x))\\ &= g(yx) + \mu(x)(g \circ \tau)(x\tau(y)) = g(yx) + \frac{\mu(x)}{\mu(x\tau(y))}g(x\tau(y))\\ &= g(yx) + \mu(y)g(x\tau(y)) \qquad \text{for all } x, y \in S,\end{aligned}$$

implies that g is central.

(c) For any $x, y, z \in S$ we get from (9.11) that $g(xyz) + \mu(z)g(xy\tau(z)) = 2g(xy)g(z)$. To get rid of the second term we apply (9.11) to it and get $g(xy\tau(z)) + \mu(y\tau(z))g(xz\tau(y)) = 2g(x)g(y\tau(z))$. That introduced another second term that we compute by $g(xz\tau(y)) + \mu(\tau(y))g(xzy) = 2g(xz)g(\tau(y))$. Combining the three identities we find

$$g(xyz) + g(xzy) = 2g(xy)g(z) - 2g(x)g(y\tau(z))\mu(z) + 2g(xz)\mu(y)g(\tau(y)),$$

from which small reformulations based on (a) and (b) reveal that g is a solution of the pre-d'Alembert functional equation.

(d) Lemma 8.9(b).

(e) Put $y = e$ in (9.11). □

We note some useful properties of classical d'Alembert functions.

Corollary 9.18. *Let g be a classical d'Alembert function on a group G.*

(a) *g is even and central.*

(b) *$g(e) = 1$.*

(c) *g is a solution of the pre-d'Alembert functional equation.*

(d) *If g is bounded, then $g(G) \subseteq [-1, 1]$.*

Proof. (a), (b) and (c) are immediate from Proposition 9.17.

(d) Let $x \in G$ be arbitrary. Let g_1 be the restriction of g to the abelian subgroup $\langle x \rangle$ of G generated by x. By Kannappan's theorem (Theorem 9.22) there exists a unitary character χ on $\langle x \rangle$ such that $g_1(y) = (\chi(y) + \check{\chi}(y))/2 = (\chi(y) + \overline{\chi(y)})/2$ for all $y \in \langle x \rangle$. Writing $\chi(x) = \exp(i\theta)$, where $\theta \in \mathbb{R}$, we find that $g(x) = g_1(x) = \cos\theta \in [-1, 1]$. □

Propositions 8.11 and 9.17(c) reveal that pre-d'Alembert functions and μ-d'Alembert functions are very closely related on groups:

Theorem 9.19. *Let $\mathcal{P}$ denote the set of pre-d'Alembert functions on a group G. For any $\mu \in \Gamma =$ the set of characters of G we let $S(\mu)$ denote the set of μ-d'Alembert functions on G. Then*

$$\mathcal{P} = \bigcup_{\mu \in \Gamma} S(\mu),$$

so the set of pre-d'Alembert functions is the disjoint union of the sets of μ-d'Alembert functions, μ ranging over the characters of G.

Definition 9.15 requires μ to be a character. On 2-divisible, abelian groups that requirement is a consequence of the normalization $g(e) = 1$ as implied by Theorem 9.20 (or Exercise 9.28).

Theorem 9.20. *Let $\mu : G \to \mathbb{C}$ where G is a group. Let $g : G \to \mathbb{C}$ be a solution of the exponential-cosine functional equation*

$$g(xy) + \mu(y)g(xy^{-1}) = 2g(x)g(y), \qquad x, y \in G, \tag{9.13}$$

such that $g(e) = 1$, e being the neutral element of G.

(a) *$\mu(x) = 2g(x)^2 - g(x^2)$ and $g(x) = \mu(x)g(x^{-1})$ for all $x \in G$, $\mu(e) = 1$ and $\mu(G) \subseteq \mathbb{C}^*$.*

(b) *If G is generated by its squares and g is central, then μ is a character on G.*

Proof. (a) Putting $y = x$ in (9.13) we get the first statement. When we take $x = e$ in that we get $\mu(e) = 1$. Taking $x = e$ in (9.13) we get the invariance property $g(y) = \mu(y)g(y^{-1})$. We prove that μ vanishes nowhere by contradiction: Assume $\mu(y_0) = 0$. From the functional equation we get $g(xy_0) = 2g(x)g(y_0)$. Taking $x = e$ we see that $g(y_0) = 0$. Then $g(xy_0) = 0$ for all $x \in G$, so that $g = 0$, contradicting $g(e) = 1$.

(b) The right hand side of (9.13) is unchanged under interchange of x and y, so we get

$$g(xy) + \mu(y)g(xy^{-1}) = g(yx) + \mu(x)g(yx^{-1}) = g(yx) + \mu(x)g((xy^{-1})^{-1}).$$

Here $g(xy)$ and $g(yx)$ cancel one another, because g is central. When we multiply the remaining identity by $\mu(xy^{-1})$, use the invariance property and replace x by xy we obtain the identity

$$[\mu(xy) - \mu(x)\mu(y)]g(x) = 0 \qquad \text{for all } x, y \in G. \tag{9.14}$$

It suffices to prove the claim $\mu(a^2y) = \mu(a^2)\mu(y)$, for all $a, y \in G$, to obtain (b). Indeed, the claim implies that μ is multiplicative, because G is generated by its squares, and we have furthermore that $\mu(e) = 1$ from (a).

If $g(a^2) \neq 0$ the claim is immediate from (9.14) with $x = a^2$, so we may assume that $g(a^2) = 0$. Since μ vanishes nowhere we see from $\mu(a) = 2g(a)^2 - g(a^2) = 2g(a)^2$ that $g(a) \neq 0$, so we may apply (9.14) with $x = a$ from which we read that $\mu(ay) = \mu(a)\mu(y)$ for all $y \in G$. In particular that $\mu(a^2) = \mu(a)^2$. Finally, $\mu(a^2y) = \mu(a(ay)) = \mu(a)\mu(ay) = \mu(a)^2\mu(y) = \mu(a^2)\mu(y)$, proving the claim. □

Let g be a d'Alembert function. At e the corresponding multiplicative function $d = d_g$ assumes the value 1 (Proposition 9.17(e)), so on groups d vanishes nowhere (Lemma 3.4(a)). This is not so on monoids: Here is an abelian d'Alembert function g such that the corresponding d vanishes everywhere except of course at the neutral element: $M = (\mathbb{R}^+ \cup \{0\}, +)$, $\tau = \text{id}$ and $g = \mathbf{1}_{\{0\}}$, which means that $d = \mathbf{1}_{\{0\}}$.

Formulas that show how the solutions of the μ-d'Alembert functional equation look like on abelian groups, are written down in Proposition 9.31. For $\mu = 1$ they coincide with Kannappan's result (Corollary 9.22).

9.4 Abelian d'Alembert functions

In this section we derive and extend Kannappan's result about the structure of abelian d'Alembert functions.

If a solution g of d'Alembert's functional equation (9.4) on a semigroup satisfies Kannappan's condition then it is an abelian function, because the requirement of centrality is satisfied automatically: Replacing y by $\tau(y)$ in (9.4) we infer that g is τ-invariant, so that we can refer to Theorem 7.1(a). So even on semigroups we may concentrate on verifying Kannappan's condition.

Let us as motivation for Kannappan's result (Corollary 9.22) note that the continuous, classical d'Alembert functions on the group $\mathbb{R}$ are the functions

$$g_\lambda(x) = \cos(\lambda x) = \frac{e^{i\lambda x} + e^{-i\lambda x}}{2}, \qquad x \in \mathbb{R},$$

where λ ranges over $\mathbb{C}$ (see Proposition 9.4). So $g_\lambda = (\chi_\lambda + \check{\chi_\lambda})/2$ where χ_λ is the character $\chi_\lambda(x) = e^{i\lambda x}$, $x \in \mathbb{R}$. This fact was extended from $\mathbb{R}$ to any abelian group $(G, +)$ by Kannappan in his seminal paper [125] from 1968 in which he proved that the non-zero functions $g : G \to \mathbb{C}$ satisfying

$$g(x+y) + g(x-y) = 2g(x)g(y), \qquad x, y \in G, \tag{9.15}$$

are the functions of the form $g = (\chi + \check{\chi})/2$, where χ ranges over the characters of G. These are the solutions (9.5) from Lemma 9.2. So Kannappan's result is the inverse of Lemma 9.2 in the case of S being an abelian group and τ the group inversion.

Theorem 9.21 below which is our inverse of Lemma 9.2, is a minor extension of Kannappan's result, in which we roughly speaking replace his abelian group G by an abelian semigroup and the group inversion by an involution. The treatment is purely algebraic. It is remarkable that neither we nor Kannappan impose any kind of regularity of the solutions such as continuity or measurability, so the methods are bound to be completely different from the ones used to prove Proposition 9.4 where we differentiated to get the solutions.

Theorem 9.21. *Let S be a semigroup, and let $\tau : S \to S$ be an involution of S. Let $g : S \to \mathbb{C}$.*

(a) *g is an abelian solution of d'Alembert's functional equation (9.4), if and only if there exists a multiplicative function $\chi : S \to \mathbb{C}$, such that*

$$g = \frac{\chi + \chi \circ \tau}{2}. \tag{9.16}$$

(b) *Given g of the form (9.16), where $\chi : S \to \mathbb{C}$ is a multiplicative function, then*

(i) *χ is unique, except that it can be replaced by $\chi \circ \tau$.*

(ii) *g is bounded if and only if χ is bounded.*

(iii) *Let S be a topological semigroup. Then g is continuous if and only if χ is continuous.*

Proof. (a). We have already seen (Lemma 9.2) that if g has the form (9.16), then g is a solution of d'Alembert's functional equation (9.4). Assume conversely that g is an abelian solution of d'Alembert's functional equation. Since g is a solution of the pre-d'Alembert functional equation (Proposition 9.17(c)), there exist according to Theorem 8.13 two multiplicative functions $\chi, \sigma : S \to \mathbb{C}$ such that $g = (\chi + \sigma)/2$.

Since g is τ-invariant, $\chi + \sigma = \chi \circ \tau + \sigma \circ \tau$. According to Corollary 3.19 either $\sigma = \chi \circ \tau$ or $\chi = \chi \circ \tau$. The first case gives us the formula we want. In the second case we get from the expression for g that $g(x\tau(y)) = g(xy)$ which via the functional equation reveals g is multiplicative. Finally, g being even with respect to τ,

$$g = \frac{g+g}{2} = \frac{g + g\circ\tau}{2},$$

finishing the proof of (a).

(b) follows from Corollary 3.19 and Theorem 3.18. □

The main content of Kannappan's paper [125] is the following Corollary 9.22, except that Kannappan only considers d'Alembert's classical functional equation and that you don't find (b) and the regularity result (d) in [125]. Corollary 9.22 may be viewed as an extension of Euler's decomposition $\cos x = (e^{ix} + e^{-ix})/2$ of $\cos x$.

Corollary 9.22 (Kannappan). *Let G be a group, and let $\tau : G \to G$ be an involution. The abelian d'Alembert functions on G are the functions*

$$g = \frac{\chi + \chi \circ \tau}{2}, \tag{9.17}$$

where χ ranges over the characters of G.

Given g of the form (9.17) where χ is a character of G, then

(a) *χ is unique, except that it can be replaced by $\chi \circ \tau$.*

(b) *g is bounded if and only if χ is unitary.*

(c) *If G is a topological group and g is continuous, then χ is also continuous.*

(d) *Let G be a locally compact group. If g is measurable with respect to the (left) Haar measure on G, then χ, $\chi \circ \tau$ and g are continuous.*

Proof. The corollary is immediate from Theorem 9.21 and Lemma 3.4(b) except for (d). For (d) we note that χ and $\chi \circ \tau$ are measurable according to Theorem 3.18(f). And then they are continuous according to Proposition D.4(b). □

For applications of Corollary 9.22 see the Examples 9.5, 9.7 and 9.9. For the more general case of abelian μ-d'Alembert functions see Proposition 9.31.

Let g be a d'Alembert function on a group G. It is of interest to have criteria in terms of g and/or G for when g is abelian, because Kannappan's result (Corollary 9.22) then tells us how g looks like.

Proposition 9.23. *Let $g : G \to \mathbb{C}$ be a non-zero solution of d'Alembert's classical functional equation $g(xy) + g(xy^{-1}) = 2g(x)g(y)$, $x, y \in G$, on a group G. If $g(z_0)^2 \neq 1$ for some $z_0 \in Z(g)$, then there exists a character $\chi : G \to \mathbb{C}^*$, $\chi \neq \check{\chi}$, such that $g = (\chi + \check{\chi})/2$.*

Proof. Since any solution of d'Alembert's equation is a solution of the pre-d'Alembert functional equation (Proposition 9.17(c)) we get from Proposition 8.14 that $g = (\chi_1 + \chi_2)/2$, where χ_1 and χ_2 are multiplicative functions. In particular g is abelian. By Corollary 9.22 there exists then a character $\chi : G \to \mathbb{C}^*$, such that $g = (\chi + \check{\chi})/2$. The rest of the proof is left to the reader. □

Proposition 9.23 comes sometimes in handy (see for example Exercise 9.18). An even more informative result holds for solutions of d'Alembert's long functional equation (Theorem 10.1) and consequently also for d'Alembert's functional equation.

Since d'Alembert functions are pre-d'Alembert functions the results of Section 8.5 apply. As a corollary we note:

Corollary 9.24. *Let G be a group which is generated by its squares. Let g be a d'Alembert function on G such that $g|_{[G,G]}$ is abelian. Then g is abelian.*

Proof. Theorem 8.16. □

Nilpotent groups are in many respects close to being abelian. This meta-statement is supported by the following result:

Theorem 9.25. *Let G be a nilpotent group which is generated by its squares. Then any classical d'Alembert function on G is abelian.*

Proof. See Proposition 10.6. □

The assumption that G is generated by its squares cannot be deleted from Theorem 9.25: The classical d'Alembert function g_0 on Q_8 (Example 9.10) is a counter-example.

9.5 Non-abelian d'Alembert functions

We have seen that non-abelian d'Alembert functions exist (Example 9.11), so a study of non-abelian d'Alembert functions is not a study of the empty set.

Theorem 9.26. *Let M be a topological monoid and $\tau : M \to M$ an involution. The non-abelian d'Alembert functions $g \in C(M)$ are the functions of the form $g = \frac{1}{2}\operatorname{tr}\rho$ such that $\rho(\tau(x)) = \operatorname{adj}(\rho(x))$ for all $x \in M$, where ρ is a continuous, algebraically irreducible representation of M on $\mathbb{C}^2$.*

Proof. Except for recalling the defining identity $A + \operatorname{adj}(A) = \operatorname{tr}(A)I$ (formula (A.2)) we leave it to the reader to check that a function g of the indicated form is a continuous d'Alembert function. g is a non-abelian function according to Theorem 8.26(c).

Let conversely $g \in C(M)$ be a non-abelian d'Alembert function. According to Proposition 9.17(c) it is a pre-d'Alembert function, so it has the desired form according to Theorem 8.26, except possibly for the identity $\rho(\tau(x)) = \operatorname{adj}(\rho(x))$ that we derive as follows: Substituting $g = \frac{1}{2}\operatorname{tr}\rho$ into d'Alembert's functional equation $g(xy) + g(x\tau(y)) = 2g(x)g(y)$ we find that

$$\operatorname{tr}(\rho(x)\rho(y) + \rho(x)\rho(\tau(y))) = \operatorname{tr}\rho(x)\operatorname{tr}\rho(y),$$

or equivalently

$$\operatorname{tr}\{\rho(x)[\rho(y) + \rho(\tau(y)) - \operatorname{tr}(\rho(y))I]\} = 0.$$

Now Corollary E.12 tells us that $\rho(y)+\rho(\tau(y)) = \operatorname{tr}(\rho(y))I$, and a comparison with the identity (A.2) yields $\rho(\tau(y)) = \operatorname{adj}(\rho(y))$ for all $y \in M$. □

For the more general functional equation $g(xy) + \mu(y)g(x\tau(y)) = 2g(x)g(y)$ (equation (9.11)) we deduce in the same way that $\operatorname{adj}(\rho(y)) = \mu(y)\rho(\tau(y))$ for all $y \in M$.

We finish this section by returning to our point of departure, d'Alembert's classical functional equation on groups.

Theorem 9.27. *Let G be a topological group and $g \in C(G)$ a non-abelian classical d'Alembert's function.*

(a) *g can be written in the form $g = \frac{1}{2}\operatorname{tr}\rho$, where ρ is a continuous, algebraically irreducible representation of G on $\mathbb{C}^2$ such that $\rho(x) \in \mathrm{SL}(2,\mathbb{C})$ for all $x \in G$.*

(b) *If g is bounded, we may choose the ρ from (a) such that $\rho(x) \in \mathrm{SU}(2)$ for all $x \in G$.*

Proof. (a) Here $d = 1$, so we need only cite Theorem 8.26(b).

(b) If g is bounded, then so is ρ (Theorem 8.24(f)). And when ρ is bounded, a result by Weil (see [111, 22.23(c)]) says that there exists an invertible matrix A such that $A^{-1}\rho(x)A$ is unitary for all $x \in G$. We may replace ρ by $x \mapsto A^{-1}\rho(x)A \in U(2)$, because that does not change neither the trace nor the determinant. □

The converse of Theorem 9.27 holds.

9.6 Compact groups

Theorem 9.28. *Let G be a compact group. Any solution $g \in C(G)$, $g \neq 0$, of d'Alembert's classical functional equation (9.3) has one of the following two forms*

(a) *$g(x) = \frac{1}{2}\operatorname{tr}\pi(x)$, $x \in G$, where $\pi : G \to \mathrm{SU}(2)$ is a continuous, irreducible representation.*

(b) *$g = (\chi + \check{\chi})/2$, where χ is a continuous, unitary character of G.*

Proof. Combine Corollary 9.22 and Theorem 9.27. □

We cite related result due to Yang [213, Theorem 9].

Theorem 9.29. *Let G be a compact connected group. If G has a direct factor isomorphic to $\mathrm{SU}(2)$, then non-abelian cosine functions on G exist and each of them factors through one of these direct factors; otherwise, every cosine function on G is abelian.*

9.7 Exercises

Exercise 9.1. Show that $g(n) := (-1)^n$, $n \in \mathbb{Z}$, is a classical d'Alembert function on $(\mathbb{Z}, +)$. Compare it with the function $x \mapsto \cos(\pi x)$ on $\mathbb{R}$.

Exercise 9.2. Let $g : G \to \mathbb{C}$ be a solution of d'Alembert's classical functional equation (9.3) on a group G. Let $u \in G$ be such that $g(u) = 1$. Show that

$$g(xu^n) - g(x) = n[g(xu) - g(x)] \qquad \text{for all } x \in G \text{ and } n \in \mathbb{Z},$$

and that $g(u^n) = 1$ for all $n \in \mathbb{Z}$.

Exercise 9.3. Consider the functional equation

$$g(x + y) + g(x + \tau(y)) = 2g(x)g(y), \qquad x, y \in \mathbb{R},$$

in which $\tau(x) = x$ for all $x \in \mathbb{R}$. Find all non-zero solutions $g \in C(\mathbb{R})$ and the corresponding multiplicative functions d.

Exercise 9.4. Let $G = (\mathbb{C}, +)$ with the involution $\tau(z) = \overline{z}$. Show that

$$g(z) := \frac{e^{z/2} + e^{\overline{z}/2}}{2}, \qquad z \in \mathbb{C},$$

is a solution of d'Alembert's functional equation (9.4) on G. Compute the corresponding multiplicative function d.

Exercise 9.5. What is the function $d : S \to \mathbb{C}$ in Proposition 9.17? How does it simplify if τ is the inversion of a group?

Exercise 9.6. Prove that the normalized trace

$$g(X) := \tfrac{1}{2}\operatorname{tr}(X), \qquad X \in \mathrm{GL}(2, \mathbb{C}),$$

is a continuous d'Alembert function on $\mathrm{GL}(2, \mathbb{C})$ with $\tau = \mathrm{adj}$.
Hint: Formula (A.2).

Exercise 9.7. Prove the following: If $\pi : G \to \mathrm{SU}(2)$ is a representation of a group G, then $g(x) := \frac{1}{2}\operatorname{tr}\pi(x)$, $x \in G$, is a bounded solution of $g(xy) + g(xy^{-1}) = 2g(x)g(y)$, $x, y \in G$.

Exercise 9.8. Prove the statements of Example 9.11.

Exercise 9.9. Let S be a semigroup and let $\tau : S \to S$ be an involution. The purpose of this exercise is to show that $f = 0$ is the only solution $f : S \to \mathbb{C}$ of the functional equation

$$f(xy) - f(x\tau(y)) = cf(x)f(y), \qquad x, y \in S, \tag{9.18}$$

where $c \in \mathbb{C} \setminus \{0\}$. The functional equation (9.18) differs from d'Alembert's functional equation by having a minus on the left hand side instead of a plus.

(a) Show that f is odd with respect to τ, i.e., that $f(\tau(x)) = -f(x)$ for all $x \in S$.

(b) Show that

$$f(x\tau(y)) = \frac{f(xy) - f(yx)}{2} \qquad \text{for all } x, y \in S.$$

Hint: Interchange x and y in (9.18).

(c) Show that

$$cf(x)f(y) = \frac{f(xy) + f(yx)}{2} \qquad \text{for all } x, y \in S.$$

(d) Deduce from Theorem 3.21 that cf is multiplicative.

(e) Show that $f = 0$.

Exercise 9.10. Let G be a group and let $\mathbb{F}$ be a field of characteristic different from 2. Let $g : G \to \mathbb{F}$ be a solution of the classical d'Alembert functional equation (9.3) (with $\mathbb{C}$ replaced by $\mathbb{F}$) such that $g(x)^2 = 1$ for all $x \in G$.

Show that $g(xy) = g(x)g(y)$ for all $x, y \in G$.

Hint: Multiply (9.3) by $g(xy) - g(xy^{-1})$ and infer that $g(xy) = g(xy^{-1})$.

The result can be compared with the one of Exercise 10.2.

Exercise 9.11. Find the continuous solutions of d'Alembert's classical functional equation (9.3) on the circle group $\mathbb{T} := \{z \in \mathbb{C} \mid |z| = 1\}$. Are they bounded? Are they real-valued? If in doubt consult Corollary 9.18.

Exercise 9.12. Find the solutions $g, f \in C(\mathbb{R})$ of the functional equation

$$g(x + y) + g(x - y) = 2f(x)f(y), \qquad x, y \in \mathbb{R}.$$

Exercise 9.13. Consider d'Alembert's functional equation (9.1) on the real line.

(a) Find its real-valued continuous solutions.

(b) Find its bounded continuous solutions. Show that they are real-valued and bounded in absolute value by 1.

Exercise 9.14. Prove the following

Proposition 9.30. *Let G be a group with an involution $\tau : G \to G$. Let $g : G \to \mathbb{C}$ be a d'Alembert function on G. If $g(z_0) = 0$ for some $z_0 \in Z(g) := \{z \in G \mid g(xyz) = g(xzy)$ for all $x, y \in G\}$, then g is abelian.*

Hint: Proposition 8.14.

Exercise 9.15. Let $g_1, g_2 \in C(G)$ be two different continuous solutions of d'Alembert's classical functional equation (9.3) on a compact group G. Show that $\int_G g_1(x)g_2(x)\,dx = 0$, where dx denotes the Haar measure on G. *Hint:* Consider the convolution $(f * g)(x) := \int_G f(y)g(y^{-1}x)\,dy$. Show that $(f * g_2)(x) = \{\int_G f(y)g_2(y)\,dy\}g_2(x)$ if f is even. Note that a central function f is central with respect to the convolution, i.e., $f * h = h * f$.

This result means that $g_1 \perp g_2$ in $L^2(G)$, because continuous, classical d'Alembert functions are real-valued on compact groups (Corollary 9.18(d)).

Exercise 9.16. Let f be a bounded, classical d'Alembert function on an abelian group G. Show that f is the real part of a unitary character of G. *Hint:* Theorem 3.18(c).

Exercise 9.17. Consider for fixed $\alpha \in \mathbb{R} \setminus \{0\}$ the functional equation

$$g(x + y + \alpha) + g(x - y + \alpha) = 2g(x)g(y), \qquad x, y \in \mathbb{R},$$

where $g : \mathbb{R} \to \mathbb{C}$ is the unknown function. The equation is d'Alembert's functional equation, except for the additive constant α. The corresponding functional equation with a minus instead of a plus between the terms on the left is Van Vleck's functional equation from Exercise 9.18.

(a) Find the non-zero solutions.

Answer: g has the form

$$g = \chi(\alpha)\frac{\chi + \check{\chi}}{2},$$

where $\chi : \mathbb{R} \to \mathbb{C}^*$ is a character on $\mathbb{R}$ such that $\chi(\alpha) = \pm 1$.

Hint: Introducing $f(x) := g(x+\alpha)$ we get $f(x+y)+f(x-y) = 2g(x)g(y)$. Taking $y=0$ we find that $f = cg$ for some constant $c \in \mathbb{C} \setminus \{0\}$. Substitute this into the just derived identity and get that g/c is a solution of d'Alembert's classical functional equation on $\mathbb{R}$, so that you may find the form of g/c from Kannappan's result (Corollary 9.22).

(b) Show that $g(x+2\alpha) = g(x)$ for all $x \in \mathbb{R}$ for any solution g.

(c) Find the continuous, non-zero solutions.

Answer:

$$g(x) = \cos\Big(\frac{2n\pi}{\alpha}x\Big) \quad \text{and} \quad g(x) = -\cos\Big(\frac{(2n+1)\pi}{\alpha}x\Big),$$

where $n = 0, 1, 2, \ldots$

Exercise 9.18. In the papers [206, 207] Van Vleck studied for fixed $z_0 \in \mathbb{R}$ the solutions $f : \mathbb{R} \to \mathbb{R}$ of the functional equation

$$f(x-y+z_0) - f(x+y+z_0) = 2f(x)f(y), \qquad x, y \in \mathbb{R}, \tag{9.19}$$

with a view to characterize the function sin on $\mathbb{R}$. The reader can check that $f(x) := \sin(\frac{\pi}{2z_0}x)$, $x \in \mathbb{R}$, is a solution.

The functional equation (9.19) is, except for the minus between the terms on the left being replaced by a plus, the topic for Exercise 9.17.

In this exercise we extend the setting in [206] from the group $\mathbb{R}$ to any group except for a bit of commutativity: We require z_0 (see below) to be in the center of G. We consider complex-valued solutions and not just real-valued. We solve the functional equation by reformulating it to d'Alembert's classical functional equation.

Let G be a group with neutral element e and let $z_0 \in Z(G)$ be given. Let $f : G \to \mathbb{C}$ be a non-zero solution of the functional equation

$$f(xy^{-1}z_0) - f(xyz_0) = 2f(x)f(y), \qquad x, y \in G. \tag{9.20}$$

(a) Show that f is odd, i.e., $f(y^{-1}) = -f(y)$ for all $y \in G$, so that $f(e) = 0$.

(b) Show that $f(z_0) = 1$.

Hint: Put $x = z_0^{-1}$ in (9.20) and use that $z_0 \in Z(G)$.

(c) Show that $f(xz_0^2) = -f(x)$ for all $x \in G$ and that $f(z_0^2) = 0$.

(d) Show that $g(x) := f(xz_0)$, $x \in G$, is a non-zero solution of d'Alembert's classical functional equation (9.3) such that $g(z_0) = 0$.

Hint: Use (c).

(e) Show that there is a character $\chi : G \to \mathbb{C}^*$ such that $g = (\chi + \check{\chi})/2$, and prove that $\chi(z_0) = \pm i$.

Hint: Proposition 9.23 or Proposition 9.30.

(f) Note that (9.20) has no non-zero solutions if $z_0^2 = e$.

(g) Conclude from (e) that

$$f(x) = \frac{\chi(x) - \check{\chi}(x)}{2i} \qquad \text{for all } x \in G,$$

where $\chi : G \to \mathbb{C}^*$ is a character such that $\chi(z_0) = i$ (compared with the formulas above you may have to replace χ by $\check{\chi}$).

(h) Prove the following converse of (g): If G is any group, z_0 any element of G and $\chi : G \to \mathbb{C}^*$ any character such that $\chi(z_0) = i$, then $f := (\chi - \check{\chi})/(2i)$ is a non-zero solution of (9.20).

(i) Find the solutions $f \in C(\mathbb{R})$, $f \neq 0$, of the original equation (9.19).

Answer: Non-zero solutions do not exist if $z_0 = 0$, but they do exist for any $z_0 \in \mathbb{R} \setminus \{0\}$ where the continuous ones are

$$f_n(x) = (-1)^n \sin\Big(\frac{(2n+1)\pi}{2z_0}x\Big), \qquad x \in \mathbb{R}, n \in \mathbb{N} \cup \{0\}.$$

We need not take $n < 0$ into account, because $f_{-n} = f_{n-1}$.

Exercise 9.19. Find the continuous solutions $g : \mathbb{R}^3 \to \mathbb{C}$ of the equation

$$\begin{aligned} g(x_1 + x_2, y_1 + y_2, z_1 + z_2 + x_1 y_2) \\ + g(x_1 - x_2, y_1 - y_2, z_1 - z_2 - (x_1 - x_2)y_2) \\ = 2g(x_1, y_1, z_1)g(x_2, y_2, z_2) \qquad \text{for all } x_1, y_1, z_1, x_2, y_2, z_2 \in \mathbb{R}. \end{aligned}$$

Hint: Is there a group behind? If yes, which one?

Hint: Corollary 9.24 and Example 3.14.

Exercise 9.20. Find the continuous solutions of d'Alembert's classical functional equation (9.3) on the $(ax + b)$-group.

Hint: Corollary 9.24.

Exercise 9.21. Prove the following result that extends Theorem 9.21 to abelian μ-d'Alembert functions.

Hint: Modify the proof of Theorem 9.21.

Proposition 9.31. *Let S be a semigroup and $\tau : S \to S$ an involution of S. Let $\mu : S \to \mathbb{C}^*$ be a multiplicative function such that $\mu(x\tau(x)) = 1$ for all $x \in S$. Let finally $g : S \to \mathbb{C}$ be a non-zero solution of the μ-d'Alembert functional equation* (9.11), *i.e., of*

$$g(xy) + \mu(y)g(x\tau(y)) = 2g(x)g(y), \qquad x, y \in S. \tag{9.21}$$

If g is abelian then there exists a non-zero multiplicative function $\chi : S \to \mathbb{C}$ such that

$$g = \frac{\chi + \mu(\chi \circ \tau)}{2}. \tag{9.22}$$

Conversely, if $\chi : S \to \mathbb{C}$ is a non-zero, multiplicative function then g, defined by (9.22), *is a non-zero, abelian solution of* (9.21).

Exercise 9.22. Find the continuous solutions of the μ-d'Alembert functional equation on the circle group $\mathbb{T}$ with τ as the group inversion, when $\mu(e^{i\theta}) := e^{i\theta}$, $\theta \in \mathbb{R}$. Show that the non-zero, continuous solutions cannot be written as a continuous character times a solution of d'Alembert's classical functional equation.

Exercise 9.23. Let $\sigma : G \to \mathbb{C}$ be a character on an abelian group G. Let $g : G \to \mathbb{C}$ be a non-zero solution of the following μ-d'Alembert functional equation

$$g(x+y) + \sigma(2y)g(x-y) = 2g(x)g(y), \qquad x, y \in G.$$

Show that g has the form $g = \sigma(\chi + \check{\chi})/2$, where χ is a character on G.

Hint: Multiply by $\sigma(-x-y)$ or use Proposition 9.31.

Exercise 9.24. Let $\alpha \in \mathbb{C}$. Find all solutions $g \in C(\mathbb{R})$ of the functional equation

$$g(x+y) + e^{\alpha y}g(x-y) = 2g(x)g(y), \qquad x, y \in \mathbb{R}.$$

Exercise 9.25. Find the solutions $g \in C(\mathbb{C})$ of

$$g(z+w) + e^{\operatorname{Re} w}g(z-w) = 2g(z)g(w), \qquad z, w \in \mathbb{C}.$$

Exercise 9.26. Generalize Theorem 9.27 to the functional equation (9.11).

Exercise 9.27. You shall here give a direct proof of the following theorem without resorting to Davison's structure theorem. The procedure is taken from Yang [214, Theorem 3.1] and [215, Theorem 2.2].

> **Theorem 9.32.** *Let G be a compact group and let $g \in C(G)$ be a non-zero solution of d'Alembert's classical functional equation* (9.3). *Then there exists a continuous homomorphism $\pi : G \to \mathrm{SU}(2)$ such that $g(x) = \frac{1}{2}\operatorname{tr}(\pi(x))$ for all $x \in G$.*

Proof.

(i) Take the Fourier transformation with respect to x of (9.3) to show that

$$[\pi(y) + \pi(y^{-1})]\,\widehat{g}(\pi) = 2g(y)\widehat{g}(\pi) \qquad \text{for all } [\pi] \in \widehat{G}.$$

(ii) Noting that g is central (Proposition 9.17(b)) deduce by Schur's lemma (Lemma E.10) that $\widehat{g}(\pi) = \lambda_\pi I_{d_\pi}$ for some constant $\lambda_\pi \in \mathbb{C}$.

(iii) Show that $\pi(y) + \pi(y^{-1}) = 2g(y)I_{d_\pi}$ whenever $\widehat{g}(\pi) \neq 0$.

(iv) Finish the proof by help of Yang's lemma (Lemma F.1). □

Exercise 9.28. Let G be a 2-divisible, abelian group. Let $\mu : G \to \mathbb{C}$. Let $g : G \to \mathbb{C}$ be a solution of the exponential-cosine functional equation (9.13) such that $g(e) \neq 0$ and $g(e) \neq \frac{1}{2}$. Prove that

(a) $1 + \mu(e) = 2g(e)$. In particular $\mu(e) \neq 0$. Furthermore

$$\frac{\mu(y)}{\mu(e)}g(y^{-1}) = g(y) \qquad \text{for all } y \in G.$$

(b) $\mu/\mu(e)$ is a character on G.

Hint: Proceed as in the proof of Theorem 9.20.

Exercise 9.29. Let G be a group and consider the exponential-cosine functional equation

$$g(xy) + [2g(y)^2 - g(y^2)]g(xy^{-1}) = 2g(x)g(y), \qquad x, y \in G,$$

where $g : G \to \mathbb{C}$ is the unknown function.

(a) Show that there are only 3 possibilities for $g(e)$, viz., 0, $\frac{1}{2}$ and 1.

(b) Show that $g(e) = 0 \implies g = 0$.

In the rest of this exercise we examine the case of $g(e) = \frac{1}{2}$, so from now on $g(e) = \frac{1}{2}$.

(c) Show that $[2g(y)^2 - g(y^2)]g(y^{-1}) = 0$ for all $y \in G$.

(d) Assume that $g(y_0) = 0$ for some $y_0 \in G$. Show that $g(xy_0) = g(y_0^2)g(xy_0^{-1})$ for all $x \in G$. Infer that $g(y_0^2) = 0$, and then that $g = 0$. Conclude that g vanishes nowhere.

(e) Show that $2g$ is a character.

Exercise 9.30. Consider the non-abelian d'Alembert function $g : G \to \mathbb{C}$ on the group G of unit quaternions from Example 9.11. (It was called g_1 there.) So

$$g(x + yi + zj + uk) = x \qquad \text{for } x, y, z, u \in \mathbb{R}.$$

For $x = x_1 + y_1 i + x_1 j + u_1 k$ and $y = x_2 + y_2 i + x_2 j + u_2 k$ compute that

$$\begin{aligned} g_x(y) &= -(y_1 y_2 + z_1 z_2 + u_1 u_2), \\ \Delta(x, y) &= (y_1 z_2 - z_1 y_2)^2 + (y_1 u_2 - u_1 y_2)^2 + (z_1 u_2 - u_1 z_2)^2. \end{aligned}$$

Choose $a = i$ and $b = j$. Show in the notation of Lemma 8.21 that

$$\begin{aligned} \Delta(a, b) &= 1, \\ w_1(x + yi + zj + uk) &= y, \\ w_2(x + yi + zj + uk) &= z, \\ w_3(x + yi + zj + uk) &= u. \end{aligned}$$

9.8 Notes and remarks

Yang's paper [213] about d'Alembert's functional equation on compact, connected groups was a precursor for Davison [64], but her methods are completely different from his.

For d'Alembert's functional equations on hypergroups see Roukbi and Zeglami [165].

For information about d'Alembert functions taking matrix or operator values, see papers by Chojnacki, Kisyński, Sinopoulos and Stetkær and their references.

Chojnacki [37] studies essentially bounded solutions of $\int_K g(x+ky)\,dk = g(x)g(y)$, $x,y \in G$, where G is a locally compact, abelian group and K a compact group such that G is a K-space.

Gajda [93] finds for a locally compact, abelian group G the solutions $f \in L^\infty(G)$ of

$$\int_G [f(x-s+y)+f(x+s-y)]\,d\mu(s) = f(x)f(y), \qquad x,y \in G,$$

where μ is a complex-valued regular Borel measure on G. Fechner [81] studies the corresponding Wilson equation.

Badora [19] and the generalization by Elqorachi and Akkouchi [76] study the solutions $f \in C_b(G)$ of the functional equation

$$\int_G \left\{ \int_K f(xtk \cdot y)\,dk \right\} d\check{\mu}(t) = f(x)f(y), \qquad x,y \in G.$$

Here G is a locally compact group, K is a compact group of morphisms of G, and $\mu \in M(G) = C_0(G)^*$.

Definition 9.1: We have in d'Alembert's functional equation (9.4) assumed that τ is an involution, in particular that τ is an anti-homomorphism. The same equation

$$g(xy) + g(x\sigma(y)) = 2g(x)g(y) \qquad \text{for all } x,y \in S,$$

where $\sigma : S \to S$ is an automorphism such that $\sigma(\sigma(x)) = x$ for all $x \in S$ has not been studied much on non-abelian semigroups. Exceptions are Stetkær [183, Example 6] (continuous solutions) and later Sinopoulos [179] (general solutions) who found that the solutions g are abelian for a special involutive automorphism σ of the Heisenberg group.

Definition 9.15: The μ-d'Alembert functional equation (9.11) occurs in the literature. See Parnami, Singh and Vasudeva [154], Davison [64, Proposition 2.11], Stetkær [193, Lemma IV.4] and Yang [216, Proposition 4.2]. However, it has not been treated systematically.

Theorem 9.20 is in Parnami, Singh and Vasudeva [154, Theorem 2.2] in the case of G being a Banach space X. They use it to find the solutions $g \in C(X)$ of the exponential-cosine functional equation

$$g(x+y) + [2g(y)^2 - g(2y)]g(x-y) = 2g(x)g(y), \quad x,y \in X,\ g(0) = 1.$$

Compare their solution formula with the result of our Proposition 9.31.

Corollary 9.22:

(1) The corollary is the main content of Kannappan's paper [125], except that Kannappan only considers d'Alembert's classical functional equation and that you do not find (b) and the regularity result (d) in [125].

(2) There are other proofs than ours in the literature. See for instance Baker [22] who has a direct proof that does not use the sine addition formula.

(3) Proposition 9.31 extends the corollary to abelian μ-d'Alembert functions.

(4) Sinopoulos [180] finds the solutions $g : S \to F$ of the functional equation

$$g(x+y) + g(x+\sigma(y)) = 2g(x)g(y), \qquad x, y \in S,$$

where $(S, +)$ is a commutative semigroup, σ an involution of S and F is a quadratically closed commutative field of characteristic different from 2. He was apparently the first to extend Kannappan's result (i.e., Corollary 9.22) from groups to semigroups.

Proposition 9.23: A special case of Proposition 9.23 is the result by Dacić [61, Theorem] who assumes there exists a $z_0 \in Z(g) := \{z \in G \mid g(xyz) = g(xzy) \text{ for all } x, y \in G\}$ such that $g(z_0) = 0$, where we require only $g(z_0)^2 \neq 1$. See also Proposition 9.30 and the comments to it. Theorem 10.1(d) and (e) extend Proposition 9.23 to d'Alembert's long functional equation.

Theorem 9.25 can be found as Davison [63, Proposition 5.5] under a stronger assumption than ours, viz., that G is 2-divisible. Our proof is a modification of Davison's. A special case of Theorem 9.25 is Corovei [50, Theorem 2] that deals with nilpotent groups, all of whose elements have (finite) odd order. Friis [90, Theorem 2.6] obtained Theorem 9.25 for connected, nilpotent Lie groups.

Section 9.5: Kannappan posed in [127, Section I] from 1971 the natural question of whether all d'Alembert functions are abelian. The answer is no. The counter-examples g_0 (Example 9.10) and g_1 (Example 9.11) were found by respectively Corovei [50, Example p. 105–106] in 1977 and Ng [5, Remark 5] in 1989, but examples of non-abelian d'Alembert functions must have appeared earlier in disguise in connection with formulas for the trace of matrices like $\operatorname{tr}(AB) + \operatorname{tr}(AB^{-1}) = (\operatorname{tr} A)(\operatorname{tr} B)$

for $A, B \in \mathrm{SL}(2, \mathbb{C})$ that you find in for instance Thurston [200, Problem 2.6.9(a)].

Theorem 9.28: A direct proof can be found in Yang [215, Theorem 2.2].

Proposition 9.30 is due to Dacić [61, Theorem], except that Dacić considered the group inversion, not a general involution τ. For more information see Propositions 8.14 and 9.23 and the comments to them.

Exercise 9.17 is adapted from Kannappan [132, Theorem 3.14].

Exercise 9.18: Consult Nagy [146] for solutions of (9.19) assuming values in a Banach algebra. The $L^\infty(\mathbb{R})$-solutions of (9.19) are listed in Gajda [93, Corollary 2], who finds them via his study of the solutions $f \in L^\infty(G)$ of the functional equation

$$\int_G f(x+y-s)\, d\mu(s) + \int_G f(x-y+s)\, d\mu(s) = f(x)f(y), \quad x, y \in G,$$

where G is a locally compact, abelian group and μ is a complex-valued, regular Borel measure on G with bounded variation.

Exercise 9.18: Van Vleck [206] discussed only the solution f_0 of (9.19) and not the remaining solutions f_n, $n \neq 0$.

Proposition 9.31: The general solution formula (9.22) differs a bit from the more special one of [154, Theorem 2.2]. It has to according to Exercise 9.22.

Exercise 9.29 transfers Parnami, Singh and Vasudeva [154, Remark p. 288] from Banach spaces to groups.

Chapter 10

D'Alembert's Long Functional Equation

10.1 Introduction

Let S be a semigroup with involution τ. We seek the complex-valued functions g on S that satisfy

$$g(xy) + g(yx) + g(x\tau(y)) + g(\tau(y)x) = 4g(x)g(y) \quad \text{for all } x, y \in S. \tag{10.1}$$

The functional equation (10.1) which is a symmetrized version of d'Alembert's functional equation (9.4), is called *d'Alembert's long functional equation* or abbreviated *d'Alembert's long equation.* It was originally encountered by Corovei [49] for $S = G$ a group and in the form

$$g(xy) + g(yx) + g(xy^{-1}) + g(y^{-1}x) = 4g(x)g(y) \quad \text{for all } x, y \in G, \tag{10.2}$$

in his studies of Wilson's functional equation (see Lemma 11.2(e)).

Any solution g of d'Alembert's equation $g(xy) + g(x\tau(y)) = 2g(x)g(y)$, $x, y \in S$, is central (Proposition 9.17(b)), so such a g satisfies d'Alembert's long functional equation (10.1) as well. The crucial difference between d'Alembert's long and short functional equations is thus that the solutions of the short one are central functions. On certain groups the converse holds, i.e., all solutions of the long equation are solutions of the short one, for example on abelian groups (trivially). As we shall see, less than commutativity can do it: It suffices that the group is step 2 nilpotent (Proposition 10.5). Another result for nilpotent groups is Proposition 10.6. The converse holds also for continuous solutions on compact groups (see Theorem 10.7). And for continuous solutions on the $(ax + b)$-group (Example 10.8). Actually, no example is known of a solution of d'Alembert's long functional equation which is not also a solution of the short one.

The most important results of the present chapter are the ones of Section 10.3. They present sufficient conditions for solutions of d'Alembert's long functional equation to be solutions of d'Alembert's functional equation. This is relevant, because we know how to handle d'Alembert's functional equation (Chapter 9).

10.2 The structure of the solutions

In this section we derive some basic properties of solutions of d'Alembert's long functional equation. We use the notation $d(x) := 2g(x)^2 - g(x^2)$, $x \in S$ (Definition B.6), when g is a complex-valued function on the semigroup S. As mentioned above, any d'Alembert function is a solution of d'Alembert's long functional equation. Thus any joint property of the solutions of d'Alembert's long functional equation is also possessed by the solutions of the short one. So what we derive about properties of solutions of the long functional equation holds also for d'Alembert functions.

Theorem 10.1. *Let M be a monoid with involution τ and neutral element $e \in M$. Let $g : M \to \mathbb{C}$ be a solution of d'Alembert's long functional equation* (10.1) *on M.*

(a) *g is τ-invariant, i.e., $g \circ \tau = g$.*

(b) *If $g \neq 0$, then $g(e) = 1$.*

(c) *For all $x \in M$ we have*

$$d(x) = \frac{g(x\tau(x)) + g(\tau(x)x)}{2}.$$

If M is a group and τ is the group inversion, then $d(x) = 1$ for all $x \in M$, if $g \neq 0$.

(d) *Let $z_0 \in Z(g)$. If $g(z_0)^2 \neq d(z_0)$, then there exists a monoid homomorphism $\chi : M \to \mathbb{C}$ such that $\chi \neq \chi \circ \tau$ and $g = (\chi + \chi \circ \tau)/2$. In particular g is abelian.*

(e) *Let $z_0 \in Z(g)$. If $g(z_0)^2 = d(z_0)$, then $g(xz_0) = g(x)g(z_0)$ for all $x \in M$.*

Proof. (a) The left hand side of (10.1), and consequently also its right hand side, is unchanged if y is replaced by $\tau(y)$.

(b) Putting $y = e$ in (10.1) we get by help of Lemma A.31 that $4g(x) = 4g(x)g(e)$ for all $x \in M$, which implies (b).

(c) Put $y = x$ in (10.1).

(d) and (e) Let $g_{z_0}(x) := g(xz_0) - g(x)g(z_0)$, $x \in M$. Using $z_0 \in Z(g)$ we prove the pair of the functions (g_{z_0}, g) satisfies the symmetrized sine addition formula (6.1), i.e., that

$$g_{z_0}(xy) + g_{z_0}(yx) = 2g_{z_0}(x)g(y) + 2g_{z_0}(y)g(x). \tag{10.3}$$

We do it as follows: Taking $y = z_0$ in (10.1) we get $2g(xz_0) + 2g(\tau(x)z_0) = 4g(x)g(z_0)$, and so $g(xz_0) - g(x)g(z_0) = -[g(\tau(x)z_0) - g(\tau(x))g(z_0)]$, which means g_{z_0} is odd with respect to τ, i.e., that $g_{z_0} \circ \tau = -g_{z_0}$. Next a small computation based on the functional equation and $z_0 \in Z(g)$ shows that

$$g_{z_0}(xy) + g_{z_0}(yx) + g_{z_0}(x\tau(y)) + g_{z_0}(\tau(y)x) = 4g_{z_0}(x)g(y). \tag{10.4}$$

We interchange x and y in (10.4) and add the result to (10.4). Four terms cancel because g_{z_0} is odd, and we are after division by 2 left with (10.3).

When $g_{z_0} \neq 0$ the conclusions of (d) and (e) are immediate from Proposition 6.5 with $w = g_{z_0}$. When $g_{z_0} = 0$ we are in case (e) and we get the conclusion from the definition of g_{z_0}. □

A special case of Theorem 10.1(d): If $M = G$ is a group and g vanishes somewhere on the centre of G, then g is abelian. (To prove this we may assume that $g \neq 0$, so that $g(e) = 1$. The restriction of g to the subgroup $\langle x \rangle$ of G is a solution of d'Alembert's functional equation on $\langle x \rangle$, so $d|_{\langle x \rangle}$ is multiplicative (Proposition 9.17(d)). Furthermore $d(e) = 1$, so d is a character on the group $\langle x \rangle$ and so vanishes nowhere on $\langle x \rangle$. We conclude that $d(x) \neq 0$ for all $x \in G$.)

(d) and (e) above look very much like the corresponding items of Proposition 8.14. Although the functional equations are different, they have the same root (Proposition 6.5). They are used in the proof of Proposition 10.5.

The proposition says that bounded solutions of d'Alembert's long functional equation (10.1) behave in some respects like the d'Alembert function $g(x) = \cos x$ on the real line.

Proposition 10.2. *Let $g : S \to \mathbb{C}$ be a bounded solution of d'Alembert's long functional equation* (10.1) *on a semigroup S with involution τ.*

(a) $|g(x)| \leq 1$ *for all* $x \in S$.

(b) *Assume furthermore that $S = G$ is a group and that τ is the group inversion. Then $g(G) \subseteq [-1, 1]$.*

Proof. We may assume that $g \neq 0$.

(a) Let $m := \sup_{x \in S}|g(x)| < \infty$. Taking $y = x$ in (10.1) we find that

$$4|g(x)|^2 \leq |g(xy)| + |g(yx)| + |g(x\tau(y))| + |g(\tau(y)x)| \leq 4m,$$

so that $|g(x)|^2 \leq m$ and hence $|g(x)| \leq \sqrt{m}$ for all $x \in S$. But then $m \leq \sqrt{m}$ and so $m \leq 1$.

(b) We may assume that $g \neq 0$. Let $x \in G$. Applying Theorem 10.1(d) and (e) (or Kannappan's result Corollary 9.22) to the abelian group $\langle x \rangle$ (which is invariant under τ, because τ is the group inversion here) we get a character $\chi : \langle x \rangle \to \mathbb{C}$ such that

$$g(y) = \frac{\chi(y) + (\chi \circ \tau)(y)}{2} = \frac{\chi(y) + \chi(y^{-1})}{2} \qquad \text{for all } y \in \langle x \rangle.$$

χ is bounded because g is bounded (Proposition 3.18(c)), so $|\chi(y)| = 1$ (Lemma 3.4(b)). Writing $\chi(y) = e^{i\theta}$ for some $\theta \in \mathbb{R}$ we finally find $g(x) = (\chi(y) + \chi(y^{-1}))/2 = (\chi(y) + \chi(y)^{-1})/2 = (e^{i\theta} + e^{-i\theta})/2 \in [-1, 1]$. □

Any solution g of d'Alembert's functional equation is abelian if and only if $\Delta = 0$. Indeed, any solution g of d'Alembert's functional equation is also a solution of the pre-d'Alembert functional equation (Proposition 9.17(c)), so we may refer to Lemma 8.10. Proposition 10.3 gives a partial result of the same nature for d'Alembert's long functional equation (10.2) on a group G. It should be compared to Proposition 8.15. We recall that the function $\widetilde{\Delta} : G \times G \to \mathbb{C}$ was defined by (B.2). Since $d = 1$ here, $\widetilde{\Delta} = 4\Delta$ and reduces to

$$\widetilde{\Delta}(x, y) := 2 + g(xy^2x) + g(yx^2y) - 4g(xy)g(yx) \qquad \text{for } x, y \in G.$$

Proposition 10.3. *Let $g : G \to \mathbb{C}$ be a non-zero solution of d'Alembert's long functional equation* (10.2) *on a group G. Then $\tilde{\Delta} = 0$ if and only if $g([x, y]) = 1$ for all $x, y \in G$.*

Proof. Replacing x by xy and y by yx in (10.2) we get that

$$g(xy^2x) + g(yx^2y) + g([x, y]) + g([x^{-1}, y^{-1}]) = 4g(xy)g(yx),$$

which means that $\tilde{\Delta}$ reduces to

$$\tilde{\Delta}(x, y) := 2 - [g([x, y]) + g([x^{-1}, y^{-1}])].$$

Clearly $g([x,y]) = 1$ for all $x, y \in G$ implies $\tilde{\Delta} = 0$. Let us conversely assume that $\tilde{\Delta} = 0$. This means that

$$g([x,y]) + g([x^{-1}, y^{-1}]) = 2 \qquad \text{for all } x, y \in G. \tag{10.5}$$

Replacing y by yx^{-1} in (10.5) we find that $g([x,y]) + g([y^{-1}, x]) = 2$, so that $g([x^{-1}, y^{-1}]) = g([y^{-1}, x])$, which means that $g([a,b]) = g([b, a^{-1}])$ for all $a, b \in G$. Using this formula with $a = y^{-1}$ and $b = x$ we get that $g([x^{-1}, y^{-1}]) = g([y^{-1}, x]) = g([x,y])$. Substituting this result into (10.5) we find that $g([x,y]) = 1$. □

Remark 10.4. If g in Proposition 10.3 is assumed bounded, then $g([x,y]) = 1$ for all $x, y \in G$ is equivalent to $g \equiv 1$ on all of the commutator subgroup $[G, G]$, not just on the generators $[x, y]$. See [195, Lemma 2.11].

10.3 Relations to d'Alembert's equation

In this section we exhibit some classes of groups on which d'Alembert's long functional equation (10.2) and d'Alembert's functional equation (9.3) have the same solutions. Any solution of d'Alembert's functional equation is a solution of d'Alembert's long functional equation, because it is central (Proposition 9.17(b)), so it is the converse that is of interest. All the results pertain to a group G and its inversion.

In Proposition 10.5 we treat the important class of step 2 nilpotent groups. The Heisenberg group H_3 and the quaternion group Q_8 are examples of such groups. See Theorem 8.16 and Remark 8.19 for results about d'Alembert's short functional equation on nilpotent groups.

Proposition 10.5. *On a step 2 nilpotent group G d'Alembert's long functional equation* (10.2) *has the same complex-valued solutions as d'Alembert's classical functional equation* (9.3), *i.e., as $g(xy) + g(xy^{-1}) = 2g(x)g(y)$, $x, y \in G$.*

Proof. As mentioned above any solution of d'Alembert's functional equation is a solution of d'Alembert's long functional equation. For the other direction it suffices to prove that any solution g of d'Alembert's long functional equation on a step 2 nilpotent group is central, i.e., that $g(xy) = g(yx)$ for all $x, y \in G$. This statement is trivial if $g = 0$, so we shall assume that $g \neq 0$, in which case $g(e) = 1$.

If there is a $z \in Z(g)$ such that $g(z)^2 \neq 1$, then there exists by Theorem 10.1(d) a homomorphism $\chi : G \to \mathbb{C}$ such that $g = (\chi + \chi \circ \tau)/2$. Obviously such a g is central. So we may assume that $g(z)^2 = 1$ for all $z \in Z(g)$. Then $g(xz) = g(x)g(z)$ for all $x \in G$ and $z \in Z(g)$ by Theorem 10.1(e).

Claim. $b^2 \in Z(g)$ for all $b \in G$.

Proof of the claim. Since $[G,G] \subseteq Z(G) \subseteq Z(g)$ we have

$$g(xyb^2) = g(xb^2y[y^{-1}, b^{-2}]) = g(xb^2y)g([y^{-1}, b^{-2}]),$$

so it suffices to prove that $g([x, a^2]) = 1$ for all $a, x \in G$. Using the formula $[a, b^2] = [a, b]b[a, b]b^{-1}$ that here reduces to $[a, b^2] = [a, b]^2$, we get $g([x, a^2]) = g([x, a]^2) = g([x, a])^2 = 1$, finishing the proof of the claim.

Replacing x first by xy and then by yx in (10.2) and subtracting the resulting identities we find after rearranging terms that

$$4[g(xy) - g(yx)]g(y) = g(xy^2) - g(y^2x) + g(y^{-1}xy) - g(yxy^{-1}).$$

Since $y^2 \in Z(g)$ the right hand side simplifies to

$$\begin{aligned} 0 + g(y^{-1}xy) - g(yxy^{-1}) &= g(y^2y^{-1}xyy^{-2}) - g(yxy^{-1}) \\ &= g(yxy^{-1}) - g(yxy^{-1}) = 0, \end{aligned}$$

so $g(y)[g(xy) - g(yx)] = 0$ for all $x, y \in G$. Interchanging x and y we get that also $g(x)[g(xy) - g(yx)] = 0$ for all $x, y \in G$. The desired conclusion $g(xy) = g(yx)$ follows for those x, y for which either $g(x) \neq 0$ or $g(y) \neq 0$. Thus it is left to prove that $g(x_0y_0) = g(y_0x_0)$ if $g(x_0) = g(y_0) = 0$. Taking $x = y = x_0$ in (10.2) we get that $g(x_0^2) = -1$. Similarly $g(y_0^2) = -1$. Finally, using that $g = \check{g}$ by Theorem 10.1(a), that $g(x)g(z) = g(xz)$ for all $x \in G$ and $z \in Z(g)$ and that $x^2 \in Z(g)$ for all $x \in G$ we get

$$\begin{aligned} g(x_0y_0) &= g(y_0^{-1}x_0^{-1}) = g(y_0^{-1}x_0^{-1})g(x_0^2)g(y_0^2) \\ &= g(y_0^{-1}x_0^{-1}x_0^2y_0^2) = g(y_0x_0) \end{aligned}$$

as desired. □

Proposition 10.6. *Let G be a nilpotent group, which is generated by its squares. Then any non-zero solution $g : G \to \mathbb{C}$ of d'Alembert's long functional equation* (10.2) *has the form $g = (\chi + \check{\chi})/2$, where χ is a character on G.*

Proof. Proceeding as in the proof of Proposition 10.5 we see that we may assume that $g(z)^2 = 1$ and that $g(xz) = g(x)g(z)$ for all $x \in G$ and $z \in Z(g)$.

Claim. If H is a subgroup of G such that $[G, H] \subseteq Z(g)$ then $g(ahy^2b) = g(ay^2hb)$ for all $a, b, y \in G$ and $h \in H$.

Proof of the claim. For ease of notation we prove that $g(ah^{-1}y^{-2}b) = g(ay^{-2}h^{-1}b)$ instead of $g(ahy^2b) = g(ay^2hb)$:

$$\begin{aligned} g(ah^{-1}y^{-2}b) &= g(ay^{-2}h^{-1}hy^2h^{-1}y^{-2}b) = g(ay^{-2}h^{-1}[h, y^2]b) \\ &= g(ay^{-2}h^{-1}b[h, y^2]) = g(ay^{-2}h^{-1}b)g([h, y^2]), \end{aligned}$$

so it suffices to prove that $g([h, y^2]) = 1$. And that follows from

$$\begin{aligned} g([h, y^2]) &= g([h, y]y[h, y]y^{-1)}) = g(yy^{-1}[h, y]^2) \\ &= g([h, y]^2) = g([h, y])^2 = 1. \end{aligned}$$

Since G is generated by its squares, it follows from the claim that $g(xhy) = g(xyh)$ for all $x, y \in G$ and $h \in H$. In other words that $H \subseteq Z(g)$ if H is a subgroup of G such that $[G, H] \subseteq Z(g)$.

Using the notation of Definition A.5 we have $[G, C^{n-1}(G)] = \{e\}$, so $C^{n-1}(G)$ satisfies the conditions on H of the claim, which implies that $C^{n-1}(G) \subseteq Z(g)$. From $[G, C^{n-2}(G)] = C^{n-1}(G)$ we see that $C^{n-2}(G)$ satisfies the condition on H, so that $C^{n-2}(G) \subseteq Z(g)$. Continuing inductively we end up with $C^0(G) = G \subseteq Z(g)$. This means g is abelian, so we may appeal to Kannappan's result (Corollary 9.22). □

We next treat the class of compact groups. It contains the class of finite groups.

Theorem 10.7.

(a) *Let $g \in C(G)$ be a solution of d'Alembert's long functional equation* (10.2) *on a locally compact group G. If g is square integrable with respect to a right or left Haar measure on G, then g is a solution of d'Alembert's short functional equation* (9.3).

(b) *Let G be a compact group. The solutions $g \in C(G)$ of d'Alembert's long functional equation* (10.2) *are the same as those of the short one* (9.3).

Proof. (a) Let $g \in C(G)$ be a solution of d'Alembert's long functional equation (10.2). The conclusion holds trivially if $g = 0$, so we may assume $g \neq 0$.

We use this assumption below in computations in the form $g(y^2) + 1 - 2g(y)^2 = 0$. For each $y \in G$ we define

$$r(y) := R(y) + R(y^{-1}) - 2g(y)I : C(G) \to C(G)$$

and

$$l(y) := L(y) + L(y^{-1}) - 2g(y)I : C(G) \to C(G),$$

where $R, L : C(G) \to C(G)$ denote the right and left regular representations, and I is the identity operator. Then d'Alembert's long functional equation (10.2) takes the form

$$[r(y) + l(y)]g = 0 \qquad \text{for all } y \in G. \tag{10.6}$$

Since R and L and hence also $r(y)$ and $l(y)$ commute, we get that $r(y)^2 g = l(y)^2 g$ (Multiply (10.6) by $r(y) - l(y)$). Simple computations reveal that $r(y)^2 = r(y^2) - 4g(y)r(y)$ and $l(y)^2 = l(y^2) - 4g(y)l(y)$, so that

$$r(y)^2 + l(y)^2 = r(y^2) + l(y^2) - 4g(y)[r(y) + l(y)].$$

Applying the last identity to g gives us that $[r(y)^2 + l(y)^2]g = 0$. As noted above $r(y)^2 g = l(y)^2 g$ so we conclude that $r(y)^2 g = 0$.

Let us assume that g is square integrable with respect to a right Haar measure dx on G. We may view $r(y) = R(y) + R(y)^{-1} - 2g(y)I$ as a normal operator on $L^2(G, dx)$, because R is a unitary reprersentation so that $R(y)^* = R(y^{-1})$. Due to the normality $r(y)^2 g = 0$ implies that $r(y)g = 0$ as an element of $L^2(G, dx)$ (Exercise 10.4), so $r(y)g$ vanishes almost everywhere. Being continuous it vanishes everywhere. And $(r(y)g)(x) = 0$ for all $x, y \in G$ means that g is a solution of d'Alembert's short functional equation (9.3).

The case of g being square integrable with respect to a left Haar measure can be treated analogously based on the result $l(y)^2 g = 0$, so we leave it out.

(b) is immediate from (a). □

Example 10.8. In this example we compute the continuous solutions of d'Alembert's long functional equation (10.2) on the $(ax + b)$-group

$$G := \left\{ \begin{pmatrix} a & b \\ 0 & 1 \end{pmatrix} \,\middle|\, a > 0,\ b \in \mathbb{R} \right\}$$

from Example A.17(i). More precisely, we show that any such solution is central. That suffices, because then it satisfies d'Alembert's functional equation (9.3) and the continuous solutions of (9.3) were found in Example 9.9.

The result of the present example will be used in Example 11.14.

Let $g \in C(G)$, $g \neq 0$, be a solution of d'Alembert's long functional equation (10.2) on G. To avoid too much empty space we use the abbreviation

$$(a,b) := \begin{pmatrix} a & b \\ 0 & 1 \end{pmatrix}.$$

Replacing x and y in the identity (10.2) by $(a,b) \in G$ and $(x,y) \in G$ it becomes

$$\begin{aligned} g(ax, b+ay) + g(ax, y+bx) + g(ax^{-1}, b - ax^{-1}y) \\ + g(ax^{-1}, bx^{-1} - x^{-1}y) \\ = 4g(a,b)g(x,y) \qquad \text{for all } (a,b),(x,y) \in G. \end{aligned} \tag{10.7}$$

Any continuous solution g is C^∞ by Example D.9, so we may use the theory of differential equations. We want to show that g is constant on cosets of the commutator group $[G,G] = \{(1,b) \mid b \in \mathbb{R}\}$, which means constant in the second variable b. This makes us differentiate (10.7) with respect to the second variable y. Doing so twice at $(x,y) = (1,0)$ we find that g as a function of b satisfies the second order ordinary differential equation

$$\frac{\partial^2 g}{\partial b^2}(a,b) = \frac{\lambda^2}{1+a^2}\, g(a,b), \qquad \text{where } \lambda^2 := 2\frac{\partial^2 g}{\partial b^2}(1,0) \in \mathbb{C}.$$

First case. $\lambda \neq 0$. From the theory of differential equations we know that there are for each fixed a constants $c_1(a), c_2(a) \in \mathbb{C}$, such that

$$g(a,b) = c_1(a) \exp\Big(\frac{\lambda}{\sqrt{1+a^2}}\, b\Big) + c_2(a) \exp\Big(-\frac{\lambda}{\sqrt{1+a^2}}\, b\Big) \quad \text{for all } b \in \mathbb{R}. \tag{10.8}$$

c_1 and c_2 are uniquely determined, smooth functions of $a \in \mathbb{R}^+$, because

$$\begin{aligned} g(a,0) &= c_1(a) + c_2(a), \\ \frac{\partial g}{\partial b}(a,0) &= \frac{\lambda}{\sqrt{1+a^2}}\,(c_1(a) - c_2(a)). \end{aligned}$$

Now g is even (Theorem 10.1(a)), so that $g(a,b) = g(a^{-1}, -a^{-1}b)$. Using this in (10.8) we find from the linear independence of exponentials (Corollary 3.20) that $c_2(a) = c_1(a^{-1})$, so that (replacing c_1 by $c/2$) we obtain for all $(a,b) \in G$ that

$$2g(a,b) = c(a) \exp\Big(\frac{\lambda}{\sqrt{1+a^2}}\, b\Big) + c(a^{-1}) \exp\Big(-\frac{\lambda}{\sqrt{1+a^2}}\, b\Big). \tag{10.9}$$

With $b = 0$ and $x = 1$ (10.7) becomes

$$g(a, ay) + g(a, y) + g(a, -ay) + g(a, -y) = 4g(a, 0)g(1, y).$$

Using the expression (10.9) for g we get after simple computations using that $c(1) = (c(1) + c(1))/2 = g(1, 0) = 1$ that:

$$\begin{aligned} c(a)\exp\Big(\frac{\lambda}{\sqrt{1+a^2}}\,ay\Big) &+ c(a^{-1})\exp\Big(\frac{-\lambda}{\sqrt{1+a^2}}\,ay\Big) \\ &+ c(a)\exp\Big(\frac{\lambda}{\sqrt{1+a^2}}\,y\Big) + c(a^{-1})\exp\Big(\frac{-\lambda}{\sqrt{1+a^2}}\,y\Big) \\ &+ c(a)\exp\Big(\frac{-\lambda}{\sqrt{1+a^2}}\,ay\Big) + c(a^{-1})\exp\Big(\frac{\lambda}{\sqrt{1+a^2}}\,ay\Big) \\ &+ c(a)\exp\Big(\frac{-\lambda}{\sqrt{1+a^2}}\,y\Big) + c(a^{-1})\exp\Big(\frac{\lambda}{\sqrt{1+a^2}}\,y\Big) \\ &= 2[c(a) + c(a^{-1})]\Big[\exp\Big(\frac{\lambda}{\sqrt{2}}\,y\Big) + \exp\Big(\frac{-\lambda}{\sqrt{2}}\,y\Big)\Big]. \end{aligned}$$

Since exponentials with different exponents are linearly independent we get for $a \neq 1$ that $c(a) + c(a^{-1}) = 0$. By the continuity of c this also holds for $a = 1$, contradicting $c(1) = 1$. Thus the case of $\lambda \neq 0$ does not occur.

Second case. $\lambda = 0$. Here $\partial^2 g/\partial b^2 = 0$, so $g(a, b) = \alpha(a) + \beta(a)b$ which we substitute into d'Alembert's long equation (10.7):

$$\begin{aligned} \alpha(ax) &+ \beta(ax)b + \beta(ax)ay + \alpha(ax) + \beta(ax)y + \beta(ax)xb \\ &+ \alpha(ax^{-1}) + \beta(ax^{-1})b - \beta(ax^{-1})ax^{-1}y \\ &+ \alpha(ax^{-1}) + \beta(ax^{-1})x^{-1}b - \beta(ax^{-1})x^{-1}y \\ &= 4[\alpha(a) + \beta(a)b][\alpha(x) + \beta(x)y]. \end{aligned}$$

The coefficients of the variable y must be the same on both sides, so

$$\beta(ax)a + \beta(ax) - \beta(ax^{-1})ax^{-1} - \beta(ax^{-1})x^{-1} = 4[\alpha(a) + \beta(a)b]\beta(x).$$

Here the variable b occurs on the right hand side only. Thus its coefficient is 0, i.e., $4\beta(a)\beta(x) = 0$ for all $a, x \in \mathbb{R}^+$. Then $\beta = 0$, so that $g(a, b) = \alpha(a)$. This implies g is central as desired.

10.4 Exercises

Exercise 10.1. Account for why the solutions of d'Alembert's long functional equation (10.2) on the quaternion group Q_8 satisfy d'Alembert's functional equation (9.3).

Exercise 10.2. Let $g : S \to \mathbb{C}$ be a solution of d'Alembert's long functional equation (10.1) on a semigroup S such that $|g(x)| = 1$ for all $x \in S$.

Show that g is a multiplicative function from S into $\mathbb{T}$ such that $g \circ \tau = g$.

Hint: The unit ball in $\mathbb{C}$ is strictly convex.

Assume furthermore that S is a group and τ is the group inversion. Show that $g(G) \subseteq \{\pm 1\}$.

The result can be compared with the one of Exercise 9.10.

Exercise 10.3. Prove that the set of non-zero solutions of d'Alembert's long functional equation (10.1) on a semigroup S is linearly independent in the vector space of all complex-valued functions on S.

Hint: Proceed as in the proof of Proposition 3.18.

Exercise 10.4. Let $N \in \mathcal{B}(\mathcal{H})$ be normal, i.e., commute with its adjoint. In other words $N^*N = NN^*$. Let $g \in \mathcal{H}$ satisfy that $N^2 g = 0$. Show that $Ng = 0$.

Hint: Show first that $(N^*N)^2 g = 0$.

10.5 Notes and remarks

Much of the material of the present chapter is taken from Stetkær [188], but Corovei [49] was the first one to study d'Alembert's long functional equation.

For d'Alembert's functional equations on hypergroups see Roukbi and Zeglami [165].

Theorem 10.1(d) is due to Corovei [49, Teorema 1] in the case of M being a group and τ the group inversion. See also the comment to Proposition 9.30.

Proposition 10.3 is due to Dilian Yang [216, Corollary 6.4].

Proposition 10.6 generalizes Friis [90, Theorem 2.6] and Theorem 9.25.

Theorem 10.7 is contained in An and Yang [16, Corollary 6.9]. Yang gave later a direct proof in [214, Theorem 5.1]. Both papers state the result in form of our Theorem 9.28. The proof we present has a reminiscence of Yang [214, Theorem 5.1], but is more elementary and even shorter. Our hypothesis that g is square integrable is arguably too strong on non-compact groups: Under weak conditions any such solution vanishes according to Penney and Rukhin [156, Theorem 1.3].

Chapter 11

Wilson's Functional Equation

Throughout this chapter we let M denote a topological monoid equipped with a continuous involution $\tau : M \to M$, and when we say that a function $f : M \to \mathbb{C}$ is even/odd, we mean even/odd with respect to τ.

We discuss Wilson's functional equation (11.3). The most important results are Corollary 11.7 (for the abelian case) and Theorem 11.8 (for the non-abelian case). A special instance is Jensen's functional equation that we study in details in Chapter 12.

11.1 Introduction

In [209] Wilson dealt with functional equations related to and generalizing the cosine functional equation $g(x+y) + g(x-y) = 2g(x)g(y)$ on the real line. He generalized the cosine equation to

$$f(x+y) + f(x-y) = 2f(x)g(y), \qquad x, y \in \mathbb{R}, \tag{11.1}$$

that contains the two unknown functions f and g. In [210] he introduced his second generalization

$$f(x+y) + f(x-y) = 2h(x)k(y), \qquad x, y \in \mathbb{R}, \tag{11.2}$$

that contains the three unknown functions f, h and k. We shall in this chapter study Wilson's first generalization, not just on $\mathbb{R}$, but extended to monoids. To whet the appetite let us note that a solution of (11.1) is the pair $f(x) = \sin x$ and $g(x) = \cos x$. Another solution is $f(x) = g(x) = \cos x$. And in the degenerate case of $g = 1$ we get a solution f of another nature (not a trigonometrical function, but an affine function): $f(x) = ax + b$, where $a, b \in \mathbb{C}$ are constants. Our generalization of (11.1) is

Definition 11.1. *Wilson's functional equation* or for brevity just *Wilson's equation* is

$$f(xy) + f(x\tau(y)) = 2f(x)g(y), \qquad x, y \in M, \tag{11.3}$$

where $f, g : M \to \mathbb{C}$ are the unknowns. By a *solution* of (11.3) we understand an ordered pair (f, g) of continuous functions $f, g \in C(M)$ satisfying (11.3).

We call the solutions $f \in C(M)$ of (11.3) the *Wilson functions* corresponding to $g \in C(M)$, or just Wilson functions when it is clear from the context what g is. For fixed $g \in C(M)$ and τ we let $\mathrm{Wil}(g)$ denote the complex vector space of Wilson functions, so

$$\mathrm{Wil}(g) := \{f \in C(M) \mid f \text{ is a Wilson function}\}.$$

Wilson's classical functional equation is the special case

$$f(xy) + f(xy^{-1}) = 2f(x)g(y), \qquad x, y \in G, \tag{11.4}$$

where G is a topological group, and $f, g \in C(G)$.

A variant of Wilson's classical functional equation is $f(xy) + f(y^{-1}x) = 2f(x)g(y)$, $x, y \in M$, where we have interchanged x and y^{-1} in the second term on the left hand side of (11.4). The variant has in general other solutions than the classical one. We discuss it in Subsection 11.8.

11.2 General properties of the solutions

We note a couple of properties of solutions of Wilson's equation that follow almost immediately from the definition.

Lemma 11.2. $\mathrm{Wil}(g)$ *is for any* $g \in C(M)$ *a subspace of* $C(M)$, *which is invariant under left translations.*

Proof. Left to the reader. □

Lemma 11.3. *Let* (f, g) *be a solution of Wilson's functional equation* (11.3) *such that* $f \neq 0$.

(a) g *is even.*

(b) *The even part* $(f + f \circ \tau)/2$ *of* f *is* $f(e)g$, *so it is proportional to* g.

(c) f *is odd (i.e.,* $f \circ \tau = -f$*) if and only if* $f(e) = 0$.

(d) *If f is odd, then (f, g) is a solution of the symmetrized sine addition formula, i.e.,*

$$f(xy) + f(yx) = 2f(x)g(y) + 2f(y)g(x) \quad \text{for all } x, y \in M. \tag{11.5}$$

(e) *g is a solution of d'Alembert's long equation (10.1) such that $g(e) = 1$. If f is central then g is a d'Alembert function.*

(f) *If g is a d'Alembert function, then both the even and the odd part of f are Wilson functions corresponding to g, so the space of Wilson functions is in this case the direct sum of the even and the odd Wilson functions.*

Proof. (a) Replacing y by $\tau(y)$ does not change the left hand side of (11.3).

(b) and (c) Taking $x = e$ in (11.3) we get $f(y) + f(\tau(y)) = 2f(e)g(y)$ which is (b). Using the same identity we get (c).

(d) Interchange x and y in 11.3 and add the two identities.

(e) Choosing $x_0 \in M$ such that $f(x_0) \neq 0$ we get from (11.3) that

$$g(x) = \frac{f(x_0x) + f(x_0\tau(x))}{2f(x_0)} \qquad \text{for all } x \in M,$$

which implies that $g(e) = 1$ and

$$\begin{aligned}
&2f(x_0)\{g(xy) + g(x\tau(y)) + g(yx) + g(\tau(y)x)\} \\
&\quad = f(x_0xy) + f(x_0\tau(y)\tau(x)) + f(x_0x\tau(y)) + f(x_0y\tau(x)) \\
&\qquad + f(x_0yx) + f(x_0\tau(x)\tau(y)) + f(x_0\tau(y)x) + f(x_0\tau(x)y)) \\
&\quad = f(x_0xy) + f(x_0x\tau(y)) + f(x_0\tau(y)\tau(x)) + f(x_0\tau(y)x) \\
&\qquad + f(x_0y\tau(x)) + f(x_0yx) + f(x_0\tau(x)\tau(y)) + f(x_0\tau(x)y) \\
&\quad = 2f(x_0x)g(y) + 2f(x_0\tau(y))g(x) + 2f(x_0y)g(x) + 2f(x_0\tau(x))g(y) \\
&\quad = 2[f(x_0x) + f(x_0\tau(x))]g(y) + 2[f(x_0y) + f(x_0\tau(y))]g(x) \\
&\quad = 4f(x_0)g(x)g(y) + 4f(x_0)g(y)g(x) = 8f(x_0)g(x)g(y).
\end{aligned}$$

The proof of the second statement consists of similar computations:

$$\begin{aligned}
&2f(x_0)[g(xy) + g(x\tau(y))] \\
&\quad = f(x_0xy) + f(x_0\tau(y)\tau(x)) + f(x_0x\tau(y)) + f(x_0y\tau(x)) \\
&\quad = f(x_0xy) + f(x_0x\tau(y)) + f(x_0y\tau(x)) + f(x_0\tau(y)\tau(x)) \\
&\quad = 2f(x_0x)g(y) + f(\tau(x)x_0y) + f(\tau(x)x_0\tau(y)) \\
&\quad = 2f(x_0x)g(y) + 2f(\tau(x)x_0)g(y) = 2[f(x_0x) + f(x_0\tau(x)]g(y) \\
&\quad = 4f(x_0)g(x)g(y).
\end{aligned}$$

(f) That g is a d'Alembert function means that it is a Wilson function corresponding to g. Now $f = f(e)g + (f - f(e)g)$ is the decomposition of f into its even and odd parts. □

Lemma 11.3(f) indicates the fact that it is often expedient to split a Wilson function into its even and odd parts. An example presenting even and odd solutions of the equation (11.1) with $g(x) = \cos x$ is $f(x) = \cos x$ and $f(x) = \sin x$ respectively.

From Lemma 11.3(e) we see that Wilson's functional equation is tied to d'Alembert's functional equations in a deeper way than the obvious, formal one that d'Alembert's functional equation is Wilson's functional equation with $f = g$. True, it is d'Alembert's long functional equation that enters in (e), but on many groups it reduces to d'Alembert's (short) equation (9.4). Let us recall the following types of groups, where we know that d'Alembert's long equation reduces to the short one: It obviously does on abelian groups, but it also does so for continuous solutions on step 2 nilpotent groups (Proposition 10.5), nilpotent connected Lie groups (Theorem 10.6), the $(ax + b)$-group (Example 10.8) and compact groups (Theorem 10.7). In Section 11.4 we discuss the important case of Wilson's functional equation in which g is a d'Alembert function.

11.3 The abelian case

Corollary 11.7 provides a complete and satisfactory description of the solutions of Wilson's functional equation (11.3) on abelian groups. According to Lemma 11.4 certain abelian solutions exist on any monoid. Corollary 11.7 shows that they on abelian groups constitute all solutions except for the exceptional case of $f = 0$ and g anything. Special cases of them are the familiar solutions

$$x \mapsto (\alpha \sin x + \beta \cos x, \cos x) \quad \text{and} \quad x \mapsto (\alpha x + \beta, 1)$$

of Wilson's original equation (11.1).

On the Heisenberg group there exists a Wilson function which is not central (Example 12.4), so the non-abelian case is more complicated.

Lemma 11.4. *Let $\chi \in C(M)$ be multiplicative and define $g := (\chi + \chi \circ \tau)/2$.*

(a) *$f := \alpha(\chi - \chi \circ \tau)/2 + \beta(\chi + \chi \circ \tau)/2$ is a Wilson function corresponding to g for any $\alpha, \beta \in \mathbb{C}$.*

(b) *If* $\chi = \chi \circ \tau$ *then* $f := a\chi + \beta\chi$ *is a Wilson function corresponding to* g *for any additive map* $a \in C(M)$ *and any* $\beta \in \mathbb{C}$. *Here* $g = \chi$.

Proof. Computations that are left to the reader. □

Proposition 11.5. *Let the pair* (f, g), *where* $f : M \to \mathbb{C}$ *and* $g \in C(M)$, *be a solution of Wilson's functional equation* (11.3) *such that* f *is central and* $\neq 0$.

Then g *is a d'Alembert function. Furthermore*

(a) *When* g *is non-abelian, then* $f = f(e)g$.

(b) *When* g *is abelian, then* g *has the form* $g = (\chi + \chi \circ \tau)/2$ *where* $\chi \in C(M)$ *is multiplicative and* $\chi(e) = 1$.

(i) *If* $\chi \neq \chi \circ \tau$, *then*

$$\mathrm{Wil}(g) = \mathrm{span}\left\{\frac{\chi - \chi \circ \tau}{2}, \frac{\chi + \chi \circ \tau}{2}\right\}.$$

So in this case f *has the form*

$$f = \alpha \frac{\chi - \chi \circ \tau}{2} + \beta \frac{\chi + \chi \circ \tau}{2}$$

for some $\alpha, \beta \in \mathbb{C}$.

(ii) *If* $\chi = \chi \circ \tau$ *and* M *is a topological group, then*

$$\mathrm{Wil}(g) = \mathrm{span}\{a\chi, \chi \mid a : M \to \mathbb{C} \text{ additive}\}.$$

So in this case f *has the form* $f = a\chi + \beta\chi$ *for some additive map* $a : M \to \mathbb{C}$ *and some* $\beta \in \mathbb{C}$.

Conversely the formulas above for g *and* f *define solutions of* (11.3).

Proof. According to Lemma 11.3(e) g is a d'Alembert function. Let f_e and f_o denote the even and the odd parts of f. We see from Lemma 11.3 that $f_e = f(e)g$, and that f_o is a solution of the symmetrized sine addition formula, so that (being central) f_o is a solution of the sine addition formula (4.3).

(a) Assume first g non-abelian. If $f_o \neq 0$ we get a contradiction to g being non-abelian from Theorem 4.1(e). Hence $f_o = 0$, so that f is even, and so $f = f_e = f(e)g$.

(b) Assume next g abelian. By Theorem 9.21 there exists a multiplicative function $\chi \in C(M)$ such that $g = (\chi + \chi \circ \tau)/2$. Since χ is unique, except that it can be interchanged by $\chi \circ \tau$, we may apply Theorem 4.1 with $\chi_1 = \chi$ and $\chi_2 = \chi \circ \tau$ to find the form of f_o. If $\chi \neq \chi \circ \tau$ we get that $f_o = \alpha(\chi - \chi \circ \tau)/(2)$ for some $\alpha \in \mathbb{C}$, so that

$$f = f_e + f_o = f(e)g + \alpha\frac{\chi - \chi \circ \tau}{2} = f(e)\frac{\chi + \chi \circ \tau}{2} + \alpha\frac{\chi - \chi \circ \tau}{2}.$$

If $\chi = \chi \circ \tau$ and M is a group, then $f_o = a\chi$, where $a := f_o/\chi : M \to \mathbb{C}$ is additive. Thus $f = f_o + f_e = a\chi + f(e)g = a\chi + f(e)\chi$.

The converse follows from Lemma 11.4. □

Remark 11.6. Proposition 11.5 assumes f to be central. This assumption is not automatically satisfied even when g is abelian: In case (i) of (b) we find a non-central odd Wilson function F on the $(ax + b)$-group by Proposition 11.15(b), and in case (ii) the Heisenberg group provides an example with $g = \chi = 1$ (Example 12.4).

Corollary 11.7. *Let G be an abelian, topological group. Let the pair (f, g), where $f : G \to \mathbb{C}$ and $g \in C(G)$, be a solution of Wilson's classical functional equation*

$$f(x + y) + f(x - y) = 2f(x)g(y), \qquad x, y \in G. \tag{11.6}$$

(A) *If $f \neq 0$, then there exists a character $\chi \in C(G)$ on G such that g has the form $g = (\chi + \check{\chi})/2$. The character χ is uniquely determined by $g = (\chi + \check{\chi})/2$ except that it can be interchanged with $\check{\chi}$.*

(B) *Given $g = (\chi + \check{\chi})/2$ where $\chi \in C(G)$ is a character we have*

(a) *When $\chi \neq \check{\chi}$ there exist constants $\alpha, \beta \in \mathbb{C}$ such that*

$$f = \alpha\,\frac{\chi - \check{\chi}}{2} + \beta\,\frac{\chi + \check{\chi}}{2}.$$

(b) *When $\chi = \check{\chi}$ there exist an additive function $a \in C(G)$ and a constant $\beta \in \mathbb{C}$ such that $f = a\chi + \beta\chi$.*

If f is bounded and $\neq 0$, then g is bounded, χ is unitary and the additive function $a : G \to \mathbb{C}$ in (b) vanishes.

Conversely, if $\chi \in C(G)$ is a character on G then $g = (\chi + \check{\chi})/2$ and the formulas for f from (a) and (b) define solutions of (11.6).

Proof. Only the statement about what happens when f is bounded needs a proof, so assume f bounded. We see from the functional equation that so is g. Then χ is bounded (Theorem 3.18(c)) and so unitary (Lemma 3.4(b)). It follows that the additive function a in (b) is bounded. But the only bounded, additive function is $a = 0$ (Exercise 2.5). □

11.4 Wilson functions when g is a d'Alembert function

Throughout Section 11.4 we let $g \in C(M)$ be a d'Alembert function. We have in Lemma 11.3(e) seen that if a solution (f, g) of Wilson's functional equation has $f \neq 0$, then g is a solution of d'Alembert's long functional equation with $g(e) = 1$. In the present section we require more, namely that g is a d'Alembert function. The end of Section 11.2 gives a partial justification.

We recall that $W(g)$ is the vector space defined in Section 8.6 by

$$\begin{aligned} W(g) := \{w \in C(M) \mid w(xy) + w(yx) \\ = 2w(x)g(y) + 2w(y)g(x) \text{ for all } x, y \in M\}, \end{aligned}$$

so it is for fixed g the set of solutions of the symmetrized sine addition formula.

As is easily verified $T(g) := \operatorname{span}\{R(x)g \,|\, x \in M\} \subseteq \operatorname{Wil}(g) \subseteq \mathbb{C}g + W(g)$, g being a d'Alembert function (Lemma 11.2 + g being central).

11.4.1 *The case of g non-abelian*

$T(g) = \operatorname{Wil}(g) = \mathbb{C}g + W(g)$ when g is non-abelian (Lemma 8.22(b)).

The solutions of Wilson's functional equation (11.3) are for g is non-abelian given in Theorem 11.8. In Proposition 11.5 we assumed $f \in \operatorname{Wil}(g)$ was central. This assumption is not enforced here, so Theorem 11.8 is more general than Proposition 11.5(a).

Theorem 11.8. *Let $g \in C(M)$ be a non-abelian d'Alembert function (a continuous solution of* (9.4) *such that $g(e) = 1$). Then there is a continuous, algebraically irreducible representation ρ of M on $\mathbb{C}^2$ for which $\rho \circ \tau = \operatorname{adj} \circ \rho$, such that*

(a) *$g(x) = \frac{1}{2} \operatorname{tr} \rho(x)$, $x \in M$. If M is a group and g is bounded, then we may choose ρ as a unitary representation.*

(b) *For any $f \in \operatorname{Wil}(g)$ there exists an $A \in M(2 \times 2, \mathbb{C})$ such that $f(x) = \frac{1}{2} \operatorname{tr}(A\rho(x))$ for all $x \in M$.*

Conversely, let ρ be a continuous, algebraically irreducible representation of M on $\mathbb{C}^2$ for which $\rho \circ \tau = \operatorname{adj} \circ \rho$ and let $A \in M(2 \times 2, \mathbb{C})$. Define $g(x) = \frac{1}{2} \operatorname{tr} \rho(x)$ and $f(x) = \operatorname{tr}(A\rho(x))$ for all $x \in M$. Then (f, g) is a solution of Wilson's functional equation (11.3).

Proof. Checking the conversely statement consists of manipulations with formulas based on the identity $A + \operatorname{adj}(A) = (\operatorname{tr} A)I$ (see Example A.33(j)). We skip it.

(a) Theorem 9.26 tells us that g has the desired form. If M is a group and g is bounded, we refer to the proof of Theorem 9.27(b).

(b) The even and odd parts of f are again Wilson functions corresponding to g (Lemma 11.3(f)), so it suffices to prove the statement about f when f is an even and when f is an odd solution. If f is even, we refer to Lemma 11.3(b). If f is odd, we refer to Lemma 11.3(d) and Theorem 8.24(e). □

11.4.2 *Discussion for g abelian*

Any abelian d'Alembert function g on the monoid M has the form $g = (\chi + \chi \circ \tau)/2$ where $\chi : M \to \mathbb{C}$ is multiplicative and $\neq 0$ (Theorem 9.21(a)). If $\chi = \chi \circ \tau$ we find that $R(x)g = \chi(x)g$, so that $T(g) = \mathbb{C}g$ has dimension 1. If $\chi \neq \chi \circ \tau$ we find that $T(g) = \operatorname{span}\{\chi, \chi \circ \tau\}$, which has dimension 2 (Exercise C.1).

The surrounding vector space $\operatorname{Wil}(g)$ of Wilson functions may be large and larger than $T(g)$. To substantiate that statement consider Wilson's classical functional equation on an abelian group G with $g = 1$. Here we get from Corollary 11.7 that the solutions are the continuous additive functions plus the constant functions, i.e., the affine functions on G. So $\dim \operatorname{Wil}(1) = n + 1$ on $G = \mathbb{R}^n$, and $\dim \operatorname{Wil}(1) = \infty$ on an infinite dimensional vector space like $G = C([0, 1])$. Of course $g = 1$ is a rather degenerate d'Alembert function, but even in the non-degenerate case of $\chi \neq \check{\chi}$ we may have $\dim \operatorname{Wil}(g) > 2$ on non-abelian groups (Proposition 11.15(b)).

So $T(g)$ is in general a proper subspace of $\operatorname{Wil}(g)$. On the positive side it may be mentioned that the spaces coincide if M is abelian and $\chi \neq \chi \circ \tau$ (Corollary 11.5(i)).

See Remark 11.6 for references to results on some non-abelian groups.

11.5 The case of a compact group

The following theorem gives a complete description of the solutions of Wilson's classical equation on a compact group. An additive function a like the one from Corollary 11.7(b) will not show up in the solution formulas here, because continuous, additive functions vanish on compact groups (Exercise 2.5(b)).

Theorem 11.9. *Let the pair $(f,g) \in C(G) \times C(G)$ where $f \neq 0$, be a solution of the classical case of Wilson's functional equation*

$$f(xy) + f(xy^{-1}) = 2f(x)g(y), \qquad x, y \in G, \tag{11.7}$$

on a compact group G.

Then g is a classical d'Alembert function, and (f,g) has one of the following two forms:

(a) *There exist a continuous, algebraically irreducible representation $\rho : G \to \mathrm{SU}(2)$ and a matrix $A \in M(2 \times 2, \mathbb{C})$ such that*

$$g(x) = \tfrac{1}{2}\chi_\rho(x) = \tfrac{1}{2}\operatorname{tr}\rho(x), \;\; f(x) = \tfrac{1}{2}\operatorname{tr}(A\rho(x)) \qquad \text{for all } x \in G.$$

In this case g is non-abelian.

(b) *There exist a continuous, unitary character χ on G and constants $c_1, c_2 \in \mathbb{C}$ such that*

$$g = \frac{\chi + \check{\chi}}{2}, \quad f(x) = c_1\chi + c_2\check{\chi}.$$

In this case g and f are abelian.

Conversely, the formulas above define solutions of Wilson's functional equation.

Proof. g is a solution of d'Alembert's long functional equation (10.1) such that $g(e) = 1$ by Lemma 11.3(e). It follows then that g is a d'Alembert function (Theorem 10.7).

If g is non-abelian we refer to Theorem 11.8.

If g is abelian there exists by Kannappan's theorem (Corollary 9.22) a continuous character χ on G such that $g = (\chi + \check{\chi})/2$, transforming (11.7) to

$$f(xy) + f(xy^{-1}) = f(x)(\chi(y) + \check{\chi}(y)) \qquad \text{for all } x, y \in G. \tag{11.8}$$

The character χ is unitary by Lemma 3.4(b). Assuming first that $\chi \neq \check{\chi}$ we multiply (11.8) by $\overline{\chi(y)}$ and integrate the result over G with respect to the normalized Haar measure dy on G. According to the orthogonality relations

(Lemma E.17) we get for any $x \in G$ (note that $\overline{\chi} = \check{\chi}$ when χ is unitary)

$$\begin{aligned} f(x) &= f(x)\left(\int_G \chi(y)\overline{\chi(y)}\,dy + \int_G \check{\chi}(y)\overline{\chi(y)}\,dy\right) \\ &= \int_G f(xy)\overline{\chi(y)}\,dy + \int_G f(xy^{-1})\overline{\chi(y)}\,dy \\ &= \int_G f(y)\overline{\chi(x^{-1}y)}\,dy + \int_G f(y^{-1})\overline{\chi(yx)}\,dy \\ &= \left(\int_G f(y)\overline{\chi(y)}\,dy\right)\chi(x) + \left(\int_G f(y^{-1})\overline{\chi(y)}\,dy\right)\check{\chi}(x), \end{aligned}$$

which is (b). A similar argument works if $\chi = \check{\chi}$. □

11.6 Examples

Example 11.10. We seek the solutions $f, g \in C(\mathbb{R})$ of Wilson's classical functional equation

$$f(x+y) + f(x-y) = 2f(x)g(y), \qquad x, y \in \mathbb{R}.$$

We discard the trivial case of $f = 0$ and $g \in C(\mathbb{R})$ arbitrary. Combining Corollary 11.7 with Example 3.7(a) we see that there exists $\lambda \in \mathbb{C}$ such that

$$g(x) = \frac{e^{i\lambda x} + e^{-i\lambda x}}{2} = \cos(\lambda x) \qquad \text{for all } x \in \mathbb{R}.$$

If $\lambda \neq 0$, then there exist $\alpha, \beta \in \mathbb{C}$ such that

$$f(x) = \alpha \sin(\lambda x) + \beta \cos(\lambda x) \qquad \text{for all } x \in \mathbb{R}. \tag{11.9}$$

If $\lambda = 0$, then $g = 1$ and there exist $\alpha, \beta \in \mathbb{C}$ such that

$$f(x) = \alpha x + \beta \qquad \text{for all } x \in \mathbb{R}. \tag{11.10}$$

The continuous solutions of Wilson's classical functional equation on $\mathbb{R}$ are the pairs of functions, that are defined by the formulas above.

You might wonder why a solitary, non-trigonometrical function like $f(x) = \alpha x$ must enter in (11.10) among all the trigonometric and hyperbolic functions: An explanation is that it comes about as the limit for $\lambda \to 0$ of the trigonometrical solution $f(x) = \alpha\lambda^{-1}\sin(\lambda x)$ from (11.9).

Example 11.11. We will here consider solutions of Wilson's classical functional equation (11.4) on the quaternion group Q_8. Theorem 11.9 applies, Q_8 being finite. In particular, if (f, g), $f \neq 0$, is a solution of (11.4) on Q_8, then g is a d'Alembert function (it satisfies (9.3)). The d'Alembert functions on Q_8 were found in Example 9.10. The Wilson functions corresponding to the unique non-abelian d'Alembert function g_0 are computed in [192, Example 11.3].

Example 11.12. Let $M := (M(2 \times 2, \mathbb{C}), \circ)$ and $\tau := \text{adj}$. For any $A \in M$ the function $f_A(X) := \frac{1}{2}\operatorname{tr}(AX)$, $X \in M$, is a Wilson function on M corresponding to the d'Alembert function $f_I = \frac{1}{2}\operatorname{tr}$.

Example 11.13. The Wilson functions on the Heisenberg group $H_3(\mathbb{Z})$ corresponding to the unique non-abelian, classical d'Alembert function on $H_3(\mathbb{Z})$ are computed in [192, Example 11.6].

Example 11.14. Let G be the $(ax + b)$-group. We will in this example exhibit all the continuous solutions $(f, g) \in C(G) \times C(G)$ of Wilson's classical functional equation (11.4), i.e., of

$$f(xy) + f(xy^{-1}) = 2f(x)g(y), \qquad x, y \in G.$$

The example is interesting, because there is a solution (F, g) such that F is non-central, even though g is a non-degenerate, abelian d'Alembert function.

To avoid too much empty space we use the abbreviation

$$(a, b) := \begin{pmatrix} a & b \\ 0 & 1 \end{pmatrix}.$$

Below we discard the trivial set of solutions $f = 0$, $g \in C(G)$ arbitrary.

We know that g is a solution of d'Alembert's long functional equation when $f \neq 0$ (Lemma 11.3(e)). In Example 10.8 we saw that g is then a d'Alembert function. Finally, in Example 9.9 we found that the continuous d'Alembert functions are

$$g_\lambda(a, b) = \frac{a^\lambda + a^{-\lambda}}{2} \qquad \text{for } (a, b) \in G, \tag{11.11}$$

where the index λ ranges over the complex numbers. We will in this example for each fixed $\lambda \in \mathbb{C}$ solve the corresponding Wilson functional equation

$$f(xy) + f(xy^{-1}) = 2f(x)g_\lambda(y), \qquad x, y \in G, \tag{11.12}$$

where $f \in C(G)$ is the unknown function. The space of Wilson functions corresponding to g_λ is the direct sum of the even and the odd Wilson functions (Lemma 11.3(f)). The subspace of the even Wilson functions is $\mathbb{C}g_\lambda$ by Lemma 11.3(b), so the even part is known. We proceed to find the odd Wilson functions.

We use the following notation:

(i) For any $\lambda \in \mathbb{C}$ we let W_λ denote the complex vector space of all odd Wilson functions on G corresponding to g_λ, i.e., we let

$$W_\lambda := \{f \in C(G) \mid f = -\check{f},\\ f(xy) + f(xy^{-1}) = 2f(x)g_\lambda(y),\ x, y \in G\}.$$

Note that $W_\lambda = W_{-\lambda}$, because $g_\lambda = g_{-\lambda}$. Furthermore we let

$$f_\lambda(a,b) := \begin{cases} (a^\lambda - a^{-\lambda})/2 & \text{for } a \in \mathbb{R}^+,\ b \in \mathbb{R},\ \text{when } \lambda \neq 0, \\ \log a & \text{for } a \in \mathbb{R}^+,\ b \in \mathbb{R},\ \text{when } \lambda = 0. \end{cases}$$

Note that $f_{-\lambda} = -f_\lambda$ for $\lambda \neq 0$.

(ii) We define

$$F(a,b) := \frac{b}{\sqrt{a}}, \qquad a \in \mathbb{R}^+, b \in \mathbb{R}.$$

Note that F is not central.

Proposition 11.15.

(a) *If* $\lambda \in \mathbb{C} \setminus \{\pm\frac{1}{2}\}$, *then* $W_\lambda = \mathbb{C}f_\lambda$, *so* $\dim W_\lambda = 1$ *and* $\dim W(g_\lambda) = 2$.

(b) $W_{\frac{1}{2}} = \operatorname{span}\{f_{\frac{1}{2}}, F\}$, *so* $\dim W_{\frac{1}{2}} = 2$ *and* $\dim W(g_{\frac{1}{2}}) = 3$.

Proof. $f_\lambda \in W_\lambda$ for any $\lambda \in \mathbb{C}$ (Lemma 11.4(a)). By easy computations we see that $F \in W_{\frac{1}{2}}$.

Let $\lambda \in \mathbb{C}$. Consider $f \in W_\lambda$. For any $a \in \mathbb{R}^+$ we put $F_a := L((a,0)^{-1})f$. From the identity

$$f((a,0)(1,b)(1,x)) + f((a,0)(1,b)(1,x)^{-1})\\ = 2f((a,0)(1,b))g_\lambda(1,-x) = 2f((a,0)(1,b))$$

we see that $F_a(1, b+x) + F_a(1, b-x) = 2F(1,b)$, which means that $F_a(1,-)$ satisfies Jensen's functional equation on $\mathbb{R}$, so by Example 11.10 it has the form $[L((a,0)^{-1})f](1,x) = A(a)x + B(a)$, $x \in \mathbb{R}$ for some $A(a), B(a) \in \mathbb{R}$. This expression we reformulate to

$$f(a,x) = \frac{A(a)}{a}x + B(a), \qquad a \in \mathbb{R}^+, x \in \mathbb{R}. \tag{11.13}$$

When we substitute this into

$$f((a,0)(a_1,b_1)) + f((a,0)(a_1,b_1)^{-1}) = 2f(a,0)\frac{a_1^\lambda + a_1^{-\lambda}}{2},$$

we get that

$$\frac{A(aa_1)}{aa_1}ab_1 + B(aa_1) + \frac{A(aa_1^{-1})}{aa_1^{-1}}(-aa_1^{-1}b_1) + B(aa_1^{-1}) = f(a,0)(a_1^\lambda + a_1^{-\lambda}).$$

Since b_1 ranges freely over $\mathbb{R}$ the coefficient of it must vanish, which means that $A(aa_1) = a_1 A(aa_1^{-1})$. Taking $a_1 = a$ we see that $A(a) = A(1)\sqrt{a}$, so that (11.13) with $c := A(1)$ reduces to

$$f(a,x) = c\frac{x}{\sqrt{a}} + B(a), \qquad a \in \mathbb{R}^+, x \in \mathbb{R}. \tag{11.14}$$

Applying the functional equation to $a \mapsto f(a,0) = B(a)$ we find that B is a continuous and odd solution of Wilson's functional equation

$$B(aa_1) + B(aa_1^{-1}) = 2B(a)\frac{a_1^\lambda + a_1^{-\lambda}}{2}, \qquad a, a_1 \in \mathbb{R}^+,$$

to which we apply Corollary 11.7. If $\lambda \neq 0$ we get

$$B(a) = \alpha\frac{a^\lambda - a^{-\lambda}}{2} + \beta\frac{a^\lambda + a^{-\lambda}}{2}.$$

But B is odd, so the last term vanishes and so $B(a) = \alpha f_\lambda(a,b)$. For $\lambda = 0$ we get by help of Exercise 2.9(a) that $B(a) = \alpha f_0(a,b)$. In any case (11.14) says that $f \in W_\lambda$ may be written as $f = cF + \alpha f_\lambda$, where $c, \alpha \in \mathbb{C}$ are constants. If $\lambda \neq \frac{1}{2}$ this formula gives the desired result. If $\lambda = \frac{1}{2}$ we note that

$$cF = f - \alpha f_\lambda \in W_{\frac{1}{2}} \cap W_\lambda = \{0\}, \quad \text{so that} \quad cF = 0. \qquad \square$$

11.7 Generalizations of Wilson's functional equations

See Section 14.3 for works by the Moroccan school.

Let G be a locally compact abelian group with a Haar measure dx, and let $M(G)$ denote the set of all bounded, regular, complex-valued Borel measures on G. For given $\mu \in M(G)$ Fechner [81] studies the functional equation

$$\int_G \{f(x+y-s)+f(x-y+s)\}\, d\mu(s) = 2f(x)g(y), \qquad x,y \in G, \quad (11.15)$$

where $f, g \in \mathcal{L}^\infty(G, dx)$ are to be determined. (11.15) becomes Wilson's equation when μ is the Dirac measure at $e \in G$. The special case of $f = g$, that gives a generalization of d'Alembert's classical functional equation, was studied earlier by Gajda [93].

In [85] Fechner studies the similarly looking functional equation

$$\int_G \{f(x+y-s)+f(x-y+s)\}\, d\mu(s) = 2g(x)f(y), \qquad x,y \in G,$$

where $f, g \in \mathcal{L}^\infty(G, dx)$ are to be determined.

Badora [20] solves the functional equation

$$\int_K f(x+k\cdot y)\, dk = f(x)g(y), \qquad x,y \in G,$$

where K is a finite group of continuous automorphisms of G and dk the normalized Haar measure on K. The solutions (f, g) are sought such that f is almost periodic and $g \in C_b(G)$.

11.8 A variant of Wilson's equation

Throughout Section 11.8 G denotes a group. We shall study the functional equation

$$f(xy)+f(y^{-1}x) = 2f(x)g(y), \qquad x,y \in G, \quad (11.16)$$

where $f, g : G \to \mathbb{C}$ are to be found. The functional equation (11.16) is very similar in appearance to Wilson's classical functional equation

$$f(xy)+f(xy^{-1}) = 2f(x)g(y), \qquad x,y \in G, \quad (11.17)$$

but differs from it on the second term, where the new equation (11.16) has $f(y^{-1}x)$, while the classical one (11.17) has $f(xy^{-1})$. The two versions of course agree if G is abelian.

Ng [150] proved (see Section 12.8) in the special case of $g = 1$ that each solution of

$$f(xy) + f(y^{-1}x) = 2f(x), \qquad x, y \in G, \tag{11.18}$$

also satisfies Jensen's functional equation

$$f(xy) + f(xy^{-1}) = 2f(x), \qquad x, y \in G. \tag{11.19}$$

On certain groups (11.19) possesses more solutions than (11.18) (e.g., the Heisenberg group, see [191, Example 5.1]). So the functional equations (11.16) and (11.17) need not have the same solutions.

We shall here study the solutions of (11.16) and their relations to the solutions of Wilson's classical equation (11.17). We are in particular interested in conditions on G ensuring that each solution of (11.16) satisfies (11.17) as well, like mentioned for Jensen's functional equation. We prove that so is the case, if G is generated by its squares (Corollary 11.18), but also that it is not true in general (Example 11.19). For more information we refer to the paper [193].

The new equation has a more aesthetic structure than the previous one: The left hand side of (11.16) is a sum of two commuting representations (the right regular representation R and the left regular representation L), while the left hand side of (11.17) is the sum of a representation and an anti-homomorphism (R and the map $y \mapsto R(y^{-1})$ respectively). We exploit the structure in the following lemma.

We recall that we associate to any function $g : G \to \mathbb{C}$ the function $d(x) := 2g(x)^2 - g(x^2)$, $x \in G$.

Lemma 11.16. *Let the pair $f, g : G \to \mathbb{C}$ be a solution of the variant (11.16) of Wilson's functional equation such that $f \neq 0$.*

(a) *$g(e) = 1$, g is central, and $g = \check{g}$.*

(b) *$f(y^{-1}xy) = d(y)f(x)$ for all $x, y \in G$.*

(c) *d is a homomorphism of G into the multiplicative group $\{\pm 1\}$.*

(d) *g is a d-d'Alembert function, i.e.,*

$$g(xy) + d(y)g(xy^{-1}) = 2g(x)g(y), \qquad x, y \in G.$$

(e) *$d = 1$ on the center $Z(G)$ and on the subgroup of G generated by the squares of G.*

Proof. (a) To get the first statement put $y = e$ in (11.16). The functional equation (11.16) may be rewritten as $[L(y) + R(y)]f = 2g(y)f$ for all $y \in G$. Applying $L(z) + R(z)$ to this we get (starting with the right hand side) that

$$\begin{aligned} 4g(y)g(z)f &= [L(z) + R(z)][L(y) + R(y)]f \\ &= [L(zy) + R(zy)]f + [L(z)R(y) + R(z)L(y)]f \\ &= 2g(zy)f + [L(z)R(y) + R(z)L(y)]f, \end{aligned}$$

so (because L and R commute)

$$2g(zy)f = 4g(y)g(z)f - [L(z)R(y) + L(y)R(z)]f, \qquad y, z \in G. \quad (11.20)$$

The value of the right hand side of (11.20) does not change if z and y are interchanged. From f being non-zero we infer that g is central.

Let $x_0 \in G$ be arbitrary. Assume first that either $f(x_0) \neq 0$ or $f(e) \neq 0$. Then the pair $\{f, g\}$ is a solution of (11.16) on the subgroup $\langle x_0 \rangle$ of G generated by x_0, and f is not identically 0 on $\langle x_0 \rangle$. The group $\langle x_0 \rangle$ is abelian, so the pair $\{f, g\}$ is a solution of Wilson's functional equation (11.17) on $\langle x_0 \rangle$, from which it follows that g is a d'Alembert function and so that $g(x_0) = g(x_0^{-1})$.

We may thus assume that $f(e) = f(x_0) = 0$. Putting $x = e$ in (11.16) we see that f is odd on G. We get for any $x \in G$ that

$$\begin{aligned} 2f(x)[g(x_0) - g(x_0^{-1})] &= f(xx_0) + f(x_0^{-1}x) - f(xx_0^{-1}) - f(x_0x) \\ &= f(xx_0) + f(x_0^{-1}x) + f(x_0x^{-1}) + f(x^{-1}x_0^{-1}) \\ &= f(x_0x^{-1}) + f(xx_0) + f(x_0^{-1}x) + f(x^{-1}x_0^{-1}) \\ &= 2f(x_0)g(x^{-1}) + 2f(x_0^{-1})g(x) = 0 - 2f(x_0)g(x) = 0 - 0 = 0, \end{aligned}$$

which implies that $g(x_0) = g(x_0^{-1})$, because $f \neq 0$.

(b) Putting $z = y$ in (11.20) we find that $[2g(y)^2 - g(y^2)]f(x) = f(y^{-1}xy)$ for all $x, y \in G$, i.e., that (b) holds.

(c) The following computation shows that d is multiplicative:

$$\begin{aligned} d(yz)f(x) &= f((yz)^{-1}xyz) = f(z^{-1}(y^{-1}xy)z) \\ &= d(z)f(y^{-1}xy) = d(z)d(y)f(x). \end{aligned}$$

Since $d(e) = 1 \neq 0$, we infer d is a character of G. From $g = \check{g}$ we get that $d = \check{d}$, so that $d(x)^2 = d(x)d(x^{-1}) = d(e) = 1$. Thus $d(G) \subseteq \{\pm 1\}$.

(d) Choosing $x_0 \in G$ such that $f(x_0) \neq 0$ we get

$$g(x) = \frac{1}{f(x_0)}[f(x_0x) + f(x^{-1}x_0)] \qquad \text{for all } x \in G.$$

From this formula and (b) we find for any $x, y \in G$ that

$$\begin{aligned}
&2f(x_0)[g(xy) + d(y)g(xy^{-1})] \\
&\quad = f(x_0xy) + f(y^{-1}x^{-1}x_0) + d(y)f(x_0xy^{-1}) + d(y)f(yx^{-1}x_0) \\
&\quad = f(x_0xy) + f(y^{-1}x^{-1}x_0) + f(y^{-1}x_0x) + f(x^{-1}x_0y) \\
&\quad = f(x_0xy) + f(y^{-1}x_0x) + f(x^{-1}x_0y) + f(y^{-1}x^{-1}x_0) \\
&\quad = 2f(x_0x)g(y) + 2f(x^{-1}x_0)g(y) = 2f(x_0)g(x)g(y).
\end{aligned}$$

(e) is left to the reader. □

Theorem 11.17. *Let the pair $f, g : G \to \mathbb{C}$ be a solution of the variant (11.16) of Wilson's functional equation such that $f \neq 0$. Then the following four statements are equivalent:*

(a) *(f, g) is a solution of Wilson's functional equation* (11.17).

(b) *f is central.*

(c) *g satisfies d'Alembert's classical functional equation* (9.3).

(d) $d = 1$.

Proof. (a) ⇔ (b) Assuming (a) we subtract (11.17) from (11.16) and get that $f(y^{-1}x) = f(xy^{-1})$, which shows (b). The converse statement is trivial.

(a) ⇒ (c) By the assumption (f, g) is a solution of Wilson's functional equation, so g is a solution of d'Alembert's long functional equation

$$g(xy) + g(yx) + g(xy^{-1}) + g(y^{-1}x) = 4g(x)g(y), \qquad x, y \in G, \qquad (11.21)$$

(Lemma 11.3(e)). Since g is central (Lemma 11.16(a)) it is also a solution of d'Alembert's (short) functional equation.

(c) ⇒ (d) $d(x) = 2g(x)^2 - g(x^2) = g(xx^{-1}) = g(e) = 1$.

(d) ⇒ (b) This is immediate from Lemma 11.16(b). □

Corollary 11.18. *Let the pair $f, g : G \to \mathbb{C}$ be a solution of the variant (11.16) of Wilson's functional equation. If G is generated by its squares, then f is central, and (f, g) is a solution of Wilson's classical functional equation.*

We may now refer to Proposition 11.5 for the solutions.

Finally we present an example in which $d \neq 1$, and hence is an example of a solution of (11.16) that is not a solution of Wilson's classical functional equation.

Example 11.19. On the Heisenberg group with integer entries

$$H_3(\mathbb{Z}) = \left\{ (x,y,z) = \begin{pmatrix} 1 & x & z \\ 0 & 1 & y \\ 0 & 0 & 1 \end{pmatrix} \,\middle|\, x,y,z \in \mathbb{Z} \right\}$$

there exists a solution of (11.16) which is not a solution of Wilson's classical equation:

Let us for $(k,m,n) \in H_3(\mathbb{Z})$ define

$$f(k,m,n) := \frac{i^k - i^{-k}}{2} \mathbf{1}_{2\mathbb{Z}}(m)(-1)^n,$$
$$g(k,m,n) := \frac{i^k + i^{-k}}{2} \mathbf{1}_{2\mathbb{Z}}(m)(-1)^n.$$

Elementary calculations show that (f,g) is a solution of (11.16) and that f is odd. Furthermore, a small computation based on the definition of d reveals that

$$d(k,m,n) = (-1)^m \qquad \text{for } k,m,n \in \mathbb{Z},$$

so $d \neq 1$. Hence (f,g) is not a solution of Wilson's classical functional equation (Theorem 11.17).

It might be added that g does not satisfy Kannappan's condition: Indeed,

$$g((1,1,0)(1,0,0)(0,1,0)) = g(2,2,2) = -1,$$

while

$$g((1,1,0)(0,1,0)(1,0,0)) = g(2,2,1) = 1.$$

11.9 Exercises

Exercise 11.1. Find the solutions $f : \mathbb{R} \to \mathbb{C}$ of

(a) $f(x+y) + f(x-y) = 2f(x)\cos y$, $x,y \in \mathbb{R}$.

(b) $f(x+y) + f(x-y) = 2\sin x \cos y$, $x,y \in \mathbb{R}$.

(c) $f(x+y) + f(x-y) = 2f(x)\sin y$, $x,y \in \mathbb{R}$.

Exercise 11.2. Find the continuous solutions of Wilson's classical functional equation (11.4) on the groups $\mathbb{R}^n$, $\mathbb{R}^+$ and $\mathbb{T}$.

Exercise 11.3. Let G be an abelian group. We will in this exercise find the solutions $f, g : G \to \mathbb{C}$ of the functional equation

$$f(x+y) + f(x-y) = 2g(x)f(y), \qquad x, y \in Gq. \tag{11.22}$$

The left hand side of (11.22) is the same as that of Wilson's functional equation (11.6), but on the right hand side the functions f and g have been interchanged. As we shall see this makes quite a difference in the set of solutions. (11.22) reduces to d'Alembert's functional equation for $f = g$ just like Wilson's functional equation does.

Let $f, g : G \to \mathbb{C}$ be a solution of (11.22).

(a) Show that f is even.

(b) Show that $f(y)g(x) = f(x)g(y)$ for all $x, y \in G$.
Hint: $f(x-y) = f(y-x)$.

(c) Show that there exists a constant $c \in \mathbb{C}$ such that $f = cg$.

(d) Show that g is a non-zero solution of d'Alembert's functional equation $g(x+y) + g(x-y) = 2g(x)g(y)$, if $f \neq 0$.

(e) Solve the functional equation (11.22) in terms of the characters of G. Kannappan's result (Corollary 9.22) can be used.

Exercise 11.4. Let G be a group and let $\mu : G \to \mathbb{C}^*$ be a character on G. Let $f, g : G \to \mathbb{C}$, where $f \neq 0$, satisfy

$$f(xy) + \mu(y)f(xy^{-1}) = 2f(x)g(y) \qquad \text{for all } x, y \in G.$$

(a) Show that

$$g(xy) + g(yx) + \mu(y)[g(xy^{-1}) + g(y^{-1}x)] = 4g(x)g(y) \quad \text{for all } x, y \in G.$$

(b) Assuming f central, show that g is a μ-d'Alembert function, i.e., that

$$g(xy) + \mu(y)g(xy^{-1}) = 2g(x)g(y) \qquad \text{for all } x, y \in G.$$

(c) Assuming f central, show that g is also central.
Hint: Proposition 9.17(b).

Exercise 11.5. Find the non-zero solutions $f \in C(\mathbb{R})$ of the functional equation

$$f(x+y) + f(x-y) = f(x)f(y) + f(x)f(-y), \qquad x, y \in \mathbb{R}.$$

Answer: The solutions are parametrized by $\lambda, \alpha \in \mathbb{C}$:

$$f_{\lambda,\alpha}(x) = \alpha\frac{e^{\lambda x} - e^{-\lambda x}}{2} + \frac{e^{\lambda x} + e^{-\lambda x}}{2} \quad \text{and} \quad f_{0,\alpha}(x) = \alpha x + 1.$$

Hint: f is a solution of Wilson's functional equation (11.1) with g as the even part of f.

Exercise 11.6. Let $x_0 \in G$, where $(G, +)$ is an abelian group. Show that the set of solutions $f, g : G \to \mathbb{C}$ of the following modification of Wilson's functional equation

$$f(x + x_0 + y) + f(x + x_0 - y) = 2f(x)g(y) \qquad \text{for all } x, y \in G,$$

consists of the possibilities below:

(a) $f = 0$ and g arbitrary.

(b) $f = \alpha\chi$ and $g = \chi(x_0)(\chi + \check{\chi})/2$, where $\alpha \in \mathbb{C} \setminus \{0\}$ is a constant and χ is a character on G.

(c) $f = \alpha\chi + \beta\check{\chi}$ and $g = (\chi + \check{\chi})/2$, where $\alpha, \beta \in \mathbb{C} \setminus \{0\}$ are constants and χ is a character on G such that $\chi(x_0) = 1$.

(d) $f = \alpha\chi + \beta\check{\chi}$ and $g = -(\chi + \check{\chi})/2$, where $\alpha, \beta \in \mathbb{C} \setminus \{0\}$ are constants and χ is a character on G such that $\chi(x_0) = -1$.

(e) $f = (a + \beta)\chi$ and $g = \chi(x_0)\chi$ where $a : G \to \mathbb{C}$ is an additive map such that $a(x_0) = 0$, $\beta \in \mathbb{C}$ is a constant and χ is a character on G such that $\chi = \check{\chi}$.

In (b), (c) and (d) you may assume $\chi \neq \check{\chi}$.

Hints: (1) Show that $f(x + x_0) = f(x)g(0)$ and that $g(0) \neq 0$. (2) Show that $(f, g/g(0))$ is a solution of Wilson's classical functional equation (11.6).

Exercise 11.7. Let H be a subgroup of a group G. Let $f : G \to \mathbb{C}$ where $f \neq 0$, and $g : H \to \mathbb{C}$ satisfy

$$f(xy) + f(xy^{-1}) = 2f(x)g(y) \qquad \text{for all } x \in G \text{ and } y \in H.$$

Show g is a solution of d'Alembert's long functional equation (10.2) on H.

Hint: The proof of Lemma 11.3(e).

11.10 Notes and remarks

Székelyhidi [197, Theorem 11.3] shows by methods from spectral synthesis that if f, g satisfy a functional equation of the form

$$f(x+y) + f(x-y) = \sum_{i=1}^{n} h_i(x)k_i(y), \qquad x, y \in G,$$

where G is a 2-divisible abelian group, then f and g are a exponential polynomials.

Fechner [81] studies on locally compact, abelian groups G the functional equation

$$\int_G \{g(x+y-s) + g(x-y+s\}d\mu(s) = g(x)f(y), \qquad x, y \in G,$$

where μ is a complex, regular, bounded Borel measure on G and $f, g \in L^\infty(G)$.

For vector- and matrix-valued solutions of Wilson's functional equation see Sinopoulos [178, 181, 182] and Stetkær [189].

Equations (11.1) and (11.2): For an elementary and thorough discussion of Wilson's generalizations (11.1) and (11.2) of the cosine-equation on the real line see Aczél [3, Sections 3.2.1 and 3.2.2].

Lemma 11.3: Davison [64] called the equation (11.5) Wilson's functional equation, in contrast to our terminology. We call it the symmetrized sine addition formula and discussed it in Chapters 6 (general case) and 8 (where g is a pre-d'Alembert function). (e) was for Wilson's classical functional equation noted for the group $\mathbb{R}$ by Kaczmarz [124] and for general groups by Corovei [49].

Proposition 11.5 extends Aczél, Chung and Ng [5, Lemma 3], because it

(1) does not assume g is a d'Alembert function satisfying Kannappan's condition and

(2) allows a general involution where [5] has merely the group inversion.

When g is non-degenerate the assumption of f being central is not needed if G is step 2 nilpotent (see Stetkær [192, Theorem 10.1]) or G is a connected, nilpotent Lie group (see Friis [90, Theorem 3.4]). Corovei [56, 57] has some results for P_3-groups and metabelian groups.

Corollary 11.7: Badora [20] solved the following generalization of Wilson's functional equation, in which G denotes a locally compact, abelian group and K a finite group of continuous automorphisms of G:

$$\frac{1}{|K|}\sum_{k\in K} f(x+k\cdot y) = f(x)g(y), \qquad x,y\in G.$$

He found the solutions $f, g \in C(G)$ such that f is almost periodic and g is bounded.

Theorem 11.9 is due to Yang [214, Theorem 3.3], but our proof is different from Yang's.

The results of Section 11.8 are from Stetkær [193].

Exercise 11.3 is inspired by Fechner [85], who discusses the following extension of (11.22) on a locally compact, abelian group G:

$$\int_G \{g(x+y-s)+g(x-y+s\}d\mu(s) = f(x)g(y), \qquad x,y\in G,$$

where μ is a complex, regular, bounded Borel measure on G, $g \in L^\infty(G)$ and f is measurable with respect to the Haar measure.

Chapter 12

Jensen's Functional Equation

Throughout Chapter 12 we let G denote a group and $(H, +)$ an abelian group with neutral element 0. $H = \mathbb{C}$ through most of the presentation. The main content of the chapter is a study of the normalized solutions $f : G \to H$ of Jensen's functional equation $f(xy) + f(xy^{-1}) = 2f(x)$, $x, y \in G$.

The most important results are Corollary 12.12 in which G is abelian, and Theorem 12.20 in which G is locally compact, but not necessarily abelian.

12.1 Introduction, definitions and set up

Jensen's original formulation of his functional equation ([118, in Danish] or [119, in French]) was that the function $f : I \to \mathbb{R}$ defined on an interval I on the real line should satisfy

$$f\Big(\frac{s+t}{2}\Big) = \frac{f(s)+f(t)}{2} \qquad \text{for all } s, t \in I. \tag{12.1}$$

This is related to the notion of convexity: A function $f : I \to \mathbb{R}$ is said to be *midpoint convex* or *Jensen convex*, if

$$f\Big(\frac{s+t}{2}\Big) \leq \frac{f(s)+f(t)}{2} \qquad \text{for all } s, t \in I,$$

and *midpoint concave*, if

$$f\Big(\frac{s+t}{2}\Big) \geq \frac{f(s)+f(t)}{2} \qquad \text{for all } s, t \in I.$$

These two inequalities taken together are equivalent to (12.1) from which we see that the real-valued solutions of Jensen's equation on I are the functions that are both midpoint convex and midpoint concave. Geometrically (assuming continuity of f) this means that the real-valued solutions of Jensen's equation are affine functions on I, i.e., functions of the form $f(x) = ax + b$, where $a, b \in \mathbb{R}$ are constants. So the affine functions form the transition from the set of convex functions to the set of concave functions, and they may thus be considered as limiting cases of both sets.

Definition 12.1. Let S be a semigroup and let $(H, +)$ an abelian group. We say that a map $f : S \to H$ is *affine* if it has the form $f(x) = a(x) + b$, where $a : S \to H$ is an additive map and $b \in H$ a constant.

An example of an affine function $f : \mathbb{R} \to \mathbb{C}$ is $f(x) = \alpha x + \beta$, $x \in \mathbb{R}$, where $\alpha, \beta \in \mathbb{C}$.

If we replace s by $x + y$ and t by $x - y$ in Jensen's original equation (12.1) it takes on $I = \mathbb{R}$ the form

$$f(x + y) + f(x - y) = 2f(x), \qquad x, y \in \mathbb{R}. \tag{12.2}$$

This version of Jensen's functional equation generalizes from the group $(\mathbb{R}, +)$ to any group as follows:

Definition 12.2. A *solution of Jensen's functional equation* is a map $f : G \to H$ satisfying *Jensen's functional equation*

$$f(xy) + f(xy^{-1}) = 2f(x), \qquad x, y \in G. \tag{12.3}$$

Jensen's original equation (12.1) is equivalent to (12.3) for $G = I = \mathbb{R}$ and $H = \mathbb{R}$.

It is Jensen's functional equation (12.3) with range space $H = \mathbb{C}$ that will interest us most.

A constant map is clearly a solution of (12.3). Hence, if f is a solution then the function $f_0 := f - f(e)$ is also a solution. It has the further properties that $f_0(e) = 0$ and $f_0(x^{-1}) = -f_0(x)$ for all $x \in G$. So f_0 is an odd function on G, and $f = f(e) + f_0$ is a sum of a constant solution and an odd one. These considerations lead to the following notation:

Definition 12.3. We let $S(G, H)$ denote the set of all solutions $f : G \to H$ of Jensen's functional equation (12.3) for which $f(e) = 0$. We say that the solution f is *normalized*, if $f(e) = 0$.

We note that:

1. $S(G, H)$ consists of odd solutions, meaning that if $f \in S(G, H)$ then $f(x^{-1}) = -f(x)$ for all $x \in G$ (take $x = e$ in (12.3)). If H is 2-torsion free then $S(G, H)$ is the set of odd solutions.

2. $S(G,H)$ is an abelian group under addition, and the set $\mathrm{Hom}(G,H)$ of additive maps is a subgroup of $S(G,H)$.
3. If $H = \mathbb{C}$ then $S(G,\mathbb{C})$ is a complex vector space, and $\mathrm{Hom}(G,\mathbb{C})$ is a subspace of it.

Any affine map $f : G \to H$ is a solution of Jensen's equation (12.3) as is easy to verify. When H is 2-torsion free (i.e., division by 2 is unique in H) then the converse is true on abelian groups (Corollary 12.12(c)), but it is not true in general for non-abelian groups as the following example reveals.

Example 12.4. The Heisenberg group $G = H_3$ (from Examples A.17(a)) provides a simple counter-example f_0, to boot with range space $H = \mathbb{R}$:

$$f_0 \begin{pmatrix} 1 & x & z \\ 0 & 1 & y \\ 0 & 0 & 1 \end{pmatrix} = 2z - xy, \qquad x, y, z \in \mathbb{R}.$$

As is easy to check, $f_0 \in S(H_3, \mathbb{R})$, but it is not even central and hence not affine. As a preview of things to come it might in passing be remarked that the scalar multiples of f_0 are the only odd continuous solutions $f : H_3 \to \mathbb{C}$ modulo the continuous homomorphisms, a fact which can be derived from Theorem 12.20 that characterizes the quotient vector space $S^{\mathrm{cont}}(G,\mathbb{C})/\mathrm{Hom}^{\mathrm{cont}}(G,\mathbb{C})$ for any group G. See Exercise 12.11 for a way of finding solutions like f_0.

Jensen's equation (12.3) is the special instance of Wilson's equation $f(xy) + f(xy^{-1}) = 2f(x)g(y)$ in which $g = 1$. But it can be and we have defined in a more general setting than the other functional equations that we have met so far: The sine addition theorem (4.3), d'Alembert's functional equation (9.4) and Wilson's functional equation (11.3) all have a product of functions on their right hand sides, which is not the case for Jensen's functional equation. That is why it has been formulated for functions assuming values in an abelian group, while the solutions of d'Alembert's functional equation, the sine addition theorem and Wilson's functional equation must assume values in an algebraic object like a ring in which a both an addition (on the left hand side) and a product (on the right hand side) are defined. Jensen's functional equation requires nothing but the addition. However, we shall only discuss the case of $H = \mathbb{C}$ as range space in depth. For a general H we refer the reader to Ng's papers [148, 149, 151]. As Ng states in [151, p. 132], but which is equally true of [148] and [149]:

"This paper is written with one bias – that we present only results valid for general abelian H." The main extra technical complications in dealing with a general abelian group H compared with $H = \mathbb{C}$ stem from the unpleasant, but unavoidable, fact that the equation $2x = 0$, $x \in H$, may have solutions $x \neq 0$.

12.2 Key formulas and relations

We collect in Proposition 12.5 below some key formulas and relations for normalized solutions of Jensen's functional equation (12.3) on a group. Exercise 12.9 has similar results for a more general version (12.7) of Jensen's functional equation. Consult [148], [149] and [151] for many more formulas and relations that the solutions of Jensen's equation (12.3) satisfy.

The identity in (a) of Proposition 12.5 is the symmetrized additive Cauchy equation from Section 2.4. There we proved that the symmetrized additive Cauchy equation and Jensen's equation are equivalent (for the precise formulation see Section 2.4), so it is no wonder that the identity is instrumental in deriving other results of the proposition. It is a special instance of the symmetrized sine addition formula (6.1) that has $g = 1$.

Although normalized solutions of Jensen's functional equation need not be additive (as revealed by Example 12.4), useful vestiges of additivity remain, as is apparent from Proposition 12.5. For example, (c) says that f is a semi-homomorphism.

A couple of statements below are expressed in terms of the Cauchy difference Cf of $f : G \to H$, defined by $(Cf)(x,y) := f(xy) - f(x) - f(y)$ for $x, y \in G$. The second Cauchy difference $C^2 f$ (defined by (2.5)) of f also enters.

Proposition 12.5. *Let $f \in S(G,H)$. Let $x, y, z, a, b \in G$ be arbitrary.*

(a) *f satisfies (2.3), i.e., $f(xy) + f(yx) = 2f(x) + 2f(y)$.*

(b) *$f(x^k) = kf(x)$ for all $k \in \mathbb{Z}$.*

(c) *$f(xyx) = 2f(x) + f(y)$.*

(d) *$2C^2 f = 0$.*

(e) *$2Cf : G \times G \to H$ is bi-additive. Furthermore $2Cf = f([-,-])$, i.e., $2f(xy) = 2f(x) + 2f(y) + f([x,y])$ for all $x, y \in G$.*

(f) *Cf is skew-symmetric, i.e., $Cf(x,y) = -Cf(y,x)$.*

(g) *$Cf(x,-)$ and $Cf(-,y)$ are abelian functions in $S(G,H)$.*

(h) *$[G,G] \subseteq Z(f)$ (the notation $Z(f)$ is defined by formula* (B.1)*).*

(i) *$f(xcy) = f(xy) + f(c)$ for all $c \in [G,G]$. In particular the restricted map $f|_{[G,G]} : [G,G] \to H$ is additive.*

(j) *f is a function on $G/[G,[G,G]]$. Actually $f \in S(G/[G,[G,G]],H)$.*

(k) *$f(axyb) - f(ayxb) = f(xy) - f(yx) = f([x,y])$.*

(l) *$f(xyx^{-1}) = f(y) + f([x,y])$.*

(m) *For $n \in \mathbb{N}$ and $x_1, x_2, \ldots, x_n \in G$ we extend the last formula in (e) to*

$$2f(x_1x_2\cdots x_n) = 2\sum_{i=1}^{n} f(x_i) + \sum_{1\le i<j\le n} f([x_i,x_j]). \tag{12.4}$$

Proof. (a) Interchange x and y in (12.3), add the two identities and note that $f(xy^{-1}) + f(yx^{-1}) = 0$ because f is normalized and hence odd.

(b) Putting $y = x$ in (12.3) we get $f(x^2) = 2f(x)$ and by induction on k that $f(x^k) = kf(x)$ for all $k \in \mathbb{N}$. And then, due to $f(x^{-1}) = -f(x)$, we get $f(x^k) = kf(x)$ for all $k \in \mathbb{Z}$.

(c) Replace y by yx in (12.3).

(d) f satisfies (2.3), so the statement is immediate from Lemma 2.20(c).

(e) The biadditivity of $2Cf$ is Lemma 2.19(a). If we in (12.3) replace x by xy and y by $x^{-1}y^{-1}$ we get $f([x,y]) + f(xy^2x) = 2f(xy)$. When we apply first (c) and then (b) to the second term on the left we get $2Cf = f([-,-])$.

(f) is a reformulation of the identity in (a).

(g) Due to (f) it suffices to prove the statements about $Cf(-,y)$.

From Jensen's functional equation we get $f(xyz) + f(xyz^{-1}) = 2f(xy)$. To get rid of the term $f(xyz^{-1})$ we use Jensen's equation once more and find $f(xyz^{-1}) + f(xzy^{-1}) = 2f(x)$. That introduces the term $f(xzy^{-1})$ which we compute by help of $f(xzy^{-1}) + f(xzy) = 2f(xz)$. Combining the three identities we find

$$f(xyz) + f(xzy) = 2f(xy) - 2f(x) + 2f(xz). \tag{12.5}$$

We obtain a similarly looking identity when we in the identity of (a) replace x by xz, viz.,

$$f(xzy) + f(yxz) = 2f(xz) + 2f(y).$$

We subtract identity no. 2 from identity no. 1 and get

$$f(xyz) - f(yxz) = 2f(xy) - 2f(x) - 2f(y) = f(xy) - f(yx),$$

which is a reformulation of $Cf(xy,z) = Cf(yx,z)$, so Cf is central in its first variable.

As is easy to check, for any map $\phi : G \to H$ we have the relation (= the cocycle equation that we study in Chapter 16)

$$C\phi(a,z) + C\phi(az,y_0) = C\phi(a,zy_0) + C\phi(z,y_0).$$

It enables us to move the factor z in the product occuring on the first coordinate of $C\phi(-,-)$ to the second coordinate, which we exploit as follows:

$$Cf(xyz,y_0) = Cf((xy)z,y_0) = -Cf(xy,z) + Cf(xy,zy_0) + Cf(z,y_0).$$

The right hand side is unchanged under interchange of x and y due to the centrality of Cf, so $Cf(xyz,y_0) = Cf(yxz,y_0)$ and hence $Cf(zxy,y_0) = Cf(zyx,y_0)$ which is Kannappan's condition.

We next verify that $Cf(-,z)$ satisfies Jensen's functional equation:

$$\begin{aligned} &Cf(xy,z) + Cf(xy^{-1},z) \\ &\quad = f(xyz) - f(xy) - f(z) + f(xy^{-1}z) - f(xy^{-1}) - f(z) \\ &\quad = f(xyz) + f(xy^{-1}z) - 2f(x) - 2f(z). \end{aligned}$$

When we apply (12.5) to each of the two first terms of the last line we get that $Cf(-,y)$ satisfies Jensen's functional equation. The normalization condition is easily seen to be satisfied, so $Cf(-,y) \in S(G,H)$.

(h) and (i) We find

$$\begin{aligned} f(x[a,b]y) &= f(x[a,b]y) - f(x[a,b]) - f(y) + f(x[a,b]) + f(y) \\ &= Cf(x[a,b],y) + f(x[a,b]) - f(x) - f([a,b]) \\ &\quad + f(x) + f([a,b]) + f(y) \\ &= Cf(x[a,b],y) + Cf(x,[a,b]) + f(x) + f([a,b]) + f(y). \end{aligned}$$

Since Cf satisfies Kannappan's condition in each variable, the computation reduces to the identity $f(x[a,b]y) = f(xy) + f([a,b])$, from which both (h) and (i) follow.

(j) Using (h) we compute for any $a,x,y,z \in G$ that

$$f(a[x,[y,z]]) = f(ax[y,z]x^{-1}[y,z]^{-1}) = f(axx^{-1}[y,z][y,z]^{-1}) = f(a).$$

It follows from this that f is a function on $G/[G,[G,G]]$, because $[G,[G,G]]$ is generated by $\{[x,[y,z]] \mid x,y,z \in G\}$ (Lemma A.7).

(k) The first identity can be deduced from (i) as follows: When we in

$$f(axyb) = f(ayx[x^{-1}, y^{-1}]b) = f(ayxb) + f([x^{-1}, y^{-1}])$$

take $a = b = e$ we see that $f([x^{-1}, y^{-1}]) = f(xy) - f(yx)$. To get the second identity take $a = e$ and $b = x^{-1}y^{-1}$ in the first one.

(l) is a special instance of (k).

(m) We know the formula (12.4) is correct for $n = 1$ and $n = 2$ (according to (e)). We proceed by induction on n: Assuming it for a fixed $n \geq 2$ we get, first using (e), then the induction hypothesis and finally (e) again that

$$\begin{aligned}
2f(x_1x_2\cdots x_nx_{n+1}) &= 2f(x_1(x_2\cdots x_nx_{n+1})) \\
&= 2f(x_1) + 2f(x_2\cdots x_nx_{n+1}) + f([x_1, x_2\cdots x_nx_{n+1}]) \\
&= 2f(x_1) + 2\sum_{i=2}^{n+1} f(x_i) + \sum_{2\leq i<j\leq n+1} f([x_i, x_j]) + \sum_{j=2}^{n+1} f([x_1, x_j]) \\
&= 2\sum_{i=1}^{n+1} f(x_i) + \sum_{1\leq i<j\leq n+1} f([x_i, x_j]).
\end{aligned}$$

□

Corollary 12.6 shows why Jensen's functional equation (12.3) is uninteresting for continuous, complex-valued solutions on compact and hence in particular on finite groups: Constants are the only such solutions.

Corollary 12.6. *If $f \in S(G, \mathbb{C})$ is bounded, then $f = 0$.*

Proof. Let $x \in G$. If $f(G)$ is a bounded subset of $\mathbb{C}$ then so is its subset $\{nf(x) = f(x^n) \mid n \in \mathbb{N}\}$. But then $f(x) = 0$. □

12.3 On central solutions

In Section 12.3 we mainly study central solutions of Jensen's functional equation both in the form (12.3) and in the general form (12.7) below. In particular we do not assume that $H = \mathbb{C}$. If a semigroup is abelian then any function on it is central, so the case of an abelian semigroup is covered by our discussion. It would be nice, if all central, normalized solutions were additive. However, that is not true for a general H, even when G is abelian (Exercise 12.3 provides a counter-example), so some extra assumptions on G and/or H are needed, if we want central, normalized solutions to be additive. The big question is: When is $S(G,H) = \mathrm{Hom}(G,H)$, i.e., when is any normalized solution of Jensen's functional equation additive?

Until Corollary 12.12 a point is the extreme generality of H. We impose no assumptions on H except for the standing one that it is an abelian group, so our assumptions will be on G. On the other hand, if H is 2-torsion free, i.e., $[h \in H \text{ and } 2h = 0] \Rightarrow h = 0$, then we need not require anything about the group G (Corollary 12.12(b) and (c)).

Proposition 12.7. *If G is a cyclic group, then $S(G,H)$ is the set of additive maps from G to H.*

Proof. Let $f \in S(G,H)$. Let $x_0 \in G$ be a generator of G. From Proposition 12.5(b) we get $f(x_0^n) = nf(x_0)$ for all $n \in \mathbb{Z}$. Any elements $x, y \in G$ may be written as $x = x_0^m$ and $y = x_0^n$ where $m, n \in \mathbb{Z}$. Thus

$$\begin{aligned} f(xy) &= f(x_0^m x_0^n) = f(x_0^{m+n}) = (m+n)f(x_0) = mf(x_0) + nf(x_0) \\ &= f(x_0^m) + f(x_0^n) = f(x) + f(y), \end{aligned}$$

which shows that $f : G \to H$ is additive.

The proof of the converse statement is left to the reader. □

In Proposition 12.8 below we derive that central, normalized solutions of Jensen's functional equation (12.3) are additive on many groups G. Examples include $(\mathbb{Q},+)$, $(\mathbb{R}^n,+)$ and $\mathbb{C}^*$, because they are 2-divisible, abelian groups, but the multiplicative group $\mathbb{R}^*$ qualifies as well by the second part of the proposition. The statement is also true on connected Lie groups, because they are generated by their squares (Lemma A.11). For more examples see [150, Section 2]. Example 12.4 shows that the assumption about centrality cannot be deleted.

Proposition 12.8. *Let $f \in S(G,H)$ be central. Let $\langle G^2 \rangle$ denote the subgroup of G generated by the squares. If $G = \langle G^2 \rangle$ or if the number of elements in the quotient group $G/\langle G^2 \rangle$ is 2 then f is additive.*

Proof. By Proposition 12.5 we get that $2f(x) = f(x^2)$ and $f(xx_0x) = f(x^2) + f(x_0)$ for all $x, x_0 \in G$. Using that f is central we get for all $x_0, x_1, x_2, \ldots, x_m \in G$ that

$$\begin{aligned} f(x_0x_1^2x_2^2\cdots x_m^2) &= f(x_mx_0x_1^2x_2^2\cdots x_{m-1}^2x_m) = f(x_m^2) + f(x_0x_1^2\cdots x_{m-1}^2) \\ &= \cdots = f(x_1^2) + f(x_2^2) + \cdots + f(x_m^2) + f(x_0), \end{aligned}$$

so that

$$f(x_0x) = f(x_0) + f(x) \qquad \text{for all } x_0 \in G \text{ and } x \in \langle G^2 \rangle, \tag{12.6}$$

which implies f is additive on squares. This proves the proposition in the case of $G = \langle G^2 \rangle$.

If the number of elements in the quotient group $G/\langle G^2 \rangle$ is 2 we choose an $x_0 \in G \setminus \langle G^2 \rangle$. Then $G = \langle G^2 \rangle \cup x_0\langle G^2 \rangle$. For $n, m \in \{0, 1\}$ and $q_1, q_2 \in \langle G^2 \rangle$ we find by help of (12.6) and the centrality of f that

$$\begin{aligned} f(x_0^m q_1 x_0^n q_2) &= f(x_0^{m+n} x_0^{-n} q_1 x_0^n q_2) = f(x_0^{m+n}) + f(x_0^{-n} q_1 x_0^n) + f(q_2) \\ &= (m+n)f(x_0) + f(q_1) + f(q_2) \\ &= mf(x_0) + f(q_1) + nf(x_0) + f(q_2) \\ &= f(x_0^m q_1) + f(x_0^n q_2). \end{aligned}$$

□

In contrast to the slick formulations of Propositions 12.7 and 12.8 the wordings in Proposition 12.10 and Lemma 12.11 below are a bit ackward, because they must take some nuisance factors 2 and 4 into account. This stems from the fact that division by 2 in H is not assumed to be unique. The nuisance factors disappear if H is 2-torsion free, which is the case if, e.g., $H = \mathbb{C}$. It is slightly surprising that we can get interesting results even if we do not assume that the functions are defined on a group. A semigroup with an involution suffices. That requires of course that we extend the definition of Jensen's functional equation.

Definition 12.9. Let S be a semigroup with an involution $\tau : S \to S$. A *solution of Jensen's extended functional equation* is a map $f : S \to H$ satisfying *Jensen's extended functional equation*

$$f(xy) + f(x\tau(y)) = 2f(x), \qquad x, y \in S. \tag{12.7}$$

We recall that a function $f : S \to H$ is said to be *odd* (with respect to τ), if $f \circ \tau = -f$.

Proposition 12.10. *Let S be a semigroup with an involution τ. If $f : S \to H$ is an odd solution of* (12.7)*, then f satisfies* (2.3)*, i.e., $f(xy) + f(yx) = 2f(x) + 2f(y)$. If f is also central, then $2f$ is additive.*

Proof. Proceed as in the proof of Proposition 12.5(a). □

Lemma 12.11. *Let S be a semigroup, and let $\tau : S \to S$ be an involution. Let $f : S \to H$ be a central map satisfying (12.7). Then f has the form $f = a + c$, where $a : S \to H$ is such that $4a$ is additive and $2a \circ \tau = -2a$, and where $c \in H$ is a constant.*

Proof. Let us first observe that $2f(xy\tau(y)) = 2f(x)$ for all $x, y \in S$ (to get this replace y by $y\tau(y)$ in (12.7)). We next replace x by $x\tau(x)$ in (12.7) and find that $f(x\tau(x)y) + f(x\tau(x)\tau(y)) = 2f(x\tau(x))$, or f being central, that $f(yx\tau(x)) + f(\tau(y)x\tau(x)) = 2f(x\tau(x))$. Combining this with the observation we get that

$$2f(y) + 2f(\tau(y)) = 4f(x\tau(x)) \qquad \text{for all } x, y \in S. \tag{12.8}$$

The left and hence also the right hand side of (12.8) is independent of x and is hence a constant. Fixing $x_0 \in S$ we get with $c := f(x_0\tau(x_0)) \in H$, that $4f(x\tau(x)) = 4c$ for all $x \in S$. (12.8) says that $2f + 2f \circ \tau = 4c$.

We continue by interchanging x and y in (12.7) and adding the resulting identity to (12.7). Since f is central we obtain

$$2f(xy) + f(x\tau(y)) + f(y\tau(x)) = 2f(x) + 2f(y).$$

Noting that $2f(x\tau(y)) + 2f(y\tau(x)) = 2(f + f \circ \tau)(x\tau(y)) = 4c$ we infer that $4f(xy) + 4c = 4f(x) + 4f(y)$. This can be reformulated to

$$4[f(xy) - c] = 4[f(x) - c] + 4[f(y) - c],$$

which reveals that $4a$ is additive where $a(x) := f(x) - c$, $x \in S$.

Furthermore, from (12.8) we get

$$\begin{aligned} 2(a \circ \tau)(x) &= 2[f(\tau(x)) - c] = 2f(\tau(x)) - 2c \\ &= 4c - 2f(x) - 2c = -2[f(x) - c] = -2a(x). \end{aligned}$$ □

Corollary 12.12. *Assume that H is 2-torsion free.*

(a) *Let S be an abelian semigroup, and let $\tau : S \to S$ be an involution. The solutions $f : S \to H$ of (12.7) are the maps of the form $f = a + c$, where $a : S \to H$ is an additive map such that $a \circ \tau = -a$, and where $c \in H$ is a constant. If f is an odd solution, then $c = 0$, so that $f = a$ is additive.*

(b) *Any central $f \in S(G, H)$ is additive.*

(c) *If G is abelian, then $S(G, H)$ is the set of additive maps $a : G \to H$.*

Lemma 12.13. *The following four statements are equivalent for any $f \in S(G,H)$, if H is 2-torsion free.*

(i) *f is additive, i.e., $f \in \mathrm{Hom}(G,H)$.*

(ii) *f is central, i.e., $f(xy) = f(yx)$ for all $x, y \in G$.*

(iii) *f vanishes on each commutator, i.e., $f([x,y]) = 0$ for all $x, y \in G$.*

(iv) *f vanishes identically on the commutator subgroup $[G,G]$.*

Proof. Clearly (i) $\Rightarrow$ (iv) $\Rightarrow$ (iii). (iii) $\Rightarrow$ (ii) by Proposition 12.5(l), so left is (ii) $\Rightarrow$ (i). And that is Corollary 12.12(b). □

Any $f \in S(G,H)$ is a function on the quotient group $G/[G,[G,G]]$ (Proposition 12.5(j)), so we would like to know when $G/[G,[G,G]]$ is abelian. An answer and some applications are given in Lemma 12.14(a).

Lemma 12.14.

(a) *$G/[G,[G,G]]$ is abelian $\Leftrightarrow$ $[G,G] \subseteq [G,[G,G]]$.*

(b) *$[G,G] \subseteq [G,[G,G]] \Rightarrow S(G,\mathbb{C}) = \mathrm{Hom}(G,\mathbb{C})$.*

(c) *$[G,G] = G \Rightarrow S(G,H) = \{0\}$.*

Proof. (a) is a special case of a result valid for any normal subgroup N of G: G/N is abelian if and only if $[G,G] \subseteq N$.

(b) Noting that $[G,[G,G]] = [G,G]$ we deduce from Proposition 12.5(j) that any $f \in S(G,\mathbb{C})$ is a function on the abelian group $G/[G,G]$ and hence is abelian. The additivity of f now follows from Lemma 12.13.

(c) $f \in S(G,H)$ is a function on $G/[G,[G,G]]$. But $G/[G,[G,G]] = G/G$ is here a one-element group, so f vanishes (being normalized). □

For a Lie group G many problems are easier to solve on the connected component G_e of the identity $e \in G$ than on all of G, because G_e is closely related to the Lie algebra of G where methods from linear algebra can be used. Other methods must then brought forward to extend the result from G_e to the remaining connected components of G. Lemma 12.15 can be applied when G has only finitely many connected components (take $G_0 = G_e$).

Lemma 12.15. *Let G_0 be a subgroup of G such that the coset space G/G_0 is finite.*

(a) *If $S(G_0,\mathbb{C}) = \mathrm{Hom}(G_0,\mathbb{C})$, then $S(G,\mathbb{C}) = \mathrm{Hom}(G,\mathbb{C})$.*

(b) *If $S(G_0,\mathbb{C}) = \{0\}$, then $S(G,\mathbb{C}) = \{0\}$.*

Proof. Let $f \in S(G,\mathbb{C})$ and $x,y \in G$ be arbitrary. Consider the map $n \mapsto x^n G_0$ of $\mathbb{N}$ into G/G_0. Since G/G_0 is finite the map cannot be injective, so there exist by the pigeon hole principle two natural numbers $p < q$ such that $x^p G_0 = x^q G_0$ and hence $x^n G_0 = G_0$ with $n := q - p \in \mathbb{N}$. Thus $x^n \in G_0$. Similarly there exists an $m \in \mathbb{N}$ such that $y^m \in G_0$.

(a) Now, f vanishes on $[G_0, G_0]$ according to Lemma 12.13 (by our assumption $S(G_0,\mathbb{C}) = \operatorname{Hom}(G_0,\mathbb{C})$), so $0 = f([x^n, y^m])$. Using Proposition 12.5(e) we then get that $0 = nmf([x,y])$, so that $f([x,y]) = 0$. We infer from Lemma 12.13 that $f \in \operatorname{Hom}(G,\mathbb{C})$.

(b) Here f vanishes on G_0 by our assumption, so $0 = f(x^n) = nf(x)$. □

12.4 The solutions modulo the homomorphisms

Except for exercises we shall in the rest of the present chapter concentrate on the complex vector space $S(G,\mathbb{C})$ of odd solutions solutions of Jensen's functional equation (12.3). So $H = \mathbb{C}$.

The set of additive maps $\operatorname{Hom}(G,\mathbb{C})$ forms a subspace of $S(G,\mathbb{C})$. This leads us to ask how much larger $S(G,\mathbb{C})$ is than $\operatorname{Hom}(G,\mathbb{C})$ or more precisely stated: How big is the quotient vector space $S(G,\mathbb{C})/\operatorname{Hom}(G,\mathbb{C})$? To answer this question we go back to the observation in Proposition 12.5(i) that the restriction $f|_{[G,G]}$ of any $f \in S(G,\mathbb{C})$ to the commutator subgroup $[G,G]$ is an additive function on $[G,G]$. f is a function on the quotient group $G/[G,[G,G]]$ (Proposition 12.5(j)), so $f|_{[G,G]} \in \operatorname{Hom}([G,G]/[G,[G,G]],\mathbb{C})$. Theorem 12.19 below shows that the restriction map $f \mapsto f|_{[G,G]}$ is a linear map of $S(G,\mathbb{C})$ onto $\operatorname{Hom}([G,G]/[G,[G,G]],\mathbb{C})$ with kernel $\operatorname{Hom}(G,\mathbb{C})$. So it induces a vector space isomorphism

$$S(G,\mathbb{C})/\operatorname{Hom}(G,\mathbb{C}) \simeq \operatorname{Hom}([G,G]/[G,[G,G]],\mathbb{C}).$$

Thus we may parametrize $S(G,\mathbb{C})/\operatorname{Hom}(G,\mathbb{C})$ by the set of those additive maps ϕ of $[G,G]$ into $\mathbb{C}$ that vanish on all of $[G,[G,G]]$; by Lemma A.7 this means by the set of those $\phi \in \operatorname{Hom}([G,G],\mathbb{C})$ for which $\phi([x,[y,z]]) = 0$ for all $x,y,z \in G$. Our main result, Theorem 12.20, which is the topological version of Theorem 12.19, says that the topology follows suit in the quotient construction above, at least when G is locally compact.

In the proof of Theorem 12.19 we use the following technical result about extending a solution of Jensen's functional equation to a larger group.

The proposition has independent interest, because it can be used to construct solutions (see Exercises 12.11 and 12.13).

Proposition 12.16. *Let G be a step 2 nilpotent group. Let G_0 be a normal subgroup and H an abelian subgroup of G. Let $f_0 \in S(G_0, \mathbb{C})$ and $\psi \in \operatorname{Hom}(H, \mathbb{C})$ coincide on $G_0 \cap H$. Let G_1 be the group generated by G_0 and H.*

Then there is exactly one $f_1 \in S(G_1, \mathbb{C})$ such that $f_1 = f_0$ on G_0 and $f_1 = \psi$ on H. It is given by

$$f_1(x_0 h) = f_0(x_0) + \psi(h) + \tfrac{1}{2} f_0([x_0, h]) \qquad \text{for } x_0 \in G_0, h \in H. \tag{12.9}$$

Proof. Since G_0 is normal, G_1 consists of the elements of the form $x_0 h$, where $x_0 \in G_0$ and $h \in H$. Any $f_1 \in S(G_1, \mathbb{C})$ extending f_0 and ψ is according to Proposition 12.5(e) given by (12.9). This proves not just the formula but also the uniqueness of f_1. It is left to show the existence of such an f_1.

We claim that a function f_1 is well-defined on G_1 by the prescription (12.9), i.e., that if $x_0 h = y_0 k$, where $x_0, y_0 \in G_0$ and $h, k \in H$, then the difference

$$\Delta := f_0(x_0) + \psi(h) + \tfrac{1}{2} f_0([x_0, h]) - f_0(y_0) - \psi(k) - \tfrac{1}{2} f_0([y_0, k]) = 0.$$

To verify the claim we reformulate Δ using Proposition 12.5(e) and that $hk^{-1} = x_0^{-1} y_0$:

$$\begin{aligned}
\Delta &= f_0(x_0) - f_0(y_0) + \psi(h) - \psi(k) + \tfrac{1}{2} f_0([x_0, h]) - \tfrac{1}{2} f_0([y_0, k]) \\
&= f_0(x_0) - f_0(y_0) + \psi(hk^{-1}) + \tfrac{1}{2} f_0([x_0, h]) + \tfrac{1}{2} f_0([k, y_0]) \\
&= f_0(x_0) - f_0(y_0) + f_0(x_0^{-1} y_0) \\
&\quad + \tfrac{1}{2} f_0([x_0, h][k, y_0]) - \tfrac{1}{4} f_0([[x_0, h], [k, y_0]]).
\end{aligned}$$

Commutators are in $Z(G)$, because G is step 2 nilpotent, so the last term on the right hand side vanishes, and we may continue as follows

$$\begin{aligned}
\Delta &= f_0(x_0) - f_0(y_0) + f_0(x_0^{-1}) + f_0(y_0) \\
&\quad + \tfrac{1}{2} f_0([x_0^{-1}, y_0]) + \tfrac{1}{2} f_0([x_0, h][k, y_0]) \\
&= -\tfrac{1}{2} f_0([x_0^{-1}, y_0^{-1}]) + \tfrac{1}{2} f_0([x_0, h][k, y_0]).
\end{aligned}$$

We reformulate the argument of the last term as follows, using that $h^{-1}k = y_0^{-1}x_0$ (H is abelian) and that $[G,G] \subseteq Z(G)$ due to the 2 step nilpotency:

$$\begin{aligned}[x_0,h][k,y_0] &= x_0hx_0^{-1}(h^{-1}k)y_0k^{-1}y_0^{-1} = x_0hx_0^{-1}y_0^{-1}x_0y_0k^{-1}y_0^{-1} \\ &= x_0h[x_0^{-1},y_0^{-1}]k^{-1}y_0^{-1} = x_0hk^{-1}y_0^{-1}[x_0^{-1},y_0^{-1}] \\ &= x_0x_0^{-1}y_0y_0^{-1}[x_0^{-1},y_0^{-1}] = [x_0^{-1},y_0^{-1}].\end{aligned}$$

This implies that Δ vanishes, so that (12.9) defines a function f_1 on G_1. We will show f_1 has the properties stated in the proposition.

Clearly $f_1 = f_0$ on G_0, $f_1 = \psi$ on H and $f_1(e) = 0$, so it is left to check that f_1 is a solution of Jensen's equation (12.3). Thus we shall prove

$$L := f_1(x_0hy_0k) + f_1(x_0h(y_0k)^{-1}) - 2f_1(x_0h) = 0$$

for all $x_0, y_0 \in G_0, h, k \in H$. Substituting the definition of f_1 into this we find

$$\begin{aligned}L &= f_1(x_0hy_0h^{-1}hk) + f_1(x_0hk^{-1}y_0^{-1}kh^{-1}hk^{-1}) - 2f_1(x_0h) \\ &= f_0(x_0hy_0h^{-1}) + \psi(hk) + \tfrac{1}{2}f_0([x_0hy_0h^{-1},hk]) \\ &\quad + f_0(x_0hk^{-1}y_0^{-1}kh^{-1}) + \psi(hk^{-1}) + \tfrac{1}{2}f_0([x_0hk^{-1}y_0^{-1}kh^{-1},hk^{-1}]) \\ &\quad - 2f_0(x_0) - 2\psi(h) - f_0([x_0,h]).\end{aligned}$$

The terms involving ψ vanish, because ψ is additive, so left is

$$\begin{aligned}L &= f_0(x_0hy_0h^{-1}) + f_0(x_0hk^{-1}y_0^{-1}kh^{-1}) \\ &\quad + \tfrac{1}{2}f_0([x_0hy_0h^{-1},hk]) + \tfrac{1}{2}f_0([x_0hk^{-1}y_0^{-1}kh^{-1},hk^{-1}]) \\ &\quad - 2f_0(x_0) - f_0([x_0,h]).\end{aligned}$$

Using that G is 2 step nilpotent (meaning that $[G,G] \subseteq Z(G)$) so that $[a[b,c]d,e] = [ad,e]$ for all $a,b,c,d \in G$ we find for the third term that

$$[x_0hy_0h^{-1},hk] = [x_0hy_0h^{-1}y_0^{-1}y_0,hk] = [x_0[h,y_0]y_0,hk] = [x_0y_0,hk].$$

Similarly for the fourth term, so L simplifies to

$$\begin{aligned}L &= f_0(x_0hy_0h^{-1}) + f_0(x_0hk^{-1}y_0^{-1}kh^{-1}) \\ &\quad + \tfrac{1}{2}f_0([x_0y_0,hk]) + \tfrac{1}{2}f_0([x_0y_0^{-1},hk^{-1}]) - 2f_0(x_0) - f_0([x_0,h]).\end{aligned}$$

Two properties of f_0 will be useful to handle the terms of L.

The first property is that $f_0([-,-])$ is bi-additive on $G_0 \times G$, i.e.,

$$f_0([a_0b_0,cd]) = f_0([a_0,c]) + f_0([a_0,d]) + f_0([b_0,c]) + f_0([b_0,d]).$$

Note in particular that $f_0([a_0^{-1}, c]) = f_0([a_0, c^{-1}]) = -f_0([a_0, c])$.

Derivation of the first property: Note that $[ab, c] = [a, c][b, c]$ and $[a, bc] = [a, b][a, c]$ for all $a, b, c, d \in G$ (use Lemma A.2 and that $[G, G] \subseteq Z(G)$). So for any $a_0, b_0 \in G_0$ and $c, d \in G$ we get

$$\begin{aligned} f_0([a_0b_0, cd]) &= f_0([a_0b_0, c][a_0b_0, d]) \\ &= f_0([a_0b_0, c]) + f_0([a_0b_0, d]) + \tfrac{1}{2} f_0([[a_0b_0, c], [a_0b_0, d]]) \\ &= f_0([a_0b_0, c]) + f_0([a_0b_0, d]), \end{aligned}$$

where the term with coefficient $\frac{1}{2}$ vanished because $[G, G] \subseteq Z(G)$. Repeating the argument we arrive at the first property.

The second property says for $a_0, b_0 \in G_0$ and $c \in G$ that

$$f_0(a_0[c, b_0]) = f_0(a_0) + f_0([c, b_0]),$$

and it follows from Proposition 12.5(e) and that $[G, [G, G]] = \{e\}$.

Continuing our computation of L we apply the bi-additivity of $f_0([-, -])$ on $G_0 \times G$ to the two terms with the coefficient $\frac{1}{2}$ and get:

$$L = f_0(x_0hy_0h^{-1}) + f_0(x_0hk^{-1}y_0^{-1}kh^{-1}) + f_0([y_0, k]) - 2f_0(x_0). \quad (12.10)$$

For the first term on the right hand side of (12.10) we find from the second property that

$$f_0(x_0hy_0h^{-1}) = f_0(x_0[h, y_0]y_0) = f_0(x_0y_0[h, y_0]) = f_0(x_0y_0) + f_0([h, y_0]),$$

and similarly for the second term (replace y_0 by y_0^{-1} and h by hk^{-1}) that

$$\begin{aligned} f_0(x_0hk^{-1}y_0^{-1}kh^{-1}) &= f_0(x_0y_0^{-1}) + f_0([hk^{-1}, y_0^{-1}]) \\ &= f_0(x_0y_0^{-1}) - f_0([h, y_0]) + f_0([k, y_0]), \end{aligned}$$

so (using f_0 is a solution of Jensen's equation on G_0) their sum is

$$\begin{aligned} f_0(x_0hy_0h^{-1}) + f_0(x_0hk^{-1}y_0^{-1}kh^{-1}) &= 2f_0(x_0) + f_0([k, y_0]) \\ = 2f_0(x_0) - f_0([k, y_0]^{-1}) &= 2f_0(x_0) - f_0([y_0, k]). \end{aligned}$$

Substituting this into (12.10) we find that $L = 0$, as desired. □

Lemma 12.17. *Let G be step 2 nilpotent. For any $\phi \in \mathrm{Hom}([G, G], \mathbb{C})$ there exists an $f \in S(G, \mathbb{C})$ such that $f = \phi$ on $[G, G]$.*

Proof. Consider the set $\mathcal{P}$ of all pairs $\{G_0, f_0\}$ where G_0 is a subgroup of G containing $[G,G]$, and where $f_0 \in S(G_0,\mathbb{C})$ is such that $f_0 = \phi$ on $[G,G]$. $\mathcal{P}$ is not empty since $\{[G,G],\phi\} \in \mathcal{P}$. We introduce a partial ordering $\geq$ in $\mathcal{P}$ by writing $\{G_1, f_1\} \geq \{G_2, f_2,\}$ if and only if $G_1 \supseteq G_2$ and $f_1 = f_2$ on G_2. Obviously each totally ordered subset of $\mathcal{P}$ has a majorant in $\mathcal{P}$ (the subgroup of the majorant can be taken as the union of the subgroups ocurring in the subset), so by Zorn's lemma there is a maximal element $\{G_0, f_0\}$ in $\mathcal{P}$. It remains to show that $G_0 = G$. This we do by contradiction, so we assume that there is an element $a \in G \setminus G_0$. Taking $H := \langle a \rangle$ in Proposition 12.16 we see that, in order to arrive at a contradiction to the maximality of $\{G_0, f_0\}$ in $\mathcal{P}$, it suffices to find a $\psi \in \operatorname{Hom}(\langle a\rangle,\mathbb{C})$ such that $\psi = f_0$ on $\langle a\rangle \cap G_0$.

If $\langle a\rangle \cap G_0 = \{e\}$ we may take $\psi = 0$. If not, let p be the smallest integer in $\{2,3,\ldots\}$ for which $a^p \in G_0$. Then $\langle a\rangle \cap G_0 = \{a^{np} \mid n \in \mathbb{Z}\}$, so we may define $\psi(x) := f_0(x^p)/p$, $x \in \langle a\rangle$. This ψ works, the crucial observation being that the group $\langle a\rangle \cap G_0$ is cyclic, so that the restriction of f_0 to it is additive (Proposition 12.7). □

In the next lemma we get rid of the assumption in Lemma 12.17 that G is step 2 nilpotent.

Lemma 12.18. *For any $\phi \in \operatorname{Hom}([G,G],\mathbb{C})$ such that $\phi = 0$ on $[G,[G,G]]$ there exists an $f \in S(G,\mathbb{C})$ such that $f = \phi$ on $[G,G]$.*

Proof. $[G,[G,G]]$ is a normal subgroup of G, so we may form the quotient group $\widetilde{G} := G/[G,[G,G]]$ which is a step 2 nilpotent group. The quotient map $\pi : G \to \widetilde{G}$, defined by $\pi(x) := x[G,[G,G]]$, $x \in G$, is a surjective homomorphism. Note that $\pi([G,G]) = [\widetilde{G},\widetilde{G}]$, so that any element of $[\widetilde{G},\widetilde{G}]$ may be written in the form $x[G,[G,G]]$ where $x \in [G,G]$.

Since $\phi = 0$ on $[G,[G,G]]$ we may define $\widetilde{\phi} \in \operatorname{Hom}([\widetilde{G},\widetilde{G}],\mathbb{C})$ by $\phi = \widetilde{\phi}\circ\pi$ on $[G,G]$. By the previous lemma there exists $\widetilde{f} \in S(\widetilde{G},\mathbb{C})$ such that $\widetilde{f} = \widetilde{\phi}$ on $[\widetilde{G},\widetilde{G}]$. Now $f := \widetilde{f}\circ\pi \in S(G,\mathbb{C})$. For any $x \in [G,G]$ we have $\pi(x) \in [\widetilde{G},\widetilde{G}]$, and so $f(x) = \widetilde{f}(\pi(x)) = \widetilde{\phi}(\pi(x)) = \phi(x)$. □

Theorem 12.19. *The restriction map $f \mapsto f|_{[G,G]}$ is a linear map of $S(G,\mathbb{C})$ onto $\operatorname{Hom}([G,G]/[G,[G,G]],\mathbb{C})$ with kernel $\operatorname{Hom}(G,\mathbb{C})$. So it induces a vector space isomorphism*

$$S(G,\mathbb{C})/\operatorname{Hom}(G,\mathbb{C}) \simeq \operatorname{Hom}([G,G]/[G,[G,G]],\mathbb{C}).$$

Proof. The linearity of the restriction map is obvious. We get from Lemma 12.13 that its kernel is $\mathrm{Hom}(G,\mathbb{C})$. The surjectivity statement is Lemma 12.18. □

Theorem 12.19 is purely algebraic. We also want a version for topological groups and continuous functions. Let us for any topological group G introduce the notations

$$S^{\mathrm{cont}}(G,\mathbb{C}) := S(G,\mathbb{C}) \cap C(G),$$
$$\mathrm{Hom}^{\mathrm{cont}}(G,\mathbb{C}) := \mathrm{Hom}(G,\mathbb{C}) \cap C(G).$$

Theorem 12.20. *Let us assume that G is a locally compact group. The restriction map $f \mapsto f|_{[G,G]}$ is a linear mapping from $S^{\mathrm{cont}}(G,\mathbb{C})$ onto $\mathrm{Hom}^{\mathrm{cont}}([G,G]/[G,[G,G]],\mathbb{C})$ with kernel $\mathrm{Hom}^{\mathrm{cont}}(G,\mathbb{C})$. So it induces a vector space isomorphism*

$$S^{\mathrm{cont}}(G,\mathbb{C})/\mathrm{Hom}^{\mathrm{cont}}(G,\mathbb{C}) \simeq \mathrm{Hom}^{\mathrm{cont}}([G,G]/[G,[G,G]],\mathbb{C}).$$

Proof. Proceeding as in the proof of Theorem 12.19 we find that $f \mapsto f|_{[G,G]}$ is a linear map of $S^{\mathrm{cont}}(G,\mathbb{C})$ into $\mathrm{Hom}^{\mathrm{cont}}([G,G]/[G,[G,G]],\mathbb{C})$ with kernel $\mathrm{Hom}^{\mathrm{cont}}(G,\mathbb{C})$. Left is the surjectivity of the restriction map. So let $\phi \in \mathrm{Hom}^{\mathrm{cont}}([G,G]/[G,[G,G]],\mathbb{C})$. In particular ϕ is continuous. According to Theorem 12.19 there is an $f \in S(G,\mathbb{C})$ such that $f|_{[G,G]} = \phi$. Regrettably we cannot use f, the stumbling block being that f may not be continuous. However, applying Proposition 12.5(e) we get that $f(xy) - f(x) - f(y) = \frac{1}{2}\phi([x,y])$ for all $x,y \in G$. The right hand side $\frac{1}{2}\phi([x,y])$ is continuous, so Corollary D.12 tells us that $f = f_0 + a$, where $f_0 : G \to \mathbb{C}$ is continuous and $a : G \to \mathbb{C}$ is additive. Substituting $f = f_0 + a$ we get that $f_0 \in S^{\mathrm{cont}}(G,\mathbb{C})$ and $f_0|_{[G,G]} = \phi$. □

Theorem 12.19 is the special case of Theorem 12.20 in which G is equipped with the discrete topology.

12.5 Examples

In this section we find the continuous complex-valued solutions of Jensen's functional equation on various groups.

Example 12.21. Let $G = (\mathbb{R}^n, +)$. We know from Corollary 12.12(c) that $S(G,\mathbb{C})$ coincides with the additive functions. The continuous functions

$f \in S(\mathbb{R}^n, \mathbb{C})$ are the ones of the form the form $f(x) = \langle c, x \rangle$, $x \in \mathbb{R}^n$, where $c \in \mathbb{C}^n$ (Corollary 2.4).

Example 12.22. Let us consider the $(ax+b)$-group (Examples A.17(i)). We leave it to the reader to compute that $[G,[G,G]] = [G,G]$. We then see from Lemma 12.14(b) that $S(G,\mathbb{C}) = \mathrm{Hom}(G,\mathbb{C})$. We found the continuous additive maps $h : G \to \mathbb{C}$ in Example 2.10. We conclude that the continuous, complex-valued solutions of Jensen's functional equation on the $(ax+b)$-group are the functions of the form

$$f\begin{pmatrix} a & b \\ 0 & 1 \end{pmatrix} = c \log a + d,$$

where $c, d \in \mathbb{C}$ are arbitrary constants.

Example 12.23. The Heisenberg group H_3 (Examples A.17(a)) is step 2 nilpotent, which means that $[H_3,[H_3,H_3]] = \{I\}$. From the formulas of Examples A.17(a) we read that

$$[H_3,H_3]/[H_3,[H_3,H_3]] = [H_3,H_3] = \left\{ \begin{pmatrix} 1 & 0 & z \\ 0 & 1 & 0 \\ 0 & 0 & 1 \end{pmatrix} \,\middle|\, z \in \mathbb{R} \right\} \simeq \mathbb{R}.$$

Combining this with Theorem 12.20 we find that

$$\dim\{S^{\mathrm{cont}}(H_3,\mathbb{C})/\mathrm{Hom}^{\mathrm{cont}}(H_3,\mathbb{C})\} = \dim\{\mathrm{Hom}^{\mathrm{cont}}(\mathbb{R},\mathbb{C})\} = 1,$$

the last equality by Corollary 2.4. One odd, continuous solution of Jensen's functional equation which is not additive, is

$$f_0\begin{pmatrix} 1 & x & z \\ 0 & 1 & y \\ 0 & 0 & 1 \end{pmatrix} = z - \tfrac{1}{2}xy, \qquad x, y, z \in \mathbb{R}.$$

We conclude that the odd, continuous solutions of Jensen's functional equation are the complex multiples of f_0 plus the additive, continuous maps from H_3 to $\mathbb{C}$. We found those in Example 2.11. Summing up: The odd, continuous solutions of Jensen's functional equation are the functions of the form

$$\begin{pmatrix} 1 & x & z \\ 0 & 1 & y \\ 0 & 0 & 1 \end{pmatrix} \mapsto c(z - \tfrac{1}{2}xy) + \alpha x + \beta y,$$

where the parameters $c, \alpha, \beta \in \mathbb{C}$ are arbitrary.

Example 12.24. The identity $[G,G] = G$ holds when G is a connected semisimple Lie group (see for instance [201, Corollary 3.18.10]), so $S(G,\mathbb{C}) = \{0\}$ according to Lemma 12.14(c). Using Lemma 12.15 we find that we may replace the requirement "connected" by "having only finitely many connected components", and still get that $S(G,\mathbb{C}) = \{0\}$ (A Lie group is semisimple if its connected identity component is semisimple, because semisimplicity by definition means that the Lie algebra is semisimple). Examples of semisimple Lie groups are listed in Helgason's monograph [108, Ch. X, 2]. Let us among them mention the two connected groups $\mathrm{SL}(n,\mathbb{R})$ and $\mathrm{SL}(n,\mathbb{C})$ for $n \geq 2$, and the Lorentz group $O(3,1)$ that has four connected components.

Example 12.25. Let $G = N \times_s H$ be the topological semidirect product of an abelian group N (the normal part) and a semisimple Lie group H with only finitely many components. An example is the Poincaré group = the inhomogeneous Lorentz group = $\mathbb{R}^4 \times_s O(3,1)$. We will find $S(G,\mathbb{C})$.

Consider the subgroup $G_0 := N \times_s H_e$, where H_e is the identity component of H. We shall use that $H_e = [H_e, H_e]$ ([201, Corollary 3.18.10]). The homomorphism $n \mapsto n[G_0,[G_0,G_0]]$ of N into $G_0/[G_0,[G_0,G_0]]$ is surjective: For any $n \in N$ and $h \in H_e$ we get that $nh[G_0,[G_0,G_0]] = n[G_0,[G_0,G_0]]$, because $h \in H_e = [H_e,H_e] = [H_e,[H_e,H_e]] \subseteq [G_0,[G_0,G_0]]$. The image $G_0/[G_0,[G_0,G_0]]$ is abelian, because so is N. Any $f_0 \in S(G_0,\mathbb{C})$ is a function on $G_0/[G_0,[G_0,G_0]]$, so we see from Corollary 12.12(c) that f_0 is additive. We have thus shown that $S(G_0,\mathbb{C}) = \mathrm{Hom}(G_0,\mathbb{C})$. From Lemma 12.15(a) we then read that $S(G,\mathbb{C}) = \mathrm{Hom}(G,\mathbb{C})$. As we saw in Example 2.12 any additive, complex-valued map vanishes on H. It follows that $S^{\mathrm{cont}}(G,\mathbb{C}) = \mathrm{Hom}^{\mathrm{cont}}(G,\mathbb{C})$ consists of the maps of the form $nh \mapsto a(n)$, where $a : N \to \mathbb{C}$ is a continuous additive map such that $a(hnh^{-1}) = a(n)$ for all $h \in H$ and $n \in N$.

The last condition, i.e., $a(hnh^{-1}) = a(n)$, will for some semidirect products imply that $S(G,\mathbb{C}) = \mathrm{Hom}(G,\mathbb{C}) = \{0\}$. It does so for the Poincaré group, but we shall not prove it here.

Example 12.26. [149, Theorem 3] determines $S(G,H)$ for any free group G and any abelian group H.

Example 12.27. Let $n \geq 3$. Ng proves in [149, Theorem 4] proves that $S(\mathrm{GL}(n,R),H) = \mathrm{Hom}(\mathrm{GL}(n,R),H)$ for $R = \mathbb{Z}$, $\mathbb{R}$, a quadratically closed field or a finite field, while $S(\mathrm{GL}(2,\mathbb{Z}),H)$ is described in [148, Theorem 11].

12.6 Other ways of formulating Jensen's equation

Jensen's functional equation occurs in various disguises. As we more or less saw above in (12.1) it may, when G is the additive group of a vector space, be formulated as

$$2f\Big(\frac{x+y}{2}\Big) = f(x) + f(y) \qquad \text{for all } x, y \in G.$$

A second formulation is the one of Lemma 12.28. A third is found in Proposition 2.17. For other formulations of Jensen's equation, for example in terms of semi-homomorphisms (cfr. the Notes to Chapter 2), see Corovei [53].

Let us mention other connections where Jensen's equation enters:

Let $(E, \|-\|)$ and $(F, \|-\|)$ be two real, normed spaces. If $f : E \to F$ is an isometry, meaning that

$$\|f(x) - f(y)\| = \|x - y\| \qquad \text{for all } x, y \in E,$$

of E onto F, then f is continuous and satisfies the following version of Jensen's equation

$$f(x+y) + f(x-y) = 2f(x), \qquad x, y \in E.$$

For a proof of this and more information see [6, Proposition 10.10].

Another connection is the following. Let $f : G \to H$. For any $y \in G$ we introduce the difference operator by $(\Delta_y f)(x) := f(xy) - f(x)$, $x \in G$. If G is abelian then the functions f satisfying $\Delta_y^{n+1} f = 0$ for all $y \in G$ are the polynomials of order at most n (see [197, Theorem 9.3]). In particular the solutions f of $\Delta_y^2 f = 0$ for all $y \in G$ are the polynomials of order at most 1. In view of Lemma 12.28 it might thus be said that we seek the polynomials of order at most 1, but in the setting of a possibly non-abelian group G.

Lemma 12.28. *Let $f : G \to H$. Then $\Delta_y^2 f = 0$ for all $y \in G$ if and only if f is a solution of Jensen's functional equation* (12.3).

Proof. The proof follows from the computation

$$\begin{aligned}(\Delta_y^2 f)(x) &= [\Delta_y(\Delta_y f)](x) = (\Delta_y f)(xy) - (\Delta_y f)(x)\\ &= f(xy^2) - f(xy) - f(xy) + f(x)\\ &= f([xy]y) + f([xy]y^{-1}) - 2f(xy).\end{aligned}$$

□

12.7 A Pexider-Jensen's functional equation

The paper [151] by Ng gives explicit formulas for the solutions $f, g, h : G \to H$ of the functional equation

$$f(xy) + g(xy^{-1}) = h(x), \qquad x, y \in G, \tag{12.11}$$

first for a free group G and then (as applications of the results from the free group case) for certain linear groups and semi-direct products. If G is generated by its squares then (12.11) reduces to Jensen's equation, i.e., $f = g$ and $h = 2f$ by [151, Corollary 3.1]. The treatment is purely algebraic. It should be emphasized that H is only assumed to be an abelian group, no further properties of H are imposed. If H is 2-torsion free, then (12.11) can be solved by means of the theory of Jensen's equation.

12.8 A variant of Jensen's equation

If we in the second term of Jensen's functional equation (12.3) change the order of x and y we arrive at the equation $f(xy) + f(y^{-1}x) = 2f(x)$. This equation and (12.3) coincide if G is abelian, and they both generalize (12.2). The new equation is much simpler than (12.3) if H is 2-torsion free (for instance if $H = \mathbb{C}$), as Proposition 12.29 reveals. We refer the reader who wishes to assume merely that H is an abelian group, to Ng's paper [150].

Section 11.8 has results about the corresponding variant of Wilson's functional equation.

Proposition 12.29. *Let H be 2-torsion free. $f : G \to H$ is a solution of*

$$f(xy) + f(y^{-1}x) = 2f(x), \qquad x, y \in G, \tag{12.12}$$

if and only if f is an affine map from G to H.

Proof. We leave it to the reader to show that any affine map is a solution of (12.12). To prove the converse let f be a solution of (12.12). The equation (12.12) says that $[L(y) + R(y)]f = 2f$ for all $y \in G$, where L and R denote the left and right regular representations of G (Definition E.15). Applying $L(y) + R(y)$ to the equation we get

$$\begin{aligned} 4f &= [L(y) + R(y)]^2 f = [L(y)^2 + R(y)^2 + 2L(y)R(y)]f \\ &= [L(y^2) + R(y^2)]f + 2L(y)R(y)f = 2f + 2L(y)R(y)f, \end{aligned}$$

so that $2L(y)R(y)f = 2f$, which means that $2f$ is central. Division by 2 being unique we infer that f is central. Then f is a central solution of (12.3) and hence affine by Corollary 12.12(b). □

12.9 Exercises

Exercise 12.1. Show that the differentiable solutions $f : \mathbb{R} \to \mathbb{C}$ of Jensen's functional equation $f(x+y) + f(x-y) = 2f(x)$, $x, y \in \mathbb{R}$, are the functions $f(x) = \alpha x + \beta$, $x \in \mathbb{R}$, where $\alpha, \beta \in \mathbb{C}$.

Hint: Differentiate the functional equation with respect to x and with respect to y. Compare the two expressions.

Exercise 12.2. Show directly without recourse to Corollary 12.6 that a solution $f \in C(G)$ of (12.3) is constant if G is a compact group.

Hint: Integrate the functional equation with respect to the Haar measure on G.

Exercise 12.3. Let # mean "the number of elements in". Show that

(a) $\#\mathrm{Hom}(\mathbb{Z}_2 \oplus Z_2, \mathbb{Z}_2) = 4$, and that $\#S(\mathbb{Z}_2 \oplus Z_2, \mathbb{Z}_2) = 8$.

(b) Conclude that $\mathrm{Hom}(\mathbb{Z}_2 \oplus Z_2, \mathbb{Z}_2) \subsetneq S(\mathbb{Z}_2 \oplus Z_2, \mathbb{Z}_2)$.

Exercise 12.4. Assume that each element of G has finite order. Show that the constant functions are the only complex-valued solutions of Jensen's functional equation (12.3).

Exercise 12.5. Let $f \in S(G, H)$. Show that cyclic permutations of the elements of a commutator $[x, y] = xyx^{-1}y^{-1}$ where $x, y \in G$, do not change the value of f at the commutator, i.e., that

$$f(xyx^{-1}y^{-1}) = f(yx^{-1}y^{-1}x) = f(x^{-1}y^{-1}xy) = f(y^{-1}xyx^{-1}).$$

Hint: $[G, G] \in Z(f)$.

Exercise 12.6. Let $f \in S(G, H)$, $x, y, z, u \in G$ and $n \in \mathbb{Z}$.

(a) Prove that $f(xyz) + f(xy^{-1}z) = 2f(xz)$, thus extending (12.3).
Hint: Proposition 12.5(g).

(b) Derive the formula $f(xy^nz) = nf(xyz) - (n-1)f(xz)$.

Hint: Use (a) to get $f(xy^2z) = 2f(xyz) - f(xz)$ and then induction on n.

(c) Show that $Cf(xy^nz, u) = nCf(xyz, u) - (n-1)Cf(xz, u)$.

Exercise 12.7. Let $f : G \to H$ be a central solution of Jensen's functional equation (12.3). Show that $f = a + f(e)$, where $a : G \to H$ is odd and $2a$ is additive.

Hint: Proposition 12.5(a).

Exercise 12.8. Find the functions $f, g, h : \mathbb{R} \to \mathbb{C}$ satisfying the following Pexiderized version of Jensen's functional equation

$$f(x+y) + g(x-y) = h(x), \qquad x, y \in \mathbb{R}.$$

Hints: Consider on an arbitrary group G the functional equation

$$f(xy) + g(xy^{-1}) = h(x), \qquad x, y \in G. \tag{12.13}$$

Replacing y by y^{-1} in (12.13) we get another equation. Show by adding and subtracting the two equations that h is a solution of Jensen's functional equation and that $k := f - g$ satisfies $k(xy) = k(xy^{-1})$ for all $x, y \in G$. Show that k is constant if G is generated by its squares.

Exercise 12.9. It is Jensen's functional equation (12.3) with range space $H = \mathbb{C}$ that interests us most. However, some of our considerations work in the wider framework of Jensen's extended functional equation. In Proposition 12.5 we collected a number of useful formulas and relations for complex-valued, normalized solutions of Jensen's functional equation (12.3). We generalize them in the present exercise. We use the notation $[x, y]_\tau := xy\tau(x)\tau(y)$ for $x, y \in G$.

Lemma 12.30. *Let S be a semigroup with an involution $\tau : S \to S$ and let H be an abelian group. Let $f : S \to H$ be a solution of Jensen's extended functional equation* (12.7) *such that $f \circ \tau = -f$. Then for all $u, x, y, z \in S$:*

(a) *f satisfies* (2.3), *i.e., $f(xy) + f(yx) = 2f(x) + 2f(y)$.*

(b) *Cf is odd with respect to τ in each variable, i.e., $Cf(\tau(x), y) = Cf(x, \tau(y)) = -Cf(x, y)$.*

(c) *$y\tau(y) \in Z(2f)$ and $2f(xy\tau(y)) = 2f(y\tau(y)x) = 2f(x)$. Furthermore $2f(x\tau(x)) = 0$.*

(d) $4f(xy) = 4f(x) + 4f(y) + 2f([x, y]_\tau)$. *Equivalently* $4Cf(x, y) = 2f([x, y]_\tau)$.

(e) *The map* $2f([-, -]_\tau) : S \times S \to H$ *is bi-additive, and it is odd with respect to* τ *in each variable, i.e.,* $2f([\tau(x), y]_\tau) = 2f([x, \tau(y)]_\tau) = -2f([x, y]_\tau)$.

(f) $2f([x, [y, z]_\tau]_\tau) = 2f([[y, z]_\tau, x]_\tau) = 0$.

(g) $4f(u\,[x, [y, z]_\tau]_\tau) = 4f(u)$.

(h) $4[f(xyz) - f(xzy)] = 4f([y, z]_\tau)$.

(i) $[x, y]_\tau \in Z(4f)$, *and* $4f(x[y, z]_\tau) = 4f([y, z]_\tau x)$.

(j) $4f(x[y, z]_\tau) = 4f(x) + 4f([y, z]_\tau)$.

(k) $2f(xy\tau(x)) = 2f(y) + 2f([x, y]_\tau)$.

Exercise 12.10. Let $G = \mathrm{GL}(n, \mathbb{R})$, where $n \geq 2$. Show that the continuous functions $f \in S(G, \mathbb{C})$ are the functions $f(A) = c\log|\det A|$, $A \in \mathrm{GL}(n, \mathbb{R})$, where $c \in \mathbb{C}$ is a constant.

Hint: $[G, [G, G]] = [G, G] = \mathrm{SL}(n, \mathbb{R})$, because $[\mathrm{SL}(n, \mathbb{R}), \mathrm{SL}(n, \mathbb{R})] = \mathrm{SL}(n, \mathbb{R})$ (Example A.17(d)). Example A.22(d) may be also useful.

Exercise 12.11. The Heisenberg group H_3 from Examples A.17(a) can be written as the topological semidirect product $H_3 = G_0 \times_s H$, where

$$G_0 := \left\{ \begin{pmatrix} 1 & x & z \\ 0 & 1 & 0 \\ 0 & 0 & 1 \end{pmatrix} \;\middle|\; x, z \in \mathbb{R} \right\} \quad \text{and} \quad H := \left\{ \begin{pmatrix} 1 & 0 & 0 \\ 0 & 1 & y \\ 0 & 0 & 1 \end{pmatrix} \;\middle|\; y \in \mathbb{R} \right\} :$$

$$\begin{pmatrix} 1 & x & z \\ 0 & 1 & y \\ 0 & 0 & 1 \end{pmatrix} = \begin{pmatrix} 1 & x & z - xy \\ 0 & 1 & 0 \\ 0 & 0 & 1 \end{pmatrix} \begin{pmatrix} 1 & 0 & 0 \\ 0 & 1 & y \\ 0 & 0 & 1 \end{pmatrix}.$$

Construct by help of Proposition 12.16 the solution

$$f_0 \begin{pmatrix} 1 & x & z \\ 0 & 1 & y \\ 0 & 0 & 1 \end{pmatrix} = 2z - xy, \qquad x, y, z \in \mathbb{R},$$

from Example 12.4 of Jensen's equation on H_3.

Exercise 12.12. Find $S(H_3(\mathbb{Z}), \mathbb{C})$.

Exercise 12.13. Find the continuous solutions of Jensen's functional equation (12.3) on the $(2n+1)$-dimensional Heisenberg group from Example A.17(b).

Exercise 12.14. We shall in this exercise find the continuous solutions of Jensen's functional equation (12.3) modulo $\mathrm{Hom}^{\mathrm{cont}}(G,\mathbb{C})$ on the group

$$G := \left\{ \begin{pmatrix} 1 & x_1 & x_2 & z \\ 0 & 1 & u & y_1 \\ 0 & 0 & 1 & y_2 \\ 0 & 0 & 0 & 1 \end{pmatrix} \;\middle|\; x_1, x_2, y_1, y_2, z, u \in \mathbb{R} \right\}.$$

We use the abbreviation

$$\{x_1, u, y_2, x_2, y_1, z\} := \begin{pmatrix} 1 & x_1 & x_2 & z \\ 0 & 1 & u & y_1 \\ 0 & 0 & 1 & y_2 \\ 0 & 0 & 0 & 1 \end{pmatrix} \qquad \text{for } x_1, u, y_2, x_2, y_1, z \in \mathbb{R}.$$

(a) Show that

$$[G,G] = \{\{0,0,0,x_2,y_1,z\} \mid x_2, y_1, z \in \mathbb{R}\},$$

and that

$$[G,[G,G]] = \{\{0,0,0,0,0,z\} \mid z \in \mathbb{R}\}.$$

(b) Check that the map $(x_2, y_1) \mapsto \{0,0,0,x_2,y_1,0\}/[G,[G,G]]$ is a continuous isomorphism of $\mathbb{R}^2$ onto the group $[G,G]/[G,[G,G]]$. Show that it is even a topological isomorphism.

Hint: The open mapping theorem for groups, i.e., Theorem D.5(c).

(c) Conclude that $\dim S^{\mathrm{cont}}(G,\mathbb{C}) = 2$ modulo the continuous, additive functions on G.

(d) Show that G is the semidirect product of the normal subgroup $N := \{\{0,u,0,x_2,y_1,z\} \mid u, x_2, y_1, z \in \mathbb{R}\}$ and the abelian subgroup $H := \{\{x_1,0,y_2,0,0,0\} \mid x_1, y_2 \in \mathbb{R}\}$. Indeed,

$$\begin{pmatrix} 1 & x_1 & x_2 & z \\ 0 & 1 & u & y_1 \\ 0 & 0 & 1 & y_2 \\ 0 & 0 & 0 & 1 \end{pmatrix} = \begin{pmatrix} 1 & 0 & x_2 & z - x_2 y_2 \\ 0 & 1 & u & y_1 - u y_2 \\ 0 & 0 & 1 & 0 \\ 0 & 0 & 0 & 1 \end{pmatrix} \begin{pmatrix} 1 & x_1 & 0 & 0 \\ 0 & 1 & 0 & 0 \\ 0 & 0 & 1 & y_2 \\ 0 & 0 & 0 & 1 \end{pmatrix}.$$

(e) Show that f_1 in formula (12.9) for $\psi = 0$ and $f_0(0, u, 0, x_2, y_1, z) = x_2$ is

$$f_{11}(x_1, u, y_2, x_2, y_1, z) = x_2 - \tfrac{1}{2}x_1 u \quad \text{for } \{x_1, u, y_2, x_2, y_1, z\} \in G.$$

Show that f_1 in formula (12.9) for $\psi = 0$ and $f_0(0, u, 0, x_2, y_1, z) = y_1$ is

$$f_{12}(x_1, u, y_2, x_2, y_1, z) = y_1 - \tfrac{1}{2}y_2 u \quad \text{for } \{x_1, u, y_2, x_2, y_1, z\} \in G.$$

(f) Show that $f_{11}, f_{12} \in S^{\text{cont}}(G, \mathbb{C})$ (use for example Proposition 2.17 to see that it suffices to verify (2.3)).

(g) Conclude that f_{11}, f_{12} span $S^{\text{cont}}(G, \mathbb{C})$ modulo $\text{Hom}^{\text{cont}}(G, \mathbb{C})$.

Exercise 12.15. The symmetric group S_n, $n \geq 1$, is generated by its elements of order 2. So are the Euclidean reflection groups and more generally Coxeter groups. In this exercise we derive the following result about such groups:

Proposition 12.31. *Let G be a group which is generated by its elements of order 2, and let H be an abelian group.*

Any $f \in S(G, H)$ is abelian, $2f = 0$ and f is a function on $G/\langle G^2 \rangle$, where $\langle G^2 \rangle$ denotes the subgroup of G generated by the squares.

Proof. We leave the proof to the reader, who can proceed as follows.

(a) Show that $f(xyx) = 2f(x) + f(y) = f(x^2) + f(y)$ for all $x, y \in G$.

(b) Noting that any $x \in G$ can be written in the form $x = x_1 x_2 \cdots x_n$, where $x_i^2 = e$ or equivalently $x_i^{-1} = x_i$, prove that

$$\begin{aligned} f(xyx^{-1}) &= f(x_1 x_2 \cdots x_n y x_n^{-1} \cdots x_2^{-1} x_1^{-1}) \\ &= f(x_1 x_2 \cdots x_n y x_n \cdots x_2 x_1) \\ &= f(x_1^2) + f(x_2^2) + \cdots + f(x_n^2) + f(y). \end{aligned}$$

(c) Show that $f(xyx^{-1}) = f(y)$ for all $x, y \in G$, so that f is central.

(d) Prove that $2f = 0$.

Hint: Show $2f$ is additive (Proposition 12.10) and use that to get $2f(x_1 x_2 \cdots x_n) = 0$, when $x_1^2 = x_2^2 = \cdots = x_n^2 = e$.

(e) Show that f is a function on $G/\langle G^2\rangle$.
Hint: Combine $f(yx^2) = f(xyx)$ and (a).

(f) Show that f satisfies Kannappan's condition.
Hint: Lemma B.5(d). □

As an application of the proposition take $G = S_n$. Show that any $f \in S(S_n, H)$ is additive and that $f = 0$ on the alternating group A_n.
Hint: Proposition 12.8 and the fact that A_n is contained in group generated by the squares, because for example $(ab)(cd) = [(ac)(cb)(bd)]^2$.

Exercise 12.16. In this exercise we derive Proposition 12.8, not for solutions of Jensen's functional equation, but of the version from Section 12.8. It turns out that we do not need the assumption about f being central. So let $f : G \to H$ be a normalized solution of (12.12), i.e., $f(e) = 0$ and

$$f(xy) + f(y^{-1}x) = 2f(x), \qquad x, y \in G.$$

(a) Show that $f(ab) + f(ba) = 2f(a) + 2f(b)$ for all $a, b \in G$.

(b) Show that $f(y^2z) = f(z) + f(y^2)$ for all $y, z \in G$.

(c) Show more generally that $f(xy^2z) = f(xz) + f(y^2)$ for all $x, y, z \in G$.
Hint: (a) and (b).

(d) Show that $f(xy_1^2y_2^2 \cdots y_n^2z) = f(xz) + f(y_1^2) + f(y_2^2) + \cdots + f(y_n^2)$ for all $x, y_1, y_2, \ldots, y_n, z \in G$.

(e) Show that f is additive on $\langle G^2\rangle$ and that $f(xqx^{-1}) = f(q)$ for all $x \in G$ and $q \in \langle G^2\rangle$.

(f) Let $x_0 \in G$. Show that f is additive on the group generated by x_0 and $\langle G^2\rangle$.

(g) Apply this to $G = S_n$. (Compare with Exercise 12.15.)

12.10 Notes and remarks

The chapter is based mainly on Ng's groundbreaking papers [148], [149] and on Stetkær [191], where more formulas, results and examples can be found. Ng [151] contains supplementary results about Jensen's functional equation.

Proposition 2.17 and Proposition 12.5(a) combined with Section 2.6 relate Jensen's functional equation to semi-homomorphisms.

Aczél and Dhombres [6, Historical notes] has a detailed account of the origin of affine maps.

The paper Ger and Kominek [101] discusses relations between Jensen's and Pexider's functional equations on commutative semigroups.

Proposition 12.5(a) is part of Proposition 2.17 and is the special case of $g = 1$ in Lemma 11.3(d).

Proposition 12.5: (e) says that $2Cf$ is bi-additive. It is not generally true for that Cf is bi-additive. (d) says that $2C^2f = 0$. It is not generally true that $C^2f = 0$. Ng [148, Example 3] has an example of both statements.

Proposition 12.7 is adopted from Ng [148, Theorem 5].

Proposition 12.8 is a partial generalization with a different proof of a result due to Parnami and Vasudeva [155, Theorem 6]. We present only sufficient conditions, while [155] has necessary conditions as well.

Lemma 12.11 is inspired by Sinopoulos [180].

Corollary 12.12 contains Aczél, Chung and Ng [5, Lemma 1] and Sinopoulos [180, Theorem 2] as special instances.

Lemma 12.28 and the preceding remarks were pointed out to me by R. Ger.

Section 12.8: We refer the reader who wishes to assume merely that H is an abelian group, to Ng's paper [150]. The special case of $G = S_n$ and H any abelian group was treated in [141, Theorem 3.1] (see our Exercise 12.16 for this special case).

Exercise 12.3 is taken from Ng [148, Example 1].

Exercise 12.10 can be generalized for $n \geq 3$. For any ring $\mathcal{R}$ with unit the subgroup $E_n(\mathcal{R})$ of $\mathrm{GL}(n, \mathcal{R})$ generated by the elementary matrices satisfies $[E_n(\mathcal{R}), E_n(\mathcal{R})] = E_n(\mathcal{R})$ (see Bass [24]). If $\mathcal{R}$ is a field, or a commutative Euclidean ring, then $E_n(\mathcal{R}) = \mathrm{SL}(n, \mathcal{R})$. We find that $S(\mathrm{GL}(n, \mathcal{R}), \mathbb{C}) = \mathrm{Hom}(\mathrm{GL}(n, \mathcal{R}), \mathbb{C})$, and that any $f \in S(\mathrm{GL}(n, \mathcal{R}), \mathbb{C})$ has the form $f(A) = \phi(\det A)$ for some additive map ϕ of the group of units of $\mathcal{R}$ into $\mathbb{C}$.

Proposition 12.31 extends Corovei [53, Lemma 2.3]. The application to S_n was derived directly in another way than ours by Công-Trình Lê and Trung-Hiêu Thái [141, Theorem 2.1].

Exercise 12.16: The application to S_n was derived directly in another way than ours by Công-Trình Lê and Trung-Hiêu Thái [141, Theorem 3.1].

Chapter 13

The Quadratic Functional Equation

This chapter gives an account of the quadratic functional equation on groups, both abelian and non-abelian, and of the properties of its solutions. A number of examples are given. The most important results of the chapter are Theorem 13.13, Lemma 13.11 and Example 13.20.

13.1 Introduction

The quadratic functional equation has root in the golden age of Greek mathematics. Although Apollonius from Perga (ca. 262–190 B.C.E.) is mainly known for his ground breaking work on conic sections, there is a result in elementary Euclidean geometry that bears his name: *Apollonius' theorem* states that the sum of the squares of the lengths of the two diagonals of a parallelogram is the sum of the squares of the lengths of the four sides. In other words it asserts that the *parallelogram identity*

$$\|x+y\|^2 + \|x-y\|^2 = 2\|x\|^2 + 2\|y\|^2 \tag{13.1}$$

holds for all $x, y \in \mathbb{R}^2$. The Pythagorean theorem is the case where the parallelogram is a rectangle.

A triangle is half a parallelogram, but in contrast to triangles parallelograms are connected to addition of vectors (just think of the parallelogram of forces). According to Bourbaki, Euclid's focusing on triangles held up the introduction of vectors by several centuries: "On voyait le triangle, mais non le parallélogramme qui aurait pu conduire aux vecteurs" [39, Introduction] [They saw the triangle, but not the parallelogram that could have led to vectors]. This observation was the motivation for Jean Dieudonné's battle cry from 1961: "A bas Euclide, plus de triangles!" [Down with Euclid, no more triangles!].

Much later (1935) than Apollonius and in another context the parallelogram identity came once more into the spotlight. Jordan and von Neumann [122] proved that the norm of a normed real or complex vector space $(V, \|-\|)$ arises from an inner product $(-,-)$ (meaning $\|x\| = \sqrt{(x,x)}$ for all $x \in V$) if and only if (13.1) holds for all $x, y \in V$. Thus the parallelogram identity characterizes those norms that originate from inner products, and so it is not just a nice identity in the Euclidean plane, but plays a more profound role as well. It might be noticed in passing that the Jordan-von Neumann result says that: A normed vector space over $\mathbb{R}$ (or $\mathbb{C}$) is an inner product space if and only if every 2-dimensional subspace is isometrically isomorphic to $\mathbb{R}^2$ (or $\mathbb{C}^2$). Earlier the same year another characterization of inner product spaces was given by Fréchet [88]. His law, which is discussed in Exercise 13.11, deals with 3-dimensional subspaces. Many natural geometric properties fail to hold in a general normed space in contrast to inner product spaces. To mention one property: The closed unit ball may not be strictly convex. Therefore it is of interest to determine when a normed linear space is isometric to an inner product space. Entire books, for example [14] and [15], have been dedicated this subject!

An indispensable property of a Hilbert space $\mathcal{H}$ is the existence of orthogonal projections of $\mathcal{H}$ onto any closed subspace. The orthogonal projection of a point $x \in \mathcal{H}$ on a convex closed subset C of $\mathcal{H}$ is the point in C which is nearest x. The standard proof of the existence and uniqueness of this nearest point combines the parallelogram identity with the completeness of $\mathcal{H}$. So this is another place where the parallelogram identity comes in handy.

That (13.1) holds means that the function $f(x) := \|x\|^2$, $x \in V$, satisfies the quadratic functional equation $f(x+y) + f(x-y) = 2f(x) + 2f(y)$. We shall in the sequel study this functional equation with the underlying vector space V replaced by a group (sometimes even just a semigroup), which need not be the additive group of a vector space. Not only do we discard the possibility of multiplying by scalars, we also allow the group to be non-abelian.

13.2 The set up and definitions

Definition 13.1. Let S be a semigroup with an involution $\tau : S \to S$, H an abelian group and $f : S \to H$. We say that f satisfies *the quadratic functional equation* or is a *solution of the quadratic functional equation* if

$$f(xy) + f(x\tau(y)) = 2f(x) + 2f(y), \qquad x, y \in S. \tag{13.2}$$

(13.2) is called *the quadratic functional equation.*

If $S = M$ is a monoid, we write $f \in Q_\tau(M, H)$, if $f : M \to H$ satisfies not just (13.2) but also the normalization condition $f(e) = 0$, i.e., $f \in Q_\tau(M, H)$ if and only if

$$f(xy) + f(x\tau(y)) = 2f(x) + 2f(y) \quad \text{for all } x, y \in M, \text{ and } f(e) = 0. \tag{13.3}$$

Let $M = G$ be a group. $f : G \to H$ is said to be a solution of the *classical quadratic functional equation* if

$$f(xy) + f(xy^{-1}) = 2f(x) + 2f(y) \qquad \text{for all } x, y \in G. \tag{13.4}$$

We use the abbreviation $Q(G, H)$ for the set of solutions f of (13.4) that satisfy the normalization condition $f(e) = 0$.

Note that $Q_\tau(M, H)$ is an abelian group in general and a complex vector space if $H = \mathbb{C}$.

The normalization condition $f(e) = 0$ is no essential restriction: Taking $x = y = e$ in (13.2) we see that $2f(e) = 0$. This implies that $f - f(e)$ satisfies (13.2) and since $f - f(e)$ clearly vanishes at e, $f - f(e)$ is a solution in $Q_\tau(M, H)$. Thus we may work with $f - f(e)$ instead of f. If H is 2-torsion free, for example if $H = \mathbb{C}$, then $2f(e) = 0$ implies $f(e) = 0$. For such H any solution f of the quadratic functional equation (13.2) is in $Q_\tau(M, H)$, because the normalization condition is automatically satisfied.

We recall that the Cauchy difference Cf of a function $f : S \to H$ from a semigroup S to an abelian group H is defined by $(Cf)(x, y) := f(xy) - f(x) - f(y)$, $x, y \in S$.

13.3 The case of $\mathbb{R}^n$

Let us first find the continuous solutions of the classical quadratic functional equation (13.4) on $G = \mathbb{R}$ with range space $H = \mathbb{C}$ in order to get a first glimpse of what phenomena and what type of solutions of (13.2) we can expect. The result is

Proposition 13.2. *The functions $f \in C(\mathbb{R})$ satisfying the classical quadratic functional equation*

$$f(x+y)+f(x-y)=2f(x)+2f(y), \qquad x,y\in\mathbb{R}, \tag{13.5}$$

are the ones of the form $f(x)=ax^2$, $x\in\mathbb{R}$, where a ranges over $\mathbb{C}$.

Proof. Any function of the form $f(x)=ax^2$, $x\in\mathbb{R}$, is a continuous solution of (13.5) as shown by a simple computation, so it is left to show that any continuous solution f has this form.

By the results in Appendix D.3 (Proposition D.7) on regularity we know that if $f\in C(\mathbb{R})$ is a solution then $f\in C^\infty(\mathbb{R})$. Differentiating of (13.5) twice with respect to y and evaluating the result at $y=0$ we get that $f''(x)=f''(0)$ for all $x\in\mathbb{R}$. But then $f(x)=\frac{1}{2}f''(0)x^2+bx+c$, where b and c are constants. Since the first term on the right is a solution of (13.5) (by the first part of the proof) so is $\phi(x):=bx+c$. Substituting ϕ into (13.5) we find that $b=c=0$. □

More generally, replacing $\mathbb{R}$ by $\mathbb{R}^n$,

Theorem 13.3. *Let $n\in\mathbb{N}$. The functions $f\in C(\mathbb{R}^n)$ satisfying the classical quadratic functional equation*

$$f(x+y)+f(x-y)=2f(x)+2f(y), \qquad x,y\in\mathbb{R}^n, \tag{13.6}$$

are the ones of the form $f(x)=\langle Ax,x\rangle$, $x\in\mathbb{R}^n$, where A ranges over the symmetric matrices in $M(n\times n,\mathbb{C})$.

Proof. We only show that every solution has the stated form, because the converse is easy to verify. Let $f\in C(\mathbb{R}^n)$ be a solution of (13.6). It suffices to prove that the Cauchy difference Cf is symmetric and bi-additive. Indeed, since $2f(x)=Cf(x,x)$ (follows from $f(2x)=4f(x)$) we are then done by the elementary Lemma 2.14.

Cf is clearly symmetric, so to show it is bi-additive it suffices to verify that it is additive in the first variable for fixed second variable $z\in\mathbb{R}^n$. Since $Cf(0,z)=0$ it suffices by Corollary 12.12(c) to show that $Cf(-,z)$ satisfies Jensen's functional equation in its first variable. And that it does, because

$$\begin{aligned} &Cf(x+y,z)+Cf(x-y,z)-2Cf(x,z)\\ &\quad = f(x+y+z)-f(x+y)-f(z)+f(x-y+z)-f(x-y)-f(z)\\ &\qquad -2f(x+z)+2f(x)+2f(z) \end{aligned}$$

vanishes: Apply (13.6) twice to it. □

Theorem 13.3 tells us that the continuous solutions of (13.6) are the symmetric quadratic forms, so the quadratic functional equation gives another way of characterizing symmetric quadratic forms. The quadratic forms need not be positive definite as suggested by the example of the parallelogram identity. The Minkowski norm $x_1^2 + x_2^2 + x_3^2 - x_0^2$ on $\mathbb{R}^4$ works equally well.

As we shall see, with the necessary simple reformulations Theorem 13.3 is still true when $\mathbb{R}^n$ is replaced by an abelian group (Theorem 13.13). However, it is not true for all groups (Example 13.20 provides a counter-example).

The proof of Theorem 13.3 hints that the Cauchy difference will play an important role in the sequel, as is indeed the case.

Also the solutions $(f_1, f_2, f_3, f_4) \in C(\mathbb{R}^n, \mathbb{C}^4)$ of the Pexiderized version

$$f_1(x+y) + f_2(x-y) = f_3(x) + f_4(y),$$

of (13.6) are polynomials of degree 2, but they may not be homogeneous. See [167, Theorem 9.6] for a proof.

13.4 General considerations and the abelian case

If τ is the identity map then (13.2) reduces (modulo a factor 2) to the additive Cauchy equation $f(xy) = f(x) + f(y)$. If τ is the group inversion then (13.2) becomes the classical quadratic functional equation (13.4) the solutions of which (at least for $\mathbb{R}^n$) contain quadratic terms. So in the general abelian case we should expect the solutions to be a mixture of additive and quadratic terms. This is indeed the case (Theorem 13.6). Let us first note the following solutions:

Lemma 13.4. *Let S be a semigroup with an involution $\tau : S \to S$ and H an abelian group. If $Q : S \times S \to H$ is a symmetric, bi-additive map such that $Q(\tau(x), y) = -Q(x, y)$ for all $x, y \in S$, then $f(x) := Q(x, x)$, $x \in S$, is a solution of the quadratic functional equation.*

A question is when the converse holds, at least for the classical quadratic functional equation. To answer that question and similar ones we derive some properties of solutions of the quadratic functional equation in Lemma 13.5. We work on a possibly non-abelian semigroup, not a monoid or a group, and we do not assume that division by 2 is possible or unique in the range space. We see in (d) that there is a close connection between the quadratic functional equation and Jensen's functional equation from Chapter 12.1. Indeed, the

Cauchy difference $Cf(-,-)$ (more precisely $2C$) of a solution f of the quadratic functional equation is in each of its two variables an odd solution of Jensen's functional equation. Comparing with Theorem 13.3 we see that $2Cf(x,x)$ replaces the quadratic expression $\langle Ax, x\rangle$. Note that the additive term A below vanishes when f is a normalized solution on a group with τ being the group inversion.

Lemma 13.5. *Let S be a semigroup with an involution $\tau : S \to S$ and H an abelian group. Let $f : S \to H$ satisfy the quadratic functional equation* (13.2).

(a) $2f \circ \tau = 2f$, *$2f$ is central and $2Cf : S \times S \to H$ is symmetric.*

(b) *Let $A : S \to H$ be $A(x) := f(x\tau(x))$, $x \in S$. Then $2A \circ \tau = 2A$ and $2A : S \to H$ is additive.*

(c) $2f(x) = Cf(x,x) + A(x)$ *for all $x \in S$.*

(d) *$2Cf(-, y_0)$ is for each fixed $y_0 \in S$ a τ-odd solution of Jensen's extended functional equation* (12.7), *that is a solution of*

$$\phi(st) + \phi(s\tau(t)) = 2\phi(s), \qquad s, t \in S,$$

such that $2Cf(\tau(x), y_0) = -2Cf(x, y_0)$ for all $x \in S$.

Similarly, $2Cf(x_0, -)$ is for each fixed $x_0 \in S$ a solution of Jensen's functional equation (12.7), *such that $2Cf(x_0, \tau(y)) = -2Cf(x_0, y)$ for all $y \in S$.*

Proof. (a) The left hand side of (13.2) does not change if y is replaced by $\tau(y)$. From the right hand side we get $2f(\tau(y)) = 2f(y)$, that is $2f \circ \tau = 2f$. Theorem 7.1(a) implies that $2f$ is central which in turn implies $2Cf$ is symmetric.

(b) The τ-invariance of $2A$ is immediate from the very definition of A and that $2f$ is central. Replacing x by $\tau(x)x$ and y by $y\tau(y)$ in (13.2) and using that $2f$ is central we get the additivity of $2A$.

(c) is the case of $y = x$ of the computation

$$\begin{aligned} Cf(x,y) + f(x\tau(y)) &= f(xy) + f(x\tau(y)) - f(x) - f(y) \\ &= 2f(x) + 2f(y) - f(x) - f(y) = f(x) + f(y). \end{aligned}$$

(d) Fix $y_0 \in S$. Using that $2f$ is central we find for all $x, y \in S$ that

$$
\begin{aligned}
&2Cf(xy, y_0) + 2Cf(x\tau(y), y_0)\\
&\quad = 2[f(xyy_0) - f(xy) - f(y_0) + f(x\tau(y)y_0) - f(x\tau(y)) - f(y_0)]\\
&\quad = 2[f(y_0xy) + f(y_0x\tau(y)) - [f(xy) + f(x\tau(y))] - 2f(y_0)]\\
&\quad = 2[2f(y_0x) + 2f(y) - 2f(x) - 2f(y) - 2f(y_0)]\\
&\quad = 4[f(xy_0) - f(x) - f(y_0)] = 4Cf(x, y_0),
\end{aligned}
$$

which shows that $2Cf(-, y_0)$ is a solution of Jensen's equation.

Using that $2f = 2f \circ \tau$ and that $2f$ is central we get that

$$
\begin{aligned}
&2Cf(\tau(x), y_0)\\
&\quad = 2[f(\tau(x)y_0) - f(\tau(x)) - f(y_0)] = 2f(y_0\tau(x)) - 2f(x) - 2f(y_0)\\
&\quad = 2[2f(x) + 2f(y_0) - f(y_0x)] - 2f(x) - 2f(y_0) = -2Cf(x, y_0),
\end{aligned}
$$

which says that $2C$ is odd in its first variable. The corresponding statements about fixed first variable $x_0 \in S$ follow from what we have just proved, because $2Cf$ is symmetric. □

A point of the next theorem is the absence of a neutral element, contrasting the classical result in Theorem 13.13.

Theorem 13.6. *Let $(S, +)$ be an abelian semigroup with an involution $\tau : S \to S$. Let H be an abelian group such that the map $\eta \mapsto 2\eta$ is a bijection of H onto H.*

The general solution $f : S \to H$ of the functional equation

$$f(x + y) + f(x + \tau(y)) = 2f(x) + 2f(y), \qquad x, y \in S,$$

is

$$f(x) = Q(x, x) + a(x), \qquad x \in S,$$

where $Q : S \times S \to H$ is an arbitrary symmetric, bi-additive map such that $Q(\tau(x), y) = -Q(x, y)$ for all $x, y \in S$, and where $a : S \to H$ is an arbitrary additive map such that $a \circ \tau = a$.

Proof. It is easy to check that the indicated functions are solutions. That all solutions have this form is a consequence of Lemma 13.5 in conjunction with the fact that odd solutions of Jensen's equation are additive on abelian semigroups (Corollary 12.12(a)). □

Things simplify in special cases of Theorem 13.6: Q vanishes if τ is the identity map, and a vanishes if S is a group and τ its inversion. The last leads us to a classical result (Theorem 13.13) due to Aczél.

13.5 The Cauchy differences of solutions

In this section we study the first and the second Cauchy differences of solutions of the quadratic functional equation. We use the general set up of subsection 13.2, except that we require that the underlying semigroup S is a monoid $S = M$. That simplifies the treatment a little bit, for example because we avoid some annoying factors of 2. We do not assume that M is abelian.

Lemma 13.7 is Lemma 13.5 exposed for monoids. We recall that solutions $f \in Q_\tau(M, H)$ of the quadratic equation on a monoid M are normalized so that $f(e) = 0$.

Lemma 13.7. *Let M be a monoid with an involution $\tau : M \to M$ and let H be an abelian group. Let $f \in Q_\tau(M, H)$.*

(a) *$f \circ \tau = f$, f is central and $Cf : M \times M \to H$ is symmetric.*

(b) *Let $A : M \to H$ be $A(x) := f(x\tau(x))$, $x \in M$. Then $A \circ \tau = A$ and $2A : M \to H$ is additive.*

(c) *$2f(x) = Cf(x, x) + A(x)$ for all $x \in M$.*

(d) *$Cf(x, -)$ is for each fixed $x \in M$ a solution of Jensen's extended functional equation (12.7) such that $Cf(x, e) = 0$ and $Cf(x, \tau(y)) = -Cf(x, y)$ for all $y \in M$. Furthermore $Cf(x, -)$ satisfies the symmetrized additive Cauchy equation (2.3), i.e.,*

$$Cf(x, yz) + Cf(x, zy) = 2Cf(x, y) + 2Cf(x, z), \quad \text{for all } x, y, z \in M. \tag{13.7}$$

Proof. As the proof of Lemma 13.5 with the necessary changes. The derivation of the identity (13.7) is the same as the proof of Proposition 12.5(a). We leave the details to the reader. □

The second Cauchy difference $C^2 f$ from Definition 2.18 is an indispensable tool for the study of the solutions f of the quadratic functional equation. Lemma 13.8 lists some properties of $C^2 f$ that we will need. The relation of abelian solutions f to $C^2 f$ is pointed out in (a), which thus motivates a study of $C^2 f$. We recall the notation $[x, y]_\tau := xy\tau(x)\tau(y)$ that occurred in the study of Jensen's functional equation (Exercise 12.9). For the classical quadratic functional equation more relations than those of Lemma 13.8 can be found in [212, Theorem 2.1].

Lemma 13.8. *Let $f \in Q_\tau(M,H)$ where M is a monoid with an involution $\tau : M \to M$ and H an abelian group. For all $x, y, z, u \in M$ we have:*

(a) $f(xyz) - f(xzy) = 2C^2 f(x,y,z)$. *In particular f is abelian if and only if $2C^2 f = 0$.*

(b) $C^2 f : M \times M \times M \to H$ *is alternating, that is*

$$C^2 f(x_{\sigma(1)}, x_{\sigma(2)}, x_{\sigma(3)}) = \operatorname{sgn}(\sigma) C^2 f(x_1, x_2, x_3)$$

for any permutation σ of 3 elements and all $x_1, x_2, x_3 \in M$.

(c) $2C^2 f : M \times M \times M \to H$ *is multi-additive, i.e., additive in any of its three variables when the two remaining ones are fixed.*

(d) $C^2 f(\tau(x), y, z) = C^2 f(x, \tau(y), z) = C^2 f(x, y, \tau(z)) = -C^2 f(x,y,z)$.

(e) $2C^2 f(x, y, [z,u]_\tau) = 0$.

(f) $2C^2 f(x,y,z) = 0$ *if two of the elements $x, y, z \in M$ commute.*

Proof. (a) Using f is central (Lemma 13.7(a)) and (13.7) we find that $f(xyz) - f(xzy) = Cf(x,yz) - Cf(x,zy) = Cf(x,yz) - [2Cf(x,y) + Cf(x,z) - Cf(x,yz)] = 2[Cf(x,yz) - Cf(x,y) - Cf(x,z)] = 2C^2 f(x,y,z)$.

(b) In Lemma 2.19(b) we observed that $C^2 f$ is invariant under cyclic permutations when f is central, which is the case here. So it suffices to show that $C^2 f(x,y,z) = -C^2 f(x,z,y)$ for all $x,y,z \in M$. Now,

$$C^2 f(x,y,z) = Cf(x,yz) - Cf(x,y) - Cf(x,z),$$
$$C^2 f(x,z,y) = Cf(x,zy) - Cf(x,y) - Cf(x,z).$$

Adding the right hand sides gives 0 by (13.7).

(c) Since $C^2 f$ is invariant under cyclic permutations it suffices to prove that $2C^2 f$ is additive in its first variable. By Lemma 13.7(d) $Cf(-, z_0)$ is a solution of the symmetrized additive Cauchy equation (2.3), and then Lemma 2.20(d) tells us that $2C^2(-,-,z_0)$ is bi-additive.

(d) We only need to derive the identity $C^2 f(\tau(x), y, z) = -C^2 f(x,y,z)$, because $C^2 f$ is invariant under cyclic permutations. We get from Lemma 13.7(d) that

$$\begin{aligned} C^2 f(\tau(x), y, z) &= Cf(\tau(x), yz) - Cf(\tau(x), y) - Cf(\tau(x), z) \\ &= -Cf(x,yz) + Cf(x,y) + Cf(x,z) = -2C^2 f(x,y,z) \end{aligned}$$

as desired.

(e) follows from (c) and (d) and the definition $[z,u]_\tau = zu\tau(z)\tau(u)$.

(f) Since $C^2 f$ is invariant under cyclic permutations we may assume that it is y and z that commute. By (a) we then get $2C^2 f(x,y,z) = f(xyz) - f(xzy) = f(xyz) - f(xyz) = 0$. □

The following proposition shows that solutions of the quadratic functional equation on certain semi-direct products like the $(ax+b)$-group and the Heisenberg group are abelian, so that Aczél's Theorem (Theorem 13.13) may be applied (see Exercise 13.5).

Proposition 13.9. *Let $f : M \to H$ be a solution of the quadratic functional equation* (13.2)*, where M is a monoid with an involution $\tau : M \to M$ and H an abelian group. If M is generated by two abelian subsets, then f is abelian.*

Proof. Possibly replacing f by $f - f(e)$ we may assume that $f \in Q_\tau(M,H)$.

Let the two abelian subsets be denoted M_1 and M_2. We may wlog add e to M_1, so that any element of M is a product with factors from $M_1 \cup M_2$. Any three elements $x, y, z \in M$ can be written as $x = x_{i_1} x_{i_2} \cdots x_{i_n}$, $y = y_{j_1} y_{j_2} \cdots y_{j_m}$, $z = z_{k_1} z_{k_2} \cdots z_{k_p}$, where each of the factors of x, y and z is in some M_i. By the multi-additivity it suffices to prove that $2C^2 f(x,y,z) = 0$ for $x, y, z \in M_1 \cup M_2$. There are only the two sets M_1 or M_2 to choose from, so at least two of the three elements x, y and z fall into the same set, say $x, y \in M_1$. Since M_1 is abelian so that x and y commute, Lemma 13.8(f) implies that $2C^2 f(x,y,z) = 0$, which in turn implies that f is abelian (Lemma 13.8(a)). □

Corollary 13.10. *Let H be an abelian group. Any solution $f : G \to H$ of the quadratic functional equation* (13.2) *on a group G with* 2 *generators is abelian.*

There is a non-abelian solution of the classical quadratic functional equation (13.4) on the free group with 3 generators (Example 13.20), so it is essential in Corollary 13.10 that G is generated by only two elements.

13.6 The classical quadratic functional equation

Throughout this section G is a group and H is an abelian group.

We shall concentrate on the classical quadratic functional equation (13.4), where the underlying space is a group G and τ is the group inversion. The main simplification in comparison with the general case is that the expression $f(xy\tau(y))$ reduces to $f(x)$. That has consequences for the formulas in Lemma 13.5 where several factors of 2 can be taken away.

Lemma 13.11. *Let $f \in Q(G, H)$.*

(a) $\check{f} = f$ *and* f *is central.*

(b) $f(x^n) = n^2 f(x)$ *for all* $x \in G$ *and* $n \in \mathbb{Z}$.

(c) *The Cauchy difference* $Cf : G \times G \to H$ *is symmetric, and* $(Cf)(-, y)$ *is for each fixed* $y \in G$ *a solution of Jensen's functional equation* (12.3) *such that* $(Cf)(e, y) = 0$.

(d) $2f(x) = Cf(x, x)$ *for all* $x \in G$.

(e) $f(x[y, z]) = f(x) + 2C^2 f(x, y, z)$ *for all* $x, y, z \in G$.

(f) f *is a function on* $G/[G, [G, G]]$.

(g) $f([G, G]) = \{0\}$.

Proof. (a) is part of Lemma 13.7(a).

(b) is for $n \geq 0$ a simple induction on n. For $n < 0$ we use the result for a positive n in conjunction with the fact from (a) that $\check{f} = f$.

(c) The first statement is Lemma 13.7(a) and the rest is Lemma 13.7(d).

(d) is Lemma 13.5(c), because $A(x) = f(x\tau(x)) = f(xx^{-1}) = f(e) = 0$.

(e) If we in the formula of Lemma 13.8(a) replace x by abc, y by b^{-1} and z by c^{-1} and use the multi-additivity of $2C^2 f$ (Lemma 13.8(c)) we get that

$$f(a[b, c]) - f(a) = 2C^2 f(a, b, c) + 2C^2 f(b, b, c) + 2C^2 f(c, b, c).$$

Each of the last two terms vanishes according to Lemma 13.8(f). The resulting formula is the one of (e).

(f) Using (e) and Lemma 13.8(e) we find for any $x, y, z, u \in G$ that

$$f(x[y, [z, u]]) = f(x) + 2C^2 f(x, y, [z, u]) = f(x) + 0 = f(x),$$

which suffices, because $[G, [G, G]]$ is generated by elements of the form $[y, [z, u]]$ (Lemma A.7).

(g) Using the multi-additivity of $2C^2 f$ (Lemma 13.8(c)) and that $2C^2 f$ vanishes on commutators (Lemma 13.8(e)) we get by induction on n starting with the identity (e) that

$$f(x[y_1, z_1] \cdots [y_n, z_n]) = f(x) + \sum_{i=1}^{n} 2C^2 f(x, y_i, z_i) \qquad \text{for } n \in \mathbb{N}.$$

When we here take $x = e$ we get (g), because $[G, G]$ consists of products of commutators. □

The condition in the next proposition is trivially satisfied for abelian groups, but it also holds for some non-abelian groups like $\mathrm{GL}(n, \mathbb{R})$, $n \geq 2$, and the $(ax + b)$-group.

Proposition 13.12. *If $[G, G] = [G, [G, G]]$, then any solution of the classical quadratic functional equation is abelian.*

Proof. Possibly replacing f by $f - f(e)$ we may assume that $f \in Q(G, H)$. Then f is a function on $G/[G, [G, G]]$ (Lemma 13.11(f)), and so by the assumption a function on the group $G/[G, G]$, which is abelian. □

As a corollary of Lemma 13.5 we display a classical result by Aczél.

Theorem 13.13 (Aczél 1965). *Let H be an abelian group such that the map $\eta \mapsto 2\eta$ is a bijection of H onto H. Assume that $f : G \to H$ is abelian.*

Then f is a solution of the classical quadratic functional equation (13.4) *if and only if f has the form*

$$f(x) = Q(x, x), \qquad x \in G,$$

where $Q : G \times G \to H$ is a symmetric, bi-additive map. Actually, $2Q = Cf$, where Cf denotes the Cauchy difference of f.

Proof. Apply Theorem 13.6 and Lemma 13.11. □

Finally we present a modification of Theorem 13.13 that applies to continuous solutions of the quadratic functional equation, in contrast to the purely algebraic solutions above:

Corollary 13.14. *Let G be a topological group. Let H be a topological vector space or a σ-compact, locally compact abelian group such that the map $\eta \mapsto 2\eta$ is a bijection of H onto H. Assume that $f : G \to H$ is abelian. Then $f : G \to H$ is a continuous solution of the classical quadratic functional equation if and only if f has the form*

$$f(x) = Q(x, x), \qquad x \in G,$$

where $Q : G \times G \to H$ is a continuous, symmetric, bi-additive map. Actually, $2Q = Cf$.

Proof. By the open mapping for locally compact groups (Theorem D.5(c)) division by 2 is continuous in H, and $2Q = Cf$. □

The formulas in Theorem 13.13 and Corollary 13.14 are not the final answer: Non-abelian solutions of the classical quadratic functional equation exist (see Example 13.20).

Proposition 13.15. *Asssume that the quotient group $G/[G,G]$ is finite and that H is torsion free. Then any solution $f : G \to H$ of the classical quadratic functional equation is* 0.

Proof. Let n be the order of $G/[G,G]$. For any $x \in G$ we then have $(x[G,G])^n = [G,G]$, so that $x^n \in [G,G]$. Noting that $f \in Q(G,H)$ we get $f(x^n) = 0$ according to Proposition 13.11(g). But $f(x^n) = n^2 f(x)$ by Proposition 13.11(b), so $f(x) = 0$. □

Corollary 13.16. *If $f : G \to H$ is a solution of the classical quadratic functional equation, where G is a semisimple Lie group with finitely many connected components, and where H is torsion free, then $f = 0$.*

Proof. The identity component G^e of G is a normal subgroup of G ([111, Theorem 7.1]). We see that G/G^e is a finite group, the order of which is the number of connected components of G. Now, $[G^e, G^e] = G^e$ (see for example [201, Corollary 3.18.10]), so

$$G/[G,G] = \{G/[G^e,G^e]\}/\{[G,G]/[G^e,G^e]\} = \{G/G^e\}/\{[G,G]/[G^e,G^e]\}$$

is a finite group. Since H is torsion free we get that $f \in Q(G,H)$, so we may refer to Proposition 13.15. □

13.7 Examples

We shall here collect some examples of solutions of the classical quadratic functional equation (13.4), including a non-abelian solution.

Example 13.17. Let G be a compact group and $H = \mathbb{C}$. If $f \in C(G)$ is a solution of the classical quadratic functional equation then $f(x^n) = n^2 f(x)$ for all $n \in \mathbb{N}$ and $x \in G$ by Lemma 13.11(b). The left hand side is bounded, so we see that the only solution is $f = 0$.

Example 13.18. Let G be a cyclic, infinite group and H an abelian group. Let $a \in G$ be a generator of G. From the formula $f(x^n) = n^2 f(x)$ for all $n \in \mathbb{N}$ and $x \in G$ (Lemma 13.11(b)) we infer that the normalized solutions of the classical quadratic functional equation are the functions of the form $f(a^n) = n^2 h$, $n \in \mathbb{Z}$, where $h \in H$ is a constant.

Example 13.19 (This is another treatment of Theorem 13.3.). We get from Corollary 13.14 that any solution $f \in C(\mathbb{R}^n)$ of the classical quadratic functional equation has the form $f(x) = Q(x,x)$, $x \in \mathbb{R}^n$, where $Q : \mathbb{R}^n \times \mathbb{R}^n \to \mathbb{C}$ is a continuous, symmetric, bi-additive map. Taking the continuity together with the bi-additivity we see that Q is bilinear (Lemma 2.4), so there exists an $n \times n$ matrix with complex entries such that $Q(x,y) = \langle Ax, y\rangle$ for all $x, y \in \mathbb{R}^n$. The symmetry of Q means that $A = A^T$ (T denoting transpose), in other words that A is symmetric.

The converse being easy to verify we conclude that the solutions $f \in C(\mathbb{R}^n)$ of the classical quadratic functional equation (13.4) on $\mathbb{R}^n$ are the functions of the form $f(x) = \langle Ax, x\rangle$, $x \in \mathbb{R}^n$, where A is a symmetric $n \times n$ matrix with complex entries.

Example 13.20. The present example exhibits a non-abelian solution of the classical quadratic functional equation (13.4) on the free group on 3 generators with range space $H = \mathbb{Z} \subseteq \mathbb{C}$. This is interesting, because it establishes that non-abelian solutions exist. Corollary 13.10 reveals that two generators do not suffice to produce non-abelian solutions. Here we see that three generators do.

Let $\mathcal{A} = \{e_1, e_2, e_3\}$, and let $G := \langle \mathcal{A} \rangle$. Define $B : \mathcal{A} \times \mathcal{A} \times \mathcal{A} \to \mathbb{Z}$ by:

$$B(e_{\pi(1)}, e_{\pi(2)}, e_{\pi(3)}) = \operatorname{sgn}(\pi) \text{ for any permutation } \pi \text{ of 3 objects,}$$
$$B(a,b,c) = 0 \text{ if two of the three elements } a,b,c \in \mathcal{A} \text{ coincide.}$$

Any $x \in G$ can be written in the form $x = a_1^{\varepsilon_1} \cdots a_n^{\varepsilon_n} \in G$, where $a_1, \ldots, a_n \in \mathcal{A}$ and $\varepsilon_1, \ldots, \varepsilon_n \in \{\pm 1\}$. The neutral element is the empty product. We define $f : G \to \mathbb{Z}$ as follows:

$$f(x) := \sum_{1 \le i < j < k \le n} \varepsilon_i \varepsilon_j \varepsilon_k B(a_i, a_j, a_k), \tag{13.8}$$

with the usual convention that the value of a sum over an empty index set is interpreted as 0. We claim that

1. f is well defined by (13.8), and

2. f is a non-abelian solution of the classical quadratic functional equation (13.4).

Proof of Claim 1. The problem with the definition of f is that the decomposition $x = a_1^{\varepsilon_1} \cdots a_n^{\varepsilon_n}$ is not unique, because by the very definition of a free group factors aa^{-1} and $a^{-1}a$, where $a \in \mathcal{A}$, may be inserted at any location in the product, so

$$a_1^{\varepsilon_1} \cdots a_n^{\varepsilon_n} = a_1^{\varepsilon_1} \cdots a_l^{\varepsilon_l}(aa^{-1})a_{l+1}^{\varepsilon_{l+1}} \cdots a_n^{\varepsilon_n} = a_1^{\varepsilon_1} \cdots a_l^{\varepsilon_l}(a^{-1}a)a_{l+1}^{\varepsilon_{l+1}} \cdots a_n^{\varepsilon_n}$$

for any $l = 0, 1, \ldots, n$. To have f well defined it suffices to show that prove that

$$f(a_1^{\varepsilon_1} \cdots a_l^{\varepsilon_l}(aa^{-1})a_{l+1}^{\varepsilon_{l+1}} \cdots a_n^{\varepsilon_n}) = f(a_1^{\varepsilon_1} \cdots a_n^{\varepsilon_n}),$$

and

$$f(a_1^{\varepsilon_1} \cdots a_l^{\varepsilon_l}(a^{-1}a)a_{l+1}^{\varepsilon_{l+1}} \cdots a_n^{\varepsilon_n}) = f(a_1^{\varepsilon_1} \cdots a_n^{\varepsilon_n})$$

for all $a, a_1, \ldots, a_n \in \mathcal{A}$, $\varepsilon_1, \ldots, \varepsilon_n \in \{\pm 1\}$ and $l = 0, 1, \ldots, n$. We only write down the proof of the first identity, because the second one can be proved in a similar way. We compute $f(a_1^{\varepsilon_1} \cdots a_l^{\varepsilon_l}(aa^{-1})a_{l+1}^{\varepsilon_{l+1}} \cdots a_n^{\varepsilon_n})$ according to the putative definition (13.8). The terms involving neither a nor a^{-1} are

$$\sum_{1 \le i < j < k \le n} \varepsilon_i \varepsilon_j \varepsilon_k B(a_i, a_j, a_k) = f(a_1^{\varepsilon_1} \cdots a_n^{\varepsilon_n}),$$

from which we see that it suffices to show that the remaining terms cancel one another. The terms involving a but not a^{-1} are

$$\sum_{1 \le i < j \le l} \varepsilon_i \varepsilon_j B(a_i, a_j, a) + \sum_{i=1}^{l} \sum_{j=l+1}^{n} \varepsilon_i \cdot 1 \cdot \varepsilon_j B(a_i, a, a_j)$$
$$+ \sum_{l+1 \le i < j \le n} \varepsilon_i \varepsilon_j B(a, a_i, a_j).$$

The terms involving a^{-1} but not a are the same as the ones just written down involving a, except that each of them has a minus, coming from the definition of f, so the terms involving a but not a^{-1} and the ones involving a^{-1} but not a cancel one another. Finally each of the terms involving both a and a^{-1}, i.e., of

$$\sum_{i=1}^{l} \varepsilon_i \cdot 1 \cdot (-1) B(a_i, a, a) + \sum_{j=l+1}^{n} 1 \cdot (-1) \cdot \varepsilon_j B(a, a, a_j),$$

vanishes by the definition of B. □

Proof of Claim 2. It turns out to be convenient in the following computations to know that f is even, so let us prove that first of all, using the notation above:

$$\begin{aligned} f(x^{-1}) &= f((a_1^{\varepsilon_1}\cdots a_n^{\varepsilon_n})^{-1}) = f(a_n^{-\varepsilon_n}\cdots a_1^{-\varepsilon_1}) = f(a_{n+1-1}^{-\varepsilon_{n+1-1}}\cdots a_{n+1-n}^{-\varepsilon_{n+1-n}}) \\ &= \sum_{1\le i<j<k\le n} (-\varepsilon_{n+1-i})(-\varepsilon_{n+1-j})(-\varepsilon_{n+1-k})B(a_{n+1-i},a_{n+1-j},a_{n+1-k}) \\ &= -\sum_{1\le i<j<k\le n} \varepsilon_{n+1-i}\,\varepsilon_{n+1-j}\,\varepsilon_{n+1-k}\,B(a_{n+1-i},a_{n+1-j},a_{n+1-k}). \end{aligned}$$

Introducing new summation variables $I := n+1-k$, $J := n+1-j$ and $K := n+1-i$ (and immediately afterwards renaming them back, I to i etc.) we get from the permutation property $B(a_k,a_j,a_i) = -B(a_i,a_j,a_k)$ that

$$\begin{aligned} f(x^{-1}) &= -\sum_{n\ge K>J>I\ge 1} \varepsilon_K\varepsilon_J\varepsilon_I B(a_K,a_J,a_I) \\ &= -\sum_{1\le i<j<k\le n} \varepsilon_i\varepsilon_j\varepsilon_k B(a_k,a_j,a_i) \\ &= -\sum_{1\le i<j<k\le n} \varepsilon_i\varepsilon_j\varepsilon_k(-B(a_i,a_j,a_k)) = f(x). \end{aligned}$$

The rest of the proof is advanced book keeping: If $y = b_1^{\eta_1}\cdots b_p^{\eta_p}$, where $b_1,\ldots,b_p \in \mathcal{A}$ and $\eta_1,\ldots,\eta_p \in \{\pm 1\}$, then

$$\begin{aligned} f(xy) &= f(a_1^{\varepsilon_1}\cdots a_n^{\varepsilon_n}b_1^{\eta_1}\cdots b_p^{\eta_p}) \\ &= \sum_{1\le i<j<k\le n} \varepsilon_i\varepsilon_j\varepsilon_k B(a_i,a_j,a_k) + \sum_{1\le r<s<t\le p} \eta_r\eta_s\eta_t B(b_r,b_s,b_t) \\ &\quad + \sum_{1\le i<j\le n}\sum_{r=1}^{p} \varepsilon_i\varepsilon_j\eta_r B(a_i,a_j,b_r) + \sum_{i=1}^{n}\sum_{1\le r<s\le p} \varepsilon_i\eta_r\eta_s B(a_i,b_r,b_s) \\ &= f(x) + f(y) + E(x,y), \end{aligned}$$

where we have introduced the abbreviation

$$E(x,y) := \sum_{1\le i<j\le n}\sum_{r=1}^{p} \varepsilon_i\varepsilon_j\eta_r B(a_i,a_j,b_r) + \sum_{i=1}^{n}\sum_{1\le r<s\le p} \varepsilon_i\eta_r\eta_s B(a_i,b_r,b_s).$$

To compute $f(xy^{-1})$ we note that

$$xy^{-1} = a_1^{\varepsilon_1}\cdots a_n^{\varepsilon_n}b_p^{-\eta_p}\cdots b_1^{-\eta_1},$$

so in comparison with xy we shall replace $b_r^{\eta_r}$ by $b_{p+1-r}^{-\eta_{p+1-r}}$. Thus

$$\begin{aligned}E(x,y^{-1}) = &\sum_{1\le i<j\le n}\sum_{r=1}^{p}\varepsilon_i\varepsilon_j(-\eta_{p+1-r})B(a_i,a_j,b_{p+1-r})\\ &+\sum_{i=1}^{n}\sum_{1\le r<s\le p}\varepsilon_i(-\eta_{p+1-r})(-\eta_{p+1-s})B(a_i,b_{p+1-r},b_{n+1-s})\\ =-&\sum_{1\le i<j\le n}\sum_{l=1}^{p}\varepsilon_i\varepsilon_j\eta_l B(a_i,a_j,b_l)\\ &+\sum_{i=1}^{n}\sum_{1\le l<m\le p}\varepsilon_i\eta_l\eta_m B(a_i,b_m,b_l).\end{aligned}$$

Invoking the permutation property of B we see that $E(x,y^{-1}) = -E(x,y)$, and since f is even, we show as follows that f satisfies (13.4):

$$\begin{aligned}f(xy)+f(xy^{-1}) &= f(x)+f(y)+E(x,y)+f(x)+f(y^{-1})+E(x,y^{-1})\\ &= 2f(x)+2f(y).\end{aligned}$$

It is left to prove that f is non-abelian. But that is immediate from $f(e_1e_2e_3) - f(e_1e_3e_2) = B(e_1,e_2,e_3) - B(e_1,e_3,e_2) = 1-(-1) = 2 \neq 0$. □

Example 13.21. If G is a connected, semisimple Lie group as for instance $\mathrm{SL}(n,\mathbb{R})$, $\mathrm{SL}(n,\mathbb{C})$, $\mathrm{Sp}(n)$, then $[G,G] = G$ (see [201, Corollary 3.18.10]). On such groups each normalized solution of the classical quadratic functional equation is 0 by Lemma 13.11(g). Certain semisimple Lie groups are not connected, but have several components. To take an example the pseudo-orthogonal group $O(p,q)$ has 4 connected components when $p,q\ge 1$. However, that does not change the result when H is torsion free as shown by Corollary 13.16.

13.8 Exercises

Exercise 13.1. This exercise presents an alternative way of doing Proposition 13.2.

Let $f:\mathbb{R}\to\mathbb{C}$ be a solution of $f(x+y)+f(x-y) = 2f(x)+2f(y)$, $x,y\in\mathbb{R}$.

(a) Show that $f(0) = 0$.

(b) Show that $f(-x) = f(x)$ for all $x \in \mathbb{R}$.

(c) Show that $f(nx) = n^2 f(x)$ for all $n \in \mathbb{Z}$ and $x \in \mathbb{R}$.

(d) Show that $f(rx) = r^2 f(x)$ for all $r \in \mathbb{Q}$ and $x \in \mathbb{R}$.

(e) Assume furthermore that f is continuous. Show that there is a constant $a \in \mathbb{C}$ such that $f(x) = ax^2$ for all $x \in \mathbb{R}$.

Exercise 13.2. We ask the reader to prove the following proposition along the guidelines below:

> **Proposition 13.22.** *If G is a group in which each commutator has finite order, then every solution $f : G \to \mathbb{C}$ of the classical quadratic functional equation is abelian.*

(a) Let $x_0 \in G$. Show that $Cf(x_0, -)$ is a normalized solution of Jensen's equation, vanishing at each commutator $[x, y]$, $x, y \in G$.

Hint: Proposition 12.5(b).

(b) Show that $Cf(x_0, -)$ is central.

Hint: Proposition 12.5(l).

(c) Show that $Cf(x_0, -)$ is additive.

Hint: Corollary 12.12(b).

(d) Show that $C^2 f = 0$. Conclude f is abelian.

Exercise 13.3. With the notation and assumptions from Lemma 13.8 we get from (f) of the lemma that $2C^2 f(x, x, y) = 2C^2 f(x, y, x) = 2C^2 f(y, x, x) = 0$ for all $x, y \in M$. The purpose of this exercise is to show the nuisance factor 2 is superfluous, when f is a solution of the classical quadratic functional equation.

Let $M = G$ be a group and let $f : G \to H$ be a solution of the classical quadratic functional equation (13.4). We shall show that

$$C^2 f(x, x, y) = C^2 f(x, y, x) = C^2 f(y, x, x) = 0 \qquad \text{for all } x, y \in G. \quad (13.9)$$

(a) Explain why it suffices to prove that $C^2 f(x, x, y) = 0$ for all $x, y \in G$.

(b) Prove that $C^2 f(x, x, y) = Cf(x^2, y) - 2Cf(x, y)$ for all $x, y \in G$.

(c) Deduce (13.9).

Exercise 13.4. Let G be a group with an involution $\tau : G \to G$ and let H be an abelian group such that $\mathrm{Hom}(G, H) = \{0\}$.

(a) Show that $2f = 0$ for all $f \in Q_\tau(G, H)$.
Hint: Show that $2C^2 f = 0$ and apply Exercise 2.33.

(b) Show that $Q_\tau(G, H) = \{0\}$ under the additional hypothesis that H is 2-torsion free.

Exercise 13.5. Find the continuous complex-valued solutions of the classical quadratic functional equation on the $(ax+b)$-group and on the Heisenberg group.

Hint: Proposition 13.9 and Exercises 2.27 and 2.28.

Exercise 13.6. Prove the following proposition:

Proposition 13.23. *Let $f : M \to H$ be a solution of the quadratic functional equation* (13.2) *on a monoid M with an involution $\tau : M \to M$. Let M_i, $i \in I$, be submonoids of M such that*

(i) *M is generated by the M_i, $i \in I$.*

(ii) *M_i is invariant under τ for each $i \in I$.*

(iii) *If $x_i \in M_i$ and $x_j \in M_j$ where $i, j \in I$ and $i \neq j$, then x_i and x_j commute.*

(iv) *The restriction of f to M_i is abelian for each $i \in I$.*

Then f is abelian.

Hint: Proceed as in the proof of Proposition 13.9.

Exercise 13.7. Let $f : \mathbb{R}^n \to \mathbb{C}$ be a solution of the classical quadratic functional equation (13.6). Let $x, y \in \mathbb{R}^n$. Show that

$$f\left(x + \frac{y}{n}\right) \to f(x) \qquad \text{for } n \to \infty.$$

Hint: Theorem 13.13.

Exercise 13.8. In this exercise we shall derive the following version of Aczél's classical result (Theorem 13.13).

Let G be a group generated by its squares and let H be an abelian, 2-torsion free group. Assume that $f : G \to H$ is abelian.

Then f is a solution of the classical quadratic functional equation (13.4) if and only if f has the form $f(x) = Q(x, x)$, $x \in G$, where $Q : G \times G \to H$ is a symmetric, bi-additive map.

Hint: Let $B_1 : G \times G \to H$ be bi-additive. Show that there exists exactly one map $B_2 : G \times G \to H$ such that $B_1 = 4B_2$.

Exercise 13.9. Drygas [66] considered what amounts to solutions $f : \mathbb{R} \to \mathbb{R}$ of the functional equation

$$f(x+y) + f(x-y) = 2f(x) + f(y) + f(-y), \qquad x, y \in \mathbb{R}. \tag{13.10}$$

We extend (13.10) from the group $\mathbb{R}$ to any group G as follows: *Drygas' functional equation* is

$$f(xy) + f(y^{-1}x) = 2f(x) + f(y) + f(y^{-1}), \qquad x, y \in G, \tag{13.11}$$

where $f : G \to \mathbb{C}$ is the unknown function.

We ask the reader to provide a proof of Proposition 13.24, where we reduce (13.11) to known functional equations by splitting $f : G \to \mathbb{C}$ into its odd and even parts. We recall from Section A.2 the notation $\check{f}(x) := f(x^{-1})$, $x \in G$, and that we may write $f = f_o + f_e$, where $f_o := (f - \check{f})/2$ and $f_e := (f + \check{f})/2$ denote the odd and even parts of f.

Proposition 13.24. *Let G be a group. The solutions $f : G \to \mathbb{C}$ of Drygas' functional equation* (13.11) *are the functions of the form $f = f_o + f_e$, where $f_o : G \to \mathbb{C}$ is additive, and $f_e : G \to \mathbb{C}$ is a solution of the classical quadratic functional equation* (13.4).

Let f be a solution of (13.11).

(a) Compute that f_e is a solution of (13.11), and conclude that it is a solution of the classical quadratic functional equation (13.4).

Hint: Theorem 7.1(b).

(b) Infer by linearity that also f_o is a solution of (13.11) so that

$$f_o(xy) + f_o(y^{-1}x) = 2f_o(x) \qquad \text{for all } x, y \in G.$$

Conclude that f_o is additive.

Hint: Proposition 12.29.

(c) Complete the proof of Proposition 13.24 by showing that any function of the form described in the proposition is a solution of (13.11).

Find the continuous solutions of the original functional equation (13.10).

Exercise 13.10. In this exercise we study *Whitehead's functional equation*

$$f(xyz) + f(x) + f(y) + f(z) = f(xy) + f(yz) + f(xz), \qquad x, y, z \in G, \tag{13.12}$$

which expresses that the second Cauchy difference $C^2 f$ (defined by (2.5)) vanishes. It is a fundamental functional equation, because it involves only the semigroup structure of G. That the (first) Cauchy difference of f vanishes, means that f is additive. The present exercise examines what it means that the second Cauchy difference vanishes, i.e., it describes the solutions $f : G \to \mathbb{C}$ of (13.12) or in other words the kernel of C^2. We have placed Whitehead's functional equation in this chapter, because it is related to the quadratic functional equation. How is revealed in (f) below. See Exercises 2.32, 2.33 and 2.34 for some elementary properties of the elements of the kernel of C^2, valid for any abelian group H. In the present exercise we take $H = \mathbb{C}$.

The special case of G or just f being abelian is the topic of Exercise 13.11 where Whitehead's functional equation becomes Fréchet's functional equation.

(a) Show that $f(e) = 0$ for all solutions f of (13.12).

(b) Let f be a solution of (13.12). Show that $\check{f}$ and consequently also the even part f_e and the odd part f_o of f are solutions of (13.12).

(c) Put $z = y^{-1}$ to get for any solution f of (13.12) the identity

$$f(xy) + f(xy^{-1}) = 2f(x) + 2f_e(y) \qquad \text{for all } x, y \in G. \tag{13.13}$$

(d) Let f be a solution of (13.12). Show that f_o is an odd solution of Jensen's functional equation (12.3). Show conversely that any odd solution of Jensen's functional equation is a solution of (13.12).

Hint: Proposition 12.5(d).

(e) Let f be a solution of (13.12). Show that f_e is an abelian solution of the classical quadratic functional equation (13.4).

Hint for abelian: Lemma 13.8(a).

Show conversely that any abelian solution of the classical quadratic functional equation satisfies (13.12).

(f) Conclude that the solutions $f : G \to \mathbb{C}$ of (13.12) are the functions of the form $f = f_o + f_e$, where f_o is an odd solution of Jensen's functional equation and f_e is an abelian solution of the classical quadratic functional equation. The form of f_e can be read from Theorem 13.13.

Exercise 13.11. Jordan and von Neumann [122] found that the norm of a normed vector space $(V, \|-\|)$ stems from an inner product if and only if the parallelogram identity (13.1) holds. Another identity that also characterizes the inner product spaces among the normed ones is the *parallelepiped law*: For all $x, y, z \in V$

$$\|x+y+z\|^2+\|x\|^2+\|y\|^2+\|z\|^2 = \|x+y\|^2+\|y+z\|^2+\|z+x\|^2. \quad (13.14)$$

Give a geometric interpretation of (13.14) in $\mathbb{R}^3$. Show that if (13.14) holds then so does the parallelogram identity.

The parallelogram identity is a requirement of the norm on 2-dimensional subspaces of V, while the parallelepiped law governs the behavior of the norm of 3-dimensional subspaces of V.

We shall in this exercise study a generalization of (13.14) in which we replace the vector space V by a group G, viz., *Fréchet's equation*

$$f(xyz)+f(x)+f(y)+f(z) = f(xy)+f(yz)+f(zx), \quad x, y, z \in G, \quad (13.15)$$

where $f : G \to \mathbb{C}$ is the unknown function.

Fréchet's equation differs from Whitehead's equation (13.12) from Exercise 13.10 only on the very last term on the right hand side, where Fréchet's equation has $f(zx)$, while Whitehead's equation has $f(xz)$. Any solution f of Fréchet's equation is a solution of Whitehead's equation, because such an f is abelian as is apparent from Proposition 13.25. The two equations coincide on abelian groups.

(A) We ask the reader to provide a proof of Proposition 13.25 below from his knowledge of the solutions of Whitehead's functional equation. The crucial fact to establish is that any solution f of Fréchet's equation is abelian.

Hint: Prove first that f is central by putting $y = e$ in (13.15).

Proposition 13.25. *Let G be a group. The solutions $f : G \to \mathbb{C}$ of Fréchet's equation* (13.15) *are the functions $f(x) = A(x) + Q(x, x)$, $x \in G$, where $A : G \to \mathbb{C}$ is additive, and $Q : G \times G \to \mathbb{C}$ is symmetric and bi-additive. A and Q are uniquely determined by f.*

If G is a topological group, then the solution f is continuous if and only if A and Q are continuous.

(B) What are the continuous solutions of Fréchet's equation when $G = \mathbb{R}$?

13.9 Notes and remarks

The chapter is based mainly on the papers Friis and Stetkær [91] and Yang [212] where more results and examples can be found. It was Yang [212] who recognized the importance of the function $C^2 f$ for the quadratic functional equation.

We work on groups, but lots of efforts have been invested in the setting of vector spaces. For an overview of results about the quadratic functional equation and related functional equations on vector spaces consult Kannappan [132, Chapter 5]. Recent results on real vector spaces X are due to Kobal [134]. He studies for $\lambda \in \mathbb{R}$ solutions $\|-\| : X \to \mathbb{R}$ of the functional equation

$$\|x + \lambda y\|^2 = \lambda\|x + y\|^2 + (1 - \lambda)\|x\|^2 + \lambda(\lambda - 1)\|y\|^2, \qquad x, y \in X,$$

that for $\lambda = -1$ is the parallelogram identity. The norm in an inner product space satisfies the equation for all $\lambda \in \mathbb{R}$. For any fixed $\lambda \neq 0, 1$ Kobal [134, Theorem 3.1] shows that the positive definite, homogeneous solutions are norms stemming from inner products on X. This extends works on how to characterize inner product spaces by Kurepa [137] and Šemrl [168].

A natural question is whether any measurable (with respect to a left Haar measure) solution $f : G \to \mathbb{C}$ of the classical quadratic functional equation (13.4) on a locally compact group G is continuous. The answer is yes, as follows from Járai [116, Theorem 8.2].

On the group $\mathbb{R}$ special results are known that do not even require measurability. See Maksa [144] for an example.

The classical quadratic functional equation on a group G was generalized in Chung, Ebanks, Ng and Sahoo [41] (and the correction [42]) to

$$f_1(xy) + f_2(xy^{-1}) = f_2(x) + f_4(y) + f_5(x)f_6(y), \qquad x, y \in G,$$

where the unknown functions $f_i : G \to K$, $i = 1, \ldots, 5$, take their values in a quadratically closed commutative field K with characteristic different from 2 and 3. [41] derives formulas for the solutions, assuming Kannappan's condition for f_1 and f_2. The special case of $f_1(x + y) + f_2(x - y) = f_3(x) + f_4(y)$, $x, y \in \mathbb{R}$, where $f_1, f_2, f_3, f_4 : \mathbb{R} \to \mathbb{R}$, is solved in [167, Theorem 9.6].

Łukasik [143] finds the solutions $f, g, h : S \to H$ of the functional equation

$$\sum_{\lambda \in \Lambda} f(x + \lambda y) = |\Lambda| g(x) + h(y), \qquad x, y \in S,$$

where S is an abelian semigroup, Λ a finite subgroup of its automorphism group and H is an abelian group.

For equations like

$$\frac{f(x+y)+f(x+\sigma(y))}{2}=f(x)g(y)+f(y)g(x), \qquad x,y\in G,$$

see Chung, Kannappan and Ng [45, 46] and Friis and Stetkær [89]. See also Stetkær [186, Corollary III.8].

For multiplicative quadratic functions on a division algebra check Heuvers and Ng [110].

Faraut and Harzallah [78] viewed the functional equation

$$\int_K f(xky)\,dk+\int_K f(xky^{-1})\,dk=2f(x)+2f(y), \qquad x,y\in G,$$

as an extension of the parallelogram law. In their paper G is a locally compact group and K is a compact subgroup with normalized Haar measure dk.

Theorem 13.6 is Sinopoulos [180, Theorem 3].

Corollary 13.10 is in the literature for the classical quadratic functional equation. Kurepa [139, Theorem 1] proved it under the condition that the range group H was 2-torsion free. Yang [212, Corollary 3.2] removed the 2-torsion free condition.

Theorem 13.13: Aczél [1, Theorem 2] assumes G abelian while we assume only that f satisfies Kannappan's condition. We need this slight generalization in our discussion of Fréchet's functional equation (Exercise 13.11) and examples like the ones in Exercise 13.5. Exercise 13.8 is a modified version of Aczél's result.

Example 13.20 is adapted from Yang [212, Section 3] that contains the complete solution of the classical quadratic functional equation on free groups with any set of generators and with no restrictions on the abelian group H.

Proposition 13.22 covers for complex-valued solutions the case of a P_3-group that was studied by Corovei [55, Lemma 4].

Exercise 13.8 is a modified version of Aczél's result (Theorem 13.13). A condition on the range H in Aczél's theorem is replaced by one on G.

In Exercise 13.9 we presented a generalization of Drygas' original functional equation (13.10). Faĭziev and Sahoo [79] have another way of generalizing it. They consider the solutions $f : G \to \mathbb{R}$ of following system of equations:

$$f(xy) + f(xy^{-1}) = 2f(x) + f(y) + f(y^{-1}),$$
$$f(yx) + f(y^{-1}x) = 2f(x) + f(y) + f(y^{-1}).$$

Exercise 13.10: (1) The functional equation (13.12), i.e., $C^2 f = 0$ for $f : G \to H$, is named after Whitehead, because it occurred in his paper Whitehead [208, Ch. II, §5], although just for G abelian and f even. (2) See Exercises 2.32, 2.33 and 2.34 for some elementary properties of the elements of the kernel of C^2, valid for any abelian group H.

Exercise 13.11: (1) In Exercise 13.11 we considered only complex-valued solutions. If the range space of the solutions f is an abelian group H, which is not 2-divisible, then there may exist solutions of (13.15) that cannot be written as a sum of an additive and a quadratic function. For a simple example take $G = H = \mathbb{Z}$ and $f(n) := n(n+1)/2$, $n \in \mathbb{Z}$, i.e., the triangular numbers extended from $\mathbb{N}$ to $\mathbb{Z}$. (2) For more about Fréchet's equation and references to the literature about it see Kannappan [132, Section 5.1].

Chapter 14

K-Spherical Functions

Throughout this chapter K denotes a compact group with normalized Haar measure dk, and G a topological group with neutral element e. K will in general not be a subgroup of G, but in Section 14.3 it is.

14.1 Introduction and notation

Examples 14.1.

(a) If K is the circle group, $K = \{e^{i\theta} \mid \theta \in \mathbb{R}\}$, then dk is the arc length measure, normalized such that the circle has total length 1. In other words,

$$\int_K f(k)\, dk = \tfrac{1}{2\pi} \int_{-\pi}^{\pi} f(e^{i\theta})\, d\theta \qquad \text{for all } f \in C(K).$$

(b) If K is a finite group, then dk is the counting measure, normalized such that K has total measure 1. In other words,

$$\int_K f(k)\, dk = \tfrac{1}{\#K} \sum_{k \in K} f(k) \qquad \text{for all } f \in C(K),$$

where $\#K$ denotes the number of elements in K. So the integral of f is the mean value of f over the finite set K.

In particular, if $K = \{e\}$, then $\int_K f(k)\, dk = f(e)$, and if $K = \mathbb{Z}_2 = \{\pm 1, \cdot\}$, then $\int_K f(k)\, dk = (f(1) + f(-1))/2$.

We recall the notion of a spherical function from the theory of harmonic analysis on groups: If K is a compact subgroup of a locally compact group G, then a function $\phi \in C(G)$, $\phi \neq 0$, is said to be spherical if

$$\int_K \phi(xky)\, dk = \phi(x)\phi(y) \qquad \text{for all } x, y \in G. \tag{14.1}$$

Any solution $\phi \in C(G)$ of (14.1) is also a solution of

$$\int_K \phi(xkyk^{-1})\,dk = \phi(x)\phi(y) \qquad \text{for all } x, y \in G, \tag{14.2}$$

because any solution of (14.1) is a function on G/K.

The functional equation (14.2) is well known in the case of $G = K$, where it is called *Weyl's functional equation for characters*. Its solutions $\phi \in C(G) \setminus \{0\}$ are the functions $\phi = \chi_U / \dim U$ where U ranges over the finite-dimensional continuous representations of the compact group G (χ_U denotes the character of U, introduced in Definition E.8(g)). See [112, (27.53)] for a proof and references.

We will study a general version of (14.2). To do so we recall the notion of a topological transformation group:

Definition 14.2. Let X a topological space. A *topological transformation group* of X is a pair consisting of

1. a topological group H and
2. a continuous map $(h, x) \mapsto h \cdot x$ af $H \times X$ into X ($H \times X$ is given the product topology)

such that

(i) $(h_1 h_2) \cdot x = h_1 \cdot (h_2 \cdot x)$ for all $h_1, h_2 \in H$ and $x \in X$.

(ii) $e \cdot x = x$ for all $x \in X$, where e denotes the neutral element of H.

We say that $h \cdot x$ is the action by $h \in H$ on $x \in X$.

When the action is given by the context we say that H is a topological transformation group of X or just that H acts on X.

Examples 14.3.

(1) $\mathrm{GL}(n, \mathbb{R})$ is a topological transformation group of $\mathbb{R}^n$.

(2) $\mathrm{SO}(3)$ is a topological transformation group of the sphere S^2.

(3) The $(ax + b)$-group acts on $\mathbb{R}$ by

$$\begin{pmatrix} a & b \\ 0 & 1 \end{pmatrix} \cdot x = ax + b.$$

(4) A subgroup H of G acts by inner automorphisms on G as follows: $h \cdot x := hxh^{-1}$ for $h \in H$ and $x \in G$.

(5) $\mathbb{Z}_2 := (\{\pm 1\}, \cdot)$ acts on G by $1 \cdot x = x$ and $(-1) \cdot x = \sigma(x)$, $x \in G$, when $\sigma : G \to G$ is a continuous homomorphism such that $\sigma^2 = I$.

(6) $\mathbb{Z}_2 := (\{\pm 1\}, \cdot)$ acts on G by $1 \cdot x = x$ and $(-1) \cdot x = x^{-1}$, $x \in G$. More generally, if $\tau : G \to G$ is a continuous, involutive automorphism, then $\mathbb{Z}_2 := (\{\pm 1\}, \cdot)$ acts on G by $1 \cdot x = x$ and $(-1) \cdot x = \tau(x)$, $x \in G$.

An interesting theory results when the X has an algebraic structure in addition to the topological one: We take $X = G$, so X is a topological group. Furthermore, the action on X will respect the algebraic structure on X as well. Our set up, enforced from now on, is the following:

Set up. K is a topological transformation group of the topological group G, acting by automorphisms.

The Examples 14.3(1), (4) and (5) satisfy our set up. So does (6), if G is abelian, for in that case it is a special instance of (5).

Definition 14.4. A *K-spherical function* is a function $\phi \in C(G)$ such that $\phi(e) = 1$ and

$$\int_K \phi(x\, k \cdot y)\, dk = \phi(x)\phi(y) \qquad \text{for all } x, y \in G, \tag{14.3}$$

where $x\, k \cdot y$ means $x(k \cdot y)$.

Definition 14.5.

(a) For any $k \in K$ and $F : G \to \mathbb{C}$ we define $(k \cdot F)(x) := F(k^{-1} \cdot x)$ for $x \in G$. We say that F is *K-invariant* if $k \cdot F = F$ for all $k \in K$.

(b) Let G be locally compact and equipped with a left Haar measure. The elements of $L_1(G)$ are in principle not functions, but equivalence classes of functions that are equal almost everywhere. Here we say that the equivalence class of $[f] \in L_1(G)$ is K-invariant if for each $k \in K$ we have that $k \cdot f = f$ almost everywhere. Below we will (as is usual) delete the brackets around f.

Lemma 14.6. *A solution $\phi \in C(G)$ of* (14.3) *is non-zero if and only if $\phi(e) = 1$, and it is K-invariant.*

Proof. Put $y = e$ for the first statement. Use the invariance of dk for the second. □

If $K = \{e\}$ then (14.3) reduces to $\phi(xy) = \phi(x)\phi(y)$, so the solutions are in this case the continuous group characters of G. If $G = (G, +)$ is abelian, then (14.3) reduces in Example 14.3(5) to

$$\frac{\phi(x+y) + \phi(x+\tau(y))}{2} = \phi(x)\phi(y) \qquad \text{for all } x, y \in G,$$

which is d'Alembert's equation (9.4) for the abelian group G. This we solved earlier (Corollary 9.22).

Corresponding to Lemma 11.3(e) about Wilson functions we obtain

Lemma 14.7. *Let $\phi, w \in C(G)$ satisfy*

$$\int_K w(x\,k \cdot y)\,dk = w(x)\phi(y) \qquad \textit{for all } x, y \in G. \tag{14.4}$$

If $w \neq 0$ then ϕ is a K-spherical function.

Proof. Changing order of integration and use dk's left invariance we find

$$\begin{aligned}
w(x_0)\int_K \phi(x\,k \cdot y)\,dk &= \int_K \left\{\int_K w(x_0\,k_1 \cdot [x\,k \cdot y])dk_1\right\} dk \\
&= \int_K \left\{\int_K w(x_0\,[k_1 \cdot x]\,[(k_1 k) \cdot y])dk_1\right\} dk \\
&= \int_K \left\{\int_K w(x_0\,[k_1 \cdot x]\,[(k_1 k) \cdot y])dk\right\} dk_1 \\
&= \int_K \left\{\int_K w(x_0\,[k_1 \cdot x]\,[k \cdot y])dk\right\} dk_1 \\
&= \int_K w(x_0\,k_1 \cdot x)\phi(y)\,dk_1 = w(x_0)\phi(x)\phi(y).
\end{aligned}$$

Choosing $x_0 \in G$ such that $w(x_0) \neq 0$ we find that ϕ is a solution of (14.3). Taking $y = e$ and $x = x_0$ in (14.4) we get that $w(x_0) = w(x_0)\phi(e)$ which implies that $\phi(e) = 1$. □

If $\phi = 1$, then (14.4) expresses that w has a mean value property with respect to K.

If G is abelian then the defining equation (14.3) for a K-spherical function ϕ on G takes the form

$$\int_K \phi(x + k \cdot y)\,dk = \phi(x)\phi(y) \qquad \text{for all } x, y \in G, \tag{14.5}$$

while (14.4) becomes

$$\int_K w(x + k \cdot y)\, dk = w(x)\phi(y) \qquad \text{for all } x, y \in G. \tag{14.6}$$

If $K = \{e\}$, then (14.5) is Cauchy's multiplicative functional equation on G. If $K = \mathbb{Z}_2$ acts by $(\pm 1) \cdot x = \pm x$, then (14.5) is d'Alembert's classical functional equation on G and (14.6) is Wilson's functional equation on G.

The equations (14.5) and (14.6) were investigated by Chojnacki [37] for bounded solutions. We cite Chojnacki's result [37, Theorem 1.2] using our notation and set up (Chojnacki is a bit more general, having a measurable action by K, not just a continuous). Roughly speaking it says that any bounded K-spherical function is the mean value over K of a unitary character of G.

Theorem 14.8. *Let G be a locally compact group. Let $w \in \mathcal{L}^\infty(G)$ be essentially non-zero with respect to the Haar measure on G. Let Ω be an open subset of G, and ϕ a complex function on Ω. Suppose that for each $y \in \Omega$ the identity*

$$\int_K w(x + k \cdot y)\, dk = w(x)\phi(y)$$

holds for almost all $x \in G$. Then there exists $\chi \in \widehat{G}$ (the dual group) such that

$$\phi(y) = \int_K (k \cdot \chi)(y)\, dk \qquad \text{for each } y \in \Omega.$$

Some of Chojnacki's results were later generalized by Badora [19] to a convolution version, but still the solutions were bounded functions on an abelian group. In [20] Badora considers continuous, bounded solutions in the case of K finite.

For unbounded, continuous solutions the best result is due to Shin'ya [170, Corollary 3.12]. In our terminology it says

Theorem 14.9. *If G is locally compact and abelian, then any continuous K-spherical function ϕ on G has the form $\phi = \int_K k \cdot \chi\, dk$ for some continuous character χ on G.*

In the general case the author has found the following simple lemma quite useful (see for instance [194]). The character χ takes signed equations into account.

Lemma 14.10. *Let $\chi \in C(G)$ be a character. Let $f, g_1, \dots, g_N, h_1, \dots, h_N \in C(G)$, and write $g = (g_1, \dots, g_N)$, $h = (h_1, \dots, h_N)^T$. If*

$$\int_K f(x\,k \cdot y)\overline{\chi(k)}\,dk = g(x)h(y) \qquad \text{for all } x, y \in G,$$

then

$$\int_K g(x\,k \cdot y)\,dk\,h(z) = g(x)\int_K h(y\,k \cdot z)\overline{\chi(k)}\,dk \qquad \text{for all } x, y, z \in G. \tag{14.7}$$

Weyl's functional equation for characters is a special case of our set up: G is compact, and $K = G$ acts on G by inner automorphisms, meaning that $k \cdot x = kxk^{-1}$ for $k, x \in K$. In Theorem 14.11 $\mathcal{L}^1(G)$ denotes the set of complex-valued functions on K that are integrable with respect to the Haar measure on K, and χ_π is the character of the representation π.

Theorem 14.11 (Weyl). *Let K be a compact group with normalized Haar measure dk. If π is a d-dimensional, continuous, unitary, irreducible representation of K, then*

$$\int_K \chi_\pi(xkyk^{-1})\,dk = \frac{1}{d}\chi_\pi(x)\chi_\pi(y) \qquad \text{for all } x, y \in K. \tag{14.8}$$

Conversely, if $f \in \mathcal{L}^1(G)$ satisfies the functional equation

$$\int_K f(xkyk^{-1})\,dk = \frac{1}{d}f(x)f(y) \qquad \text{for all } x, y \in K,$$

then either $f = 0$ or $f = \chi_\pi/d$ for some continuous, unitary, irreducible representation π of K.

Proof. See [112, §27.53]. □

14.2 From where do K-spherical functions originate?

Let G be a locally compact group with modular function Δ. Let dx denote a left Haar measure on G. An object under study in the harmonic analysis on G is the convolution algebra $L_1(G)$ where the convolution for $f, g \in C_c(G)$ is defined by

$$(f * g)(x) := \int_G f(y)g(y^{-1}x)\,dy, \qquad x \in G.$$

$L_1(G)$ is a Banach -$*$-algebra with norm $\|f\| := \int_G |f(x)|\, dx$ and involution $f^*(x) := \Delta(x)^{-1}\overline{f(x^{-1})}$.

Let $L^1(G)^K$ be the subset of $L_1(G)$, consisting of the K-invariant elements of $L_1(G)$. It is a closed $*$-subalgebra of $L_1(G)$.

The next two theorems give an interpretation of the K-spherical functions, and in doing so tell us that the K-spherical functions are not spurious generalizations of d'Alembert functions, made up for the sake of generalization, but that they actually play a role in harmonic analysis. They replace the exponential functions in the classical Fourier analysis, and the dual group in the Fourier analysis on locally compact abelian groups. We do not need the theorems later.

Theorem 14.12. *The algebra homomorphisms of $L^1(G)^K$ onto $\mathbb{C}$ are the mappings of the form*

$$f \mapsto \int_G f(x)\phi(x)\, dx, \qquad f \in L^1(G)^K, \tag{14.9}$$

where ϕ ranges over the bounded K-spherical functions.

Theorem 14.13. *Consider the convolution algebra $C_c(G)$. Let $C_c(G)^K$ be the subalgebra of $C_c(G)$, consisting of the K-invariant functions in $C_c(G)$. Equip $C_c(G)$ with the inductive limit topology and $C_c(G)^K$ with the topology inherited from $C_c(G)$.*

The continuous homomorphisms of the algebra $C_c(G)^K$ onto $\mathbb{C}$ are the mappings of the form

$$f \mapsto \int_G f(x)\phi(x)\, dx, \qquad f \in C_c(G)^K,$$

where ϕ ranges over the K-spherical functions.

Proof. See [109, Exercise B.16, p. 486 and p. 586]. We refrain from giving the details, except for noting that the dual space of $C_c(G)$ is the set of (unbounded) complex Radon measures on G. □

14.3 The Moroccan school

Throughout this section G denotes a locally compact group, $\tau : G \to G$ a continuous involution and μ a regular, bounded, complex measure on G.

Unless otherwise specified K is a compact subgroup of G. We present a list of some of the functional equations on groups that Moroccan mathematicians have studied. The presentation is brief, and we do not specify all assumptions that the papers mentioned impose. For that and more information the reader should consult their papers.

The solutions $f : G \to \mathbb{C}$ are usually assumed to be continuous and bounded. In many studies there is a vestige of commutativity: It could be that K is a Gelfand subgroup of G, for example that (G, K) is a Riemannian pair. Or that the functions should satisfy Kannappan's condition or a modification of it. Or that a certain measure μ on G is a Gelfand measure.

Consider for G compact the commutative Banach-algebra

$$\mathfrak{A} := \{f \in L^1(G) \mid f = \check{f} \text{ and } f \text{ is central}\},$$

which is a subalgebra of the convolution algebra $L^1(G)$. Its character space $\widehat{\mathfrak{A}}$ consists of those non-zero $\phi \in C(G)$ that satisfy

$$\int_G \phi(xzyz^{-1})\,dz + \int_G \phi(xzy^{-1}z^{-1})\,dz = 2\phi(x)\phi(y) \qquad \text{for all } x, y \in G, \tag{14.10}$$

in the sense that $\phi \in \widehat{\mathfrak{A}} \subseteq C(G)$ acts on $f \in \mathfrak{A}$ by $\phi(f) = \int_G f(x)\phi(x)\,dx$. We will not derive this characterization of $\widehat{\mathfrak{A}}$, since we do not need it here. We have included it, because it motivates a study of some of the functional equations below.

Moroccan mathematicians have studied various extensions of (14.1), (14.3) and (14.10). We describe these extensions below.

Bouikhalene and Kabbaj [30] study solutions $f \in C_b(G)$ of the functional equation

$$\sum_{\phi\in\Phi} \int_K f(xk\phi(y))\,dk = |\Phi| f(x)f(y), \qquad x, y \in G,$$

where Φ is a group of automorphisms of G with $|\Phi| < \infty$ elements.

Replacing the Haar measure on K by a measure μ on G [9] and [8] generalize (14.1) to

$$\int_G \phi(xty)\,d\mu(t) = \phi(x)\phi(y) \qquad \text{for all } x, y \in G.$$

This is in [75] extended further to the Levi-Civita's type functional equation

$$\int_G f(xty)\,d\mu(t) = \sum_{l=1}^{N} g_l(x)h_l(y) \qquad \text{for all } x, y \in G.$$

A similar type of functional equation can be found in [76]:

$$\int_G \left\{ \int_K f(xtk \cdot y)\, dk \right\} d\mu(t) = f(x)f(y), \qquad x, y \in G,$$

where K is a compact subgroup of the morphisms (i.e., homeomorphisms that are either automorphisms or antiautomorphisms) of G.

In [11] we meet for (G, K) a Gelfand pair a generalization of d'Alembert's functional equation, viz.,

$$\int_K (\phi(xky) + \phi(xky^{-1}))\, dk = 2\phi(x)\phi(y), \qquad x, y \in G.$$

[74] considers

$$\int_G f(xty)\, d\mu(t) \pm \int_G f(xt\tau(y))\, d\mu(t) = 2\sum_{l=1}^{N} g_l(x)h_l(y), \qquad x, y \in G.$$

The special cases

$$\int_G f(xty)\, d\mu(t) \pm \int_G f(xt\tau(y))\, d\mu(t) = 2f(x)f(y), \qquad x, y \in G,$$

and

$$\int_G f(xty)\, d\mu(t) + \int_G f(xt\tau(y))\, d\mu(t) = 2f(x)f(y), \qquad x, y \in G,$$

$$\int_G f(xty)\, d\mu(t) + \int_G f(xt\tau(y))\, d\mu(t) = 2f(x)g(y), \qquad x, y \in G,$$

are considered in [12] and [73], respectively.

For G compact [29] investigates the functional equation

$$\int_G f(xtyt^{-1})\, dt + \int_G f(xt\tau(y)t^{-1})\, dt = 2g(x)h(y), \qquad x, y \in G.$$

Under a Kannappan type condition [21] considers the functional equation

$$\int_K f(xkyk^{-1})\, dk + \int_K f(xk\tau(y)k^{-1})\, dk = 2f(x)f(y), \qquad x, y \in G.$$

Also signed equations like

$$\int_G \int_K f(xtk \cdot y)\, \overline{\chi(k)}\, dk\, d\mu(t) = g(x)h(y), \qquad x, y \in G,$$

and

$$\int_G \int_K f(xtk \cdot y)\, \overline{\chi(k)}\, dk\, d\mu(t) = \sum_{l=1}^{N} g_l(x) h_l(y), \qquad x, y \in G,$$

where χ is a continuous character on K, have been studied. See [31] and [10].

Finally the three addition formulas

$$\int_G f(xt\sigma(y))\, d\mu(t) = \begin{cases} f(x)g(y) + f(y)g(x) \\ f(x)g(y) - f(y)g(x) \\ g(x)g(y) + f(x)f(y) \end{cases}$$

in which $\sigma : G \to G$ is a continuous homomorphism such that $\sigma \circ \sigma = \mathrm{id}$, are discussed in [77].

14.4 Exercises

Exercise 14.1. Derive Lemma 14.7 from formula (14.7)

14.5 Notes and remarks

For more about K-spherical functions consult Benson, Jenkins and Ratcliff [25, 26].

Chapter 15

The Sine Functional Equation

In this chapter we discuss complex-valued solutions of the sine functional equation (15.3) on groups that may be non-abelian. The most important result is Theorem 15.3.

15.1 Introduction

Our point of departure is the trigonometrical identity

$$\sin(x+y)\sin(x-y) = \sin^2 x - \sin^2 y, \qquad x, y \in \mathbb{R}. \tag{15.1}$$

The functional equation corresponding to this identity, i.e.,

$$f(x+y)f(x-y) = f(x)^2 - f(y)^2, \qquad x, y \in \mathbb{R}, \tag{15.2}$$

is remarkable, because it characterizes the sine functions $x \mapsto \sin(\lambda x)$ in the sense that its continuous solutions f are the complex multiples of the sine functions

$$f(x) = \sin(\lambda x) = \frac{e^{i\lambda x} - e^{-i\lambda x}}{2i}, \qquad x \in \mathbb{R},$$

where λ ranges over $\mathbb{C}$, and the complex multiples of the limiting case

$$f(x) = \lim_{\lambda \to 0} \frac{\sin(\lambda x)}{\lambda} = x, \qquad x \in \mathbb{R},$$

in which f becomes additive (Corollary 15.7).

We generalize the domain of definition of the function f from $\mathbb{R}$ to a group G by considering the *sine functional equation*

$$f(xy)f(xy^{-1}) = f(x)^2 - f(y)^2, \qquad x, y \in G, \tag{15.3}$$

where $f : G \to \mathbb{C}$ is the unknown function to be determined. Note that if f is a solution, then so is cf for any constant $c \in \mathbb{C}$.

If G is abelian then (15.3) takes the form

$$f(x+y)f(x-y) = f(x)^2 - f(y)^2, \qquad x, y \in G, \tag{15.4}$$

which has as a consequence (take $x = a + b$ and $y = a - b$) that

$$f(2a)f(2b) = f(a+b)^2 - f(a-b)^2 \qquad \text{for all } a, b \in G. \tag{15.5}$$

Lemma 15.1 states there are solutions of (15.3) corresponding to the two types of solutions above (the sine functions and the additive functions) on any group, abelian or not. On some groups the two possibilities of Lemma 15.1 exhaust all the solutions, for example on $G = \mathbb{R}$ and $G = \mathbb{Z}$ (Theorem 15.3 and Example 15.9).

We recall that a function F on G is odd if $\check{F} = -F$, where $\check{F}(x) := F(x^{-1})$, $x \in G$.

Lemma 15.1. *Let G be a group. The following functions $f : G \to \mathbb{C}$ are solutions of the sine functional equation* (15.3) *on G:*

(i) $f = c(\chi - \check{\chi})$, *where $c \in \mathbb{C}$ and χ is a character on G, and*

(ii) $f = \chi a$, *where χ is a character on G such that $\chi = \check{\chi}$ and where a is an additive function on G.*

On $G = \mathbb{R}$ we get the solution $f(x) = \sin(\lambda x)$ out of Lemma 15.1(i) by taking $\chi(x) = e^{i\lambda x}$ and $c = 1/(2i)$, while the solution $f(x) = x$ is (ii) with $\chi = 1$ and $a(x) = x$.

On groups like $\mathbb{R}$ that are generated by their squares the character $\chi = \check{\chi}$ from (ii) has to be the trivial one $\chi = 1$ (by Lemma 3.5), so it disappears from sight on such groups. On $\mathbb{Z}$ the character $\chi(n) := (-1)^n$, $n \in \mathbb{Z}$, has the property that $\chi = \check{\chi}$ although $\chi \neq 1$.

The sine functional equation has not yet been solved on general groups. Results lack on groups that are not generated by their squares.

15.2 General results about the solutions

We derive in Lemma 15.2 properties of solutions of the sine functional equation on any group. In particular we do not assume that the group is generated by its squares, in contrast to the chapter's main result Theorem 15.3.

Most of the properties concern the zeros of the solutions. The reader could check what the results say in the illuminating instance of $f(x) = \sin x$, $x \in \mathbb{R}$.

Lemma 15.2. *Let $f : G \to \mathbb{C}$ be a solution of the sine functional equation* (15.3) *on a group G, and let $N_f := \{u \in G \mid f(u) = 0\}$ denote the set of its zeros.*

(a) $f(e) = 0$.

(b) *f is odd.*

(c) *N_f is a normal subgroup of G.*

(d) *We have the following two characterizations of N_f:*

$$N_f = \{u \in G \mid f(xu)^2 = f(x)^2 \text{ for all } x \in G\}$$
$$= \{u \in G \mid f(xu^2) = f(x) \text{ for all } x \in G\}.$$

In particular, if $f(u) = 0$, then u is a period of f^2 and u^2 a period of f.

(e) *For any $x, y \in G$ there are only the following two possibilities:*

$$[x, y] = xyx^{-1}y^{-1} \in N_f \quad \text{or} \quad x^2y^2 \in N_f.$$

(f) *The commutator $[x_1^2x_2^2 \cdots x_m^2, y_1^2y_2^2 \cdots y_n^2] \in N_f$ for all $x_1, x_2, \ldots, x_m \in G$, $y_1, y_2, \ldots, y_n \in G$ and $m, n \in \mathbb{N}$.*

Proof. (a) Take $y = e$ in (15.3).

(b) Taking $x = e$ in (15.3) we get $f(y)f(y^{-1}) = -f(y)^2$, and so $f(y)[f(y) + f(y^{-1})] = 0$ for all $y \in G$. To this we add what we get by interchanging y and y^{-1} and we obtain that $[f(y) + f(y^{-1})]^2 = 0$, implying (b).

(c) We first show that N_f is a subgroup of G: From (a) we see that $e \in N_f$, and from (b) that $x \in N_f \Rightarrow x^{-1} \in N_f$. So it is left to show that $x, y \in N_f \Rightarrow xy \in N_f$. But that is immediate from the identity $f(xy^2)f(x) = f(xy)^2 - f(y)^2$, that we obtain by replacing x by xy in (15.3).

It is left to show that the subgroup N_f is normal. By elementary group theory it suffices to prove for all $x, y \in G$ that $xy \in N_f \Rightarrow yx \in N_f$. So assume $xy \in N_f$. We then get from (15.3) that $f(x)^2 - f(y)^2 = f(xy)f(xy^{-1}) = 0$, so $f(x)^2 = f(y)^2$. Next we note by (15.3) and the fact that f is odd (according to (b)) that

$$f(yx)f(yx^{-1}) = f(y)^2 - f(x)^2 = 0,$$

and

$$f(yx)f(y^{-1}x) = f(x^{-1}y^{-1})f(x^{-1}y) = f(x^{-1})^2 - f(y^{-1})^2 = 0.$$

The desired conclusion $f(yx) = 0$ follows from these two formulas if $f(yx^{-1}) \neq 0$ or $f(y^{-1}x) \neq 0$. Left is the case of $f(yx^{-1}) = f(y^{-1}x) = 0$, i.e., $yx^{-1} \in N_f$ and $y^{-1}x \in N_f$. And here $yx = (yx^{-1})(xy)(y^{-1}x) \in N_f N_f N_f \subseteq N_f$.

(d) The first characterization: Let $u \in N_f$. N_f is normal, so (15.3) implies $f(xu)^2 - f(x)^2 = f(xux)f(xux^{-1}) = f(xux)0 = 0$. The converse inclusion is trivial (take $x = u^{-1}$).

The second characterization: Let $u \in N_f$. By the just proved result $f(xu^2)f(x) = f(xuu)f(xuu^{-1}) = f(xu)^2 - 0 = f(x)^2$. If $f(x) \neq 0$ we may divide by it and get $f(xu^2) = f(x)$ as desired. If $f(x) = 0$ then $x \in N_f$. Since also $u \in N_f$ we have $xu^2 \in N_f$, so $f(xu^2) = 0$. Thus $f(xu^2) = 0 = f(x)$.

About the converse implication: If $f(xu^2) = f(x)$ for all $x \in G$ we get with $x = u^{-1}$ that $f(u) = f(u^{-1}) = -f(u)$, which implies $f(u) = 0$.

(e) We get $f(yx)f(yx^{-1}) = f(y)^2 - f(x)^2$ by interchanging x and y in (15.3). This implies, f being odd, that $f(yx)f(xy^{-1}) = f(x)^2 - f(y)^2$. Comparing with (15.3) we get $[f(xy) - f(yx)]f(xy^{-1}) = 0$. When we multiply this by $f(xy) + f(yx)$ we get

$$\begin{aligned} 0 &= [f(xy) + f(yx)][f(xy) - f(yx)]f(xy^{-1}) \\ &= (f(xy)^2 - f(yx)^2)f(xy^{-1}) = f(xy^2x)f(xyx^{-1}y^{-1})f(xy^{-1}), \end{aligned}$$

which has (e) as a consequence: If the first factor $f(xy^2x) = 0$ then $xy^2x \in N_f$, which implies $x^2y^2 \in N_f$ by the normality of N_f. If $f(xyx^{-1}y^{-1}) = 0$, then $xyx^{-1}y^{-1} \in N_f$. If $f(xy^{-1}) = 0$, then $xy^{-1} \in N_f$. From that we get, N_f being a normal subgroup, that $xyx^{-1}y^{-1} = [x(xy^{-1})^{-1}x^{-1}]xy^{-1} \in N_f$.

(f) We first show that the commutator $[x^2, y^2] \in N_f$ for all $x, y \in G$: To do so we work our way through the two possibilities of (e), using N_f is normal:

(i) Assume first that $[x, y] \in N_f$. In this case the result follows from the formula $[x^2, y^2] = x[x, y]x^{-1}[x, y]yx[x, y]x^{-1}[x, y]y^{-1}$.

(ii) Assume next that $x^2y^2 \in N_f$. Since N_f is normal $ab \in N_f \Rightarrow ba \in N_f$. Thus $y^2x^2 \in N_f$ and so $x^{-2}y^{-2} \in N_f$. Finally $[x^2, y^2] = (x^2y^2)(x^{-2}y^{-2}) \in N_f$.

We continue by showing that the commutator $[x_1^2x_2^2 \cdots x_m^2, y_1^2y_2^2 \cdots y_n^2] \in N_f$ for all $x_1, x_2, \ldots, x_m, y_1, y_2, \ldots, y_n \in G$ and $m, n \in \mathbb{N}$: From the formulas

$[a, bc] = [a, b][bab^{-1}, bcb^{-1}]$ and $[bc, a] = [bcb^{-1}, bab^{-1}][b, a]$ it follows by induction that $[x_1^2 x_2^2 \cdots x_m^2, y_1^2 y_2^2 \cdots y_n^2]$ is a product of factors of the form $[x^2, y^2]$ where $x, y \in G$. Since such factors belong to the group N_f so does their product. □

The assumptions of Theorem 15.3 (that G and $[G, G]$ are generated by their squares) are according to Lemma A.11 satisfied for any connected Lie group G. Indeed, if G is a connected Lie group, then so is $[G, G]$ by [201, Theorem 3.18.8]. The assumptions can be verified directly for examples like the Heisenberg group and the $(ax + b)$-group.

Theorem 15.3. *If G is a group such that G and $[G, G]$ are generated by their squares, then the functions $f : G \to \mathbb{C}$ satisfying* (15.3) *are the ones of the forms*

(a) $f = c(\chi - \check{\chi})$, *where χ is a character on G and $c \in \mathbb{C} \setminus \{0\}$ a constant.*

(b) $f : G \to (\mathbb{C}, +)$ *additive.*

The character χ in (a) is unique, except that it can be interchanged with $\check{\chi}$.

If G is a topological group and f is continuous then the character χ from (a) is continuous.

If G is a locally compact group and f is measurable with respect to a (left) Haar measure on G then f is continuous.

Proof. We observed in Lemma 15.1 that the functions of the forms in (a) and (b) are solutions of (15.3), so it remains to prove that any solution f of (15.3) has one of the indicated forms. To do so we proceed as follows:

Step 1. We show that $[G, G] \subseteq N_f$, i.e., that f vanishes on the commutator subgroup $[G, G]$: This is immediate from Lemma 15.2(f) since by hypothesis G is generated by its squares.

Step 2. We show that f is a function on $G/[G, G]$: From Lemma 15.2(d) we read that $f(xu^2) = f(x)$ for all $x \in G$ and $u \in N_f$. In particular for all $u \in [G, G]$, because $[G, G] \subseteq N_f$. It follows that $f(xu_1^2 u_2^2 \cdots u_n^2) = f(x)$ for all $x \in G$ and $u_1, u_2, \ldots, u_n \in [G, G]$. By hypothesis $[G, G]$ is generated by its squares so $f(xy) = f(x)$ for all $x \in G$ and $y \in [G, G]$.

Step 3. Since G is generated by its squares, so is the abelian group $G/[G, G]$, which means it is 2-divisible.

Step 4. From now on we may assume that G is a 2-divisible, abelian group, because f is a function on the abelian group $G/[G,G]$.

The case of $f = 0$ falls under (b), so we may assume that $f \neq 0$. Since G is 2-divisible there is an $x_0 \in G$ such that $f(x_0^2) \neq 0$.

We may obtain cos from sin by the identity $2 \sin a \cos x = \sin(x+a) - \sin(x-a)$. Thinking of f as sin we are lead to consider the analogue $g(x) := f(xx_0^2) - f(xx_0^{-2})$, $x \in G$, of cosine. Having the sine addition formula (4.3) in mind we compute its right hand side by help of (15.5) applied to the individual terms:

$$\begin{aligned}
&f(x^2)g(y^2) + f(y^2)g(x^2) \\
&\quad = f(x^2)f(y^2x_0^2) - f(x^2)f(y^2x_0^{-2}) + f(y^2)f(x^2x_0^2) - f(y^2)f(x^2x_0^{-2}) \\
&\quad = f(xyx_0)^2 - f(xy^{-1}x_0^{-1})^2 - f(xyx_0^{-1})^2 + f(xy^{-1}x_0)^2 \\
&\qquad + f(xyx_0)^2 - f(xy^{-1}x_0)^2 - f(xyx_0^{-1})^2 + f(xy^{-1}x_0^{-1})^2 \\
&\quad = 2f(xyx_0)^2 - 2f(xyx_0^{-1})^2 = 2f(x^2y^2)f(x_0^2),
\end{aligned}$$

which means, G being 2-divisible, that

$$f(x)g(y) + f(y)g(x) = 2f(xy)f(x_0^2) \qquad \text{for all } x, y \in G.$$

Hence the pair $(f, g/[2f(x_0^2)])$ is a solution of the sine addition formula (4.3). By Corollary 4.4 only three possibilities occur when $f \neq 0$:

(i) $f = c\chi$, where χ is a character on G and $c \in \mathbb{C} \setminus \{0\}$ a constant.

(ii) There exist two different characters χ_1 and χ_2 on G and a constant $c \in \mathbb{C}$ such that $f = c(\chi_1 - \chi_2)$.

(iii) There exist a character χ and an additive function a on G such that $f = \chi a$.

The first case is not realized, because $f(e) = 0$ by Lemma 15.2(a).

In the second case we find because f is odd (Lemma 15.2(b)) that $\chi_1 + \check{\chi}_1 = \chi_2 + \check{\chi}_2$. From Corollary 3.19 we infer that either $\chi_1 = \chi_2$ or $\chi_1 = \check{\chi}_2$. The first possibility gives $f = 0$, contradicting our hypothesis, so left is $f = c(\chi_1 - \check{\chi}_1)$, i.e., (a) holds.

In the third case a substitution of $f = a\chi \neq 0$ into (15.3) reveals that $\chi(x)^2$ is a constant. The constant must be 1, because $\chi(e) = 1$. Since G is 2-divisible we can to any $y \in G$ find an $x \in G$ such that $y = x^2$. Now $\chi(y) = \chi(x^2) = \chi(x)^2 = 1$. Thus $f = a\chi = a$, i.e., (b) holds.

The statement about the essential uniqueness of χ is immediate from Corollary 3.20. The statement about the continuity of χ comes from

Theorem 3.18(d). To get the statement about the continuity of f combine Theorem 3.18(f) and Proposition D.4(b) in case (a), and in case (b) refer to Proposition D.4(a). □

Corollary 15.4. *If G is a connected Lie group, then the functions $f : G \to \mathbb{C}$ satisfying* (15.3) *are the ones of the forms*

(a) $f = c(\chi - \check{\chi})$, *where χ is a character on G and $c \in \mathbb{C} \setminus \{0\}$ a constant.*

(b) $f : G \to (\mathbb{C}, +)$ *additive.*

The character χ in (a) is unique, except that it can be interchanged with $\check{\chi}$.

If f is measurable with respect to a (left) Haar measure on G then $f \in C^\infty(G)$.

Proof. The assumptions that G and $[G, G]$ are generated by their squares are according to Lemma A.11 satisfied for any connected Lie group G. Indeed, if G is a connected Lie group, then so is $[G, G]$ by [201, Theorem 3.18.8].

If f is measurable then the character χ is continuous by Theorem 15.3. And any continuous character on a Lie group G is in $C^\infty(G)$ by [33, Proposition 3.12]. □

For abelian groups Theorem 15.3 reduces to the following:

Corollary 15.5. *If G is a 2-divisible, abelian group, then the functions $f : G \to \mathbb{C}$ satisfying* (15.4) *are the ones of the forms*

(a) $f = c(\chi - \check{\chi})$, *where χ is a character on G and $c \in \mathbb{C} \setminus \{0\}$ a constant.*

(b) $f : G \to (\mathbb{C}, +)$ *additive.*

The character χ in (a) is unique, except that it can be interchanged with $\check{\chi}$.

If G is a topological group and f is continuous then the character χ from (a) is continuous.

Remark 15.6. The condition in Corollary 15.5 that G is abelian may be replaced by the condition that f is abelian.

Example 15.7. Combining Corollary 15.4 or Corollary 15.5 with Example 3.7(a) we see that the solutions $f \in C(\mathbb{R})$ of the functional equation (15.2) are

(a) $f(x) = c\sin(\lambda x)$, $x \in \mathbb{R}$, where c and λ are complex constants.

(b) $f(x) = cx$, $x \in \mathbb{R}$, where c is a complex constant.

Example 15.8. As an example of a non-abelian group we consider

$$G := \mathbb{R}^2 \times_s \mathrm{SO}(2) = \{(x, R) \mid x \in \mathbb{R}^2,\ R \in \mathrm{SO}(2)\},$$

where

$$(x_1, R_1)(x_2, R_2) := (x_1 + R_1 x_2, R_1 R_2) \text{ for } (x_i, R_i) \in G, \qquad i = 1, 2.$$

It acts on $\mathbb{R}^2$ by the prescription $(x, R) \cdot y = x + Ry$, and it is isomorphic to the group of rigid, orientation preserving maps of $\mathbb{R}^2$ onto $\mathbb{R}^2$. G being a connected Lie group, we may use Theorem 15.3 to find the continuous solutions f of the sine functional equation (15.3) on it:

In Examples 3.17 and 2.13 we computed the continuous characters and the continuous additive functions on G. Using that we find that the continuous solutions of the sine functional equation on G are the functions f of the form

$$f(x, \begin{pmatrix} \cos\theta & -\sin\theta \\ \sin\theta & \cos\theta \end{pmatrix}) = c(e^{in\theta} - e^{-in\theta}) = 2ic\sin(n\theta) \qquad \text{for } \theta \in \mathbb{R},$$

where $c \in \mathbb{C}$ and $n \in \mathbb{Z}$ are constants.

15.3 The sine equation on cyclic groups

In this section we find the solutions of the sine functional equation on the group $(\mathbb{Z}, +)$. This example is interesting, because $\mathbb{Z}$ is not 2-divisible, so that Corollary 15.5 does not apply. Thus we must resort to ad hoc arguments

Note that if f is a solution of the sine functional equation on $\mathbb{Z}$ such that $f(1) = 0$, then $f = 0$ (follows from Lemma 15.2(c)). That is why $f(1) \neq 0$ in Example 15.9.

Example 15.9. For any $a_1 \in \mathbb{C} \setminus \{0\}$ and $a_2 \in \mathbb{C}$ there exists exactly one solution $f : \mathbb{Z} \to \mathbb{C}$ of the sine functional equation (15.4) on the group $G = \mathbb{Z}$ such that $f(1) = a_1$ and $f(2) = a_2$.

(a) If $f(2)/f(1) \neq \pm 2$ then there exist constants $c, \lambda \in \mathbb{C} \setminus \{0\}$ such that $f(n) = c\sin(\lambda n)$ for all $n \in \mathbb{Z}$.

(b) If $f(2)/f(1) = 2$ then there exists a constant $c \in \mathbb{C} \setminus \{0\}$ such that $f(n) = cn$ for all $n \in \mathbb{Z}$.

(c) If $f(2)/f(1) = -2$ then there exists a constant $c \in \mathbb{C} \setminus \{0\}$ such that $f(n) = c(-1)^{n-1}n$ for all $n \in \mathbb{Z}$.

The functions defined by the formulas in (a), (b) and (c) are solutions of the sine functional equation (15.4) on $\mathbb{Z}$.

The cases (a) and (b) are the restrictions from $\mathbb{R}$ to $\mathbb{Z}$ of the continuous solutions on $\mathbb{R}$ described in the introduction to the present chapter. The function in (c) is for $c \neq 0$ of a type that is not listed in Corollary 15.5, although it can be found in Lemma 15.1(ii).

Proof of the statements of the example. We take the last statement first: According to Lemma 15.1 and easy computations the functions listed in (a), (b) and (c) above are solutions of the sine functional equation on $\mathbb{Z}$ for any $c, \lambda \in \mathbb{C}$.

We next prove uniqueness of a solution $f : \mathbb{Z} \to \mathbb{C}$ of the sine functional equation such that $f(1) = a_1$ and $f(2) = a_2$: We see from the sine functional equation with $y = 1$, i.e.,

$$f(n+1)f(n-1) = f(n)^2 - f(1)^2 \qquad \text{for all } n \in \mathbb{Z}, \tag{15.6}$$

that $f(3)f(1) = f(2)^2 - f(1)^2$ so that $f(3)$ is determined uniquely from $a_1 = f(1)$ and $a_2 = f(2)$. Indeed, $f(1) \neq 0$, so that we can divide by $f(1)$. Next we find $f(4)$ from $f(4)f(2) = f(3)^2 - f(1)^2$, unless of course $f(2) = 0$. More generally, proceeding by successive applications of (15.6) we see that $f(n+1)$ is unique, given $f(1) = a_1$ and $f(2) = a_2$, if $f(i) \neq 0$ for $i = 2, 3, \ldots, n-1$.

If a_1 and a_2 are so that this procedure of successive applications of (15.6) never produces an $i \in \mathbb{N}$ such that $f(i) = 0$ then $f(n)$ is uniquely determined for any $n \in \mathbb{N}$ from $f(1) = a_1$ and $f(2) = a_2$. Since f is odd (Lemma 15.2(b)) $f(n)$ is also determined for negative integers and 0. Thus the uniqueness is established in this case. The other possibility is that the procedure produces an $i \in \mathbb{N}$ such that $f(i) = 0$. Let $n \geq 2$ denote the smallest natural number such that $f(n) = 0$. The procedure shows that $f(1), f(2), \ldots, f(n+1)$ are unique, given $f(1) = a_1$ and $f(2) = a_2$. Since f is odd (Lemma 15.2(b)) $f(i)$ is actually uniquely determined from $f(1) = a_1$ and $f(2) = a_2$ for all $i \in \mathbb{Z}$ in the interval $[-(n+1), n+1]$. The rest of the argument follows from the fact that f is periodic of period $2n$, because $f(n) = 0$ (Lemma 15.2(d)).

This means that f is determined by its values on any interval of length $2n$, in particular by its values on $[-(n+1), n+1]$. This proves the uniqueness.

We continue by proving existence of a solution $f : \mathbb{Z} \to \mathbb{C}$ of the sine functional equation such that $f(1) = a_1$ and $f(2) = a_2$, where $a_1 \neq 0$. During that process we obtain (a), (b) and (c) as well.

If $a_2/a_1 \neq \pm 2$, we write $a_2/a_1 = a + a^{-1}$ where $a \in \mathbb{C} \setminus \{0\}$. As a we may take

$$a = \frac{a_2/a_1 + \sqrt{(a_2/a_1)^2 - 4}}{2}$$

for any choice of the square root. Noting that $a \neq a^{-1}$ because $a_2/a_1 \neq \pm 2$, we define

$$f(n) := \frac{a_1}{a - a^{-1}}(a^n - a^{-n}), \qquad n \in \mathbb{Z}.$$

f is a solution of the sine functional equation such that $f(1) = a_1$ and $f(2) = a_2$. To see that f is a solution of the sine functional equation we observe that $n \mapsto a^n$ is a character on $\mathbb{Z}$ and apply Lemma 15.1(i). We get (a) when we write $a = \exp(i\lambda)$.

If $a_2/a_1 = 2$ we may take $f(n) := a_1 n$, $n \in \mathbb{Z}$. And finally, if $a_2/a_1 = -2$ we may take $f(n) := a_1(-1)^{n-1}n$, $n \in \mathbb{Z}$. $\square$

Example 15.10. Let G be the finite cyclic group $\mathbb{Z}_N$ where $N \in \mathbb{N}$ is the order of the group. The solutions of the sine equation on $\mathbb{Z}_N$ are the functions of the form

$$f(n + N\mathbb{Z}) = c \sin\Big(2\pi \frac{M}{N} n\Big), \qquad n \in \mathbb{Z}, \tag{15.7}$$

where $c \in \mathbb{C}$ and $M \in \mathbb{N}$ are constants.

To see this we let $\rho : \mathbb{Z} \to \mathbb{Z}_N$ be the map $\rho(n) := n + N\mathbb{Z}$. If $f = 0$ then f has the desired form. If f is a non-zero solution of the sine equation on $\mathbb{Z}_N$ then $f \circ \rho$ is a solution of the sine equation on $\mathbb{Z}$. The last two possibilities for $f \circ \rho$ in Example 15.9 are unbounded, but f is bounded, being a function on a finite set. So left is only the first possibility $f \circ \rho(n) = c \sin(\lambda n)$ for all $n \in \mathbb{Z}$. $f \circ \rho(N) = f \circ \rho(0)) = 0$ implies that $\lambda N = \pi p$ for some $p \in \mathbb{Z}$. However, if n is replaced by $n + N$ in (15.7) the left hand side does not change. This implies that p is even, say $p = 2M$. The remaining arguments are left to the reader.

15.4 A more general functional equation

Throughout this section we let G be a group and $z_0 \in Z(G)$ a fixed element. We are interested in the solutions $f : G \to \mathbb{C}$ of the functional equation (15.8) below. It reduces to the sine functional equation (15.3) when $z_0 = e$. On $\mathbb{R}$ it characterizes sine functions (Corollary 15.13).

Proposition 15.11. *Let G be a group and let $z_0 \in Z(G)$. The solutions $f : G \to \mathbb{C}$ of the functional equation*

$$f(xyz_0)f(xy^{-1}z_0) = f(x)^2 - f(y)^2, \qquad x, y \in G, \tag{15.8}$$

are the solutions of the sine functional equation (15.3) *for which* $f(z_0) = 0$.

Proof. Let f be a solution of the sine functional equation (15.3) such that $f(z_0) = 0$. Now $z_0 \in Z(G)$ and z_0^2 is a period of f (Lemma 15.2(d)), so

$$\begin{aligned} f(xyz_0)f(xy^{-1}z_0) &= f(xyz_0)f(xy^{-1}z_0^{-1}) = f(xyz_0)f(x(yz_0)^{-1}) \\ &= f(x)^2 - f(yz_0)^2 = f(x)^2 - f(y)^2, \end{aligned}$$

where the last equality sign is justified by Lemma 15.2(d).

Let conversely $f : G \to \mathbb{C}$ be a solution of the functional equation (15.8). Putting $x = y = e$ in (15.8) we get that $f(z_0)^2 = 0$ and so $f(z_0) = 0$. Thus it is left to show that f is a solution of (15.3).

We also get that $f(z_0^{-1}) = f(e) = 0$: The left hand side of (15.8) does not change if y is replaced by y^{-1}, so we find that $f(y)^2 = f(y^{-1})^2$. In particular $f(z_0^{-1})^2 = f(z_0)^2 = 0$, so that also $f(z_0^{-1}) = 0$. We obtain $f(e) = 0$ by taking $x = z_0^{-1}$ and $y = e$ in (15.8).

We prove next that $f(xz_0^2) = f(x)$ for all $x \in G$: Putting $y = z_0$ in (15.8) gives $f(xz_0^2)f(x) = f(x)^2 - f(z_0)^2 = f(x)^2$, so $f(xz_0^2) = f(x)$ if $f(x) \neq 0$. If $f(x) = 0$ we replace y by xz_0^2 in (15.8) and get $0 = 0 - f(xz_0^2)^2$, so $f(xz_0^2) = 0$.

Finally we compute, using $f(xz_0^2) = f(x)$ and $z_0 \in Z(G)$, that

$$\begin{aligned} f(xy)f(xy^{-1}) &= f(xyz_0z_0)f(x(yz_0)^{-1}z_0) = f(x)^2 - f(yz_0)^2 \\ &= f(x)^2 - f(yez_0)f(ye^{-1}z_0) = f(x)^2 - (f(y)^2 - f(e)^2) \\ &= f(x)^2 - f(y)^2 \qquad \text{for all } x, y \in G. \end{aligned}$$ □

Combining Proposition 15.11 and Theorem 15.3 we get an extension of Theorem 15.3. It reduces to Theorem 15.3, if $z_0 = e$.

Theorem 15.12. *If G is a group such that G and $[G,G]$ are generated by their squares, then the functions $f : G \to \mathbb{C}$ satisfying (15.8), where $z_0 \in Z(G)$, are the ones of the forms*

(a) $f = c(\chi - \check{\chi})$, *where χ is a character on G such that $\chi(z_0)^2 = 1$, and $c \in \mathbb{C} \setminus \{0\}$ a constant.*

(b) $f : G \to (\mathbb{C}, +)$ *an additive function such that $f(z_0) = 0$.*

The character χ in (a) is unique, except that it can be interchanged with $\check{\chi}$.

If G is a topological group and f is continuous then the character χ from (a) is continuous.

Corollary 15.13. *Let $z_0 \in \mathbb{R} \setminus \{0\}$. The solutions $f \in C(\mathbb{R})$ of the functional equation*

$$f(x + y + z_0)f(x - y + z_0) = f(x)^2 - f(y)^2, \qquad x, y \in \mathbb{R},$$

are the functions

$$f(x) = c\sin\left(\frac{n\pi}{z_0}x\right), \qquad x \in \mathbb{R},$$

where $n \in \mathbb{N}$ and $c \in \mathbb{C}$.

Example 15.7 has the result for the excluded case of $z_0 = 0$.

15.5 Exercises

Exercise 15.1. Derive the identity (15.1). What is your candid opinion of the following derivation based on the formula $(a + b)(a - b) = a^2 - b^2$?

$$\sin(x + y)\sin(x - y) = (\sin x + \sin y)(\sin x - \sin y) = (\sin x)^2 - (\sin y)^2.$$

Exercise 15.2. Find the solutions $f : \mathbb{R} \to \mathbb{C}$ of

$$f(x + y)f(x - y) = \sin^2 x - \sin^2 y, \qquad x, y \in \mathbb{R}.$$

Hint: The trigonometric identity (15.1).

Exercise 15.3. In the set up of Lemma 15.2 show that $f(u^2x) = f(x)$ for all $x \in G$ and $u \in N_f$.

Exercise 15.4. Find the continuous solutions of the sine equation on the Heisenberg group and on the $(ax+b)$-group.

Exercise 15.5. Prove the result stated in Example 15.7.

Exercise 15.6. Prove that every Lebesgue-measurable solution $f : \mathbb{R}^n \to \mathbb{C}$ of the functional equation

$$f(x+y)f(x-y) = f(x)^2 - f(y)^2, \qquad x, y \in \mathbb{R}^n,$$

has one of the following two forms

1. $f(x) = c\sin\langle\lambda, x\rangle$, $x \in \mathbb{R}^n$, where $c \in \mathbb{C}$ and $\lambda \in \mathbb{C}^n$ are constants.
2. $f(x) = \langle\lambda, x\rangle$, $x \in \mathbb{R}^n$, where $\lambda \in \mathbb{C}^n$ is a constant.

Hint: Combine Corollary 15.5, Theorem 3.18 and Proposition D.4.

Exercise 15.7. Let $G = (M(n \times n, \mathbb{R}), +)$, where $n \geq 2$. Let $f \in C(G)$ be a central solution of the sine functional equation on G. Show that f has one of the following two forms:

(i) $f(x) = \alpha \sin(\beta \operatorname{tr} x)$, $x \in G$, where $\alpha, \beta \in \mathbb{C}$ are constants.

(ii) $f(x) = \alpha \operatorname{tr} x$, $x \in G$, where $\alpha \in \mathbb{C}$ is a constant.

Hint: Apply Exercise 3.22 to $M(n\times n, \mathbb{R})$ equipped with the Hilbert-Schmidt norm.

Exercise 15.8. Let $\mathcal{H}$ be a real Hilbert space with inner product $\langle -, - \rangle$. Let $f : \mathcal{H} \to \mathbb{R}$ be a continuous solution of the sine functional equation (15.4) on $\mathcal{H}$, viewed as an additive group. Show f has one of the following forms where $c \in \mathbb{R}$ and $x_0 \in \mathcal{H}$ denote constants:

$$f(x) = c\sinh\langle x, x_0\rangle, \quad f(x) = c\sin\langle x, x_0\rangle \quad \text{or} \quad f(x) = \langle x, x_0\rangle, \qquad x \in \mathcal{H}.$$

Proceed as follows: Write by help of Exercise 3.22 the continuous character χ from the first part of Corollary 15.5 in the form

$$\chi(x) = e^{a(x)} = e^{\alpha(x)+i\beta(x)}, \qquad x \in \mathcal{H},$$

where $\alpha, \beta : \mathcal{H} \to \mathbb{R}$ are two continuous linear functionals on $\mathcal{H}$, and so for some $x_0, y_0 \in \mathcal{H}$ can be written in the form $\alpha(x) = \langle x, x_0\rangle$ and $\beta(x) = \langle x, y_0\rangle$

for all $x \in \mathcal{H}$. Now, for all $x \in \mathcal{H}$:

$$\begin{aligned} f(x) &= c\big(e^{\langle x,x_0\rangle}e^{i\langle x,y_0\rangle} - e^{-\langle x,x_0\rangle}e^{-i\langle x,y_0\rangle}\big) \\ &= 2c\big(\sinh\langle x,x_0\rangle\cos\langle x,y_0\rangle + i\sin\langle x,y_0\rangle\cosh\langle x,x_0\rangle\big). \end{aligned} \tag{15.9}$$

We may assume $c \neq 0$, because the statement is true in this case.

Get a contradiction assuming that x_0 and y_0 are linearly independent, using f is real-valued:

Choose $\alpha \in \mathbb{R}$ such that $x_2 := x_0 - \alpha y_0 \perp y_0$. Note that $\langle x_2, x_0\rangle \neq 0$. We infer from (15.9) that $f(x_2) = 2c\sinh\langle x_2, x_0\rangle$ so that c is real. Thus $f(x)/c \in \mathbb{R}$ so we get from

$$\frac{f(x)}{2c} = \sinh\langle x,x_0\rangle\cos\langle x,y_0\rangle + i\sin\langle x,y_0\rangle\cosh\langle x,x_0\rangle$$

that $\sin\langle x,y_0\rangle = 0$ for all $x \in \mathcal{H}$. But then $y_0 = 0$ contradicting the linear independence of x_0 and y_0.

Thus x_0 and y_0 are linearly dependent. If $x_0 = 0$ or $y_0 = 0$ we get the desired result from (15.9), so we may assume that $x_0 \neq 0$ and that $y_0 = rx_0$ for some $r \in \mathbb{R} \setminus \{0\}$. Writing $c = c_1 + ic_2$ where $c_1, c_2 \in \mathbb{R}$ we find that for all $x \in \mathcal{H}$, f being real-valued,

$$c_1\sin(r\langle x,x_0\rangle)\cosh\langle x,x_0\rangle + c_2\sinh\langle x,x_0\rangle\cos(r\langle x,x_0\rangle) = 0,$$

which implies that $c_1 = c_2 = 0$.

Exercise 15.9. Find the real-valued solutions $f \in C(\mathbb{R})$ of (15.2).

Exercise 15.10. Extend Corollary 15.13 from $\mathbb{R}$ to $\mathbb{R}^n$.

Exercise 15.11. We will in this exercise study a generalization of the sine functional equation (15.2) on the real line, viz.,

$$g(x+y)f(x-y) = g(x)f(x) - g(y)f(y), \qquad x,y \in \mathbb{R}, \tag{15.10}$$

with two functions $f, g : \mathbb{R} \to \mathbb{C}$ to be determined, not just one as in (15.2). If $g = f$ we get the sine equation, but the following identity reveals a solution such that $g \neq f$:

$$\cos(x+y)\sin(x-y) = \cos x\sin x - \cos y\sin y, \qquad x,y \in \mathbb{R}.$$

Theorem 15.14. *The solutions* $f, g : \mathbb{R} \to \mathbb{C}$ *of* (15.10) *are the following, where* $\chi \neq 1$ *denotes a character of* $\mathbb{R}$ *and* $c \in \mathbb{C} \setminus \{0\}$ *a constant,* $\alpha, \beta \in \mathbb{C}$ *constants and* $a : \mathbb{R} \to \mathbb{C}$ *an additive function:*

(a) $g = 0$ *and* f *arbitrary, or* $f = 0$ *and* g *arbitrary.*

(b) $f = c(\chi - \check{\chi})$ *and* $g = \alpha(\chi - \check{\chi}) + \beta(\chi + \check{\chi})$.

(c) $f = a$ *and* $g = \alpha a + \beta$.

We ask the reader to provide a proof of the theorem along the following lines:

Proof. The cases $f = 0$ and $g = 0$ are trivial, so we will assume from now on that $f \neq 0$ and $g \neq 0$.

(a) Prove f is odd.

Hint: Interchange x and y in (15.10) and add the result to (15.10).

(b) Prove that both the odd part g_o and the even part g_e of g satisfy (15.10) with f unchanged, i.e., for all $x, y \in \mathbb{R}$:

$$g_o(x + y)f(x - y) = g_o(x)f(x) - g_o(y)f(y), \tag{15.11}$$

and

$$g_e(x + y)f(x - y) = g_e(x)f(x) - g_e(y)f(y). \tag{15.12}$$

Hint: By linearity it suffices to prove the statements for $\check{g}$.

(c) Prove that $g_o = \alpha f$, where $\alpha \in \mathbb{C}$ is a constant.

Hint: Replace y by $-y$ in (15.11) and get $g_o(u)f(v) = g_o(v)f(u)$.

(d) Prove that $g_e(u)f(v) = g_e(0)[f(u + v) - f(u - v)]/2$ for all $u, v \in \mathbb{R}$.

Hint: Apply $g_e(0)f(2x) = 2g_e(x)f(x)$ to the right hand side of (15.12).

(e) Prove that f is a solution of the sine functional equation (15.2).

Hint: Since $g = g_o + g_e \neq 0$ either $g_o \neq 0$ or $g_e \neq 0$. Treat each case separately: When $g_o \neq 0$ use that $g_o = \alpha f$. When $g_e \neq 0$ proceed as follows. Show that $g_e(0) = 0$ implies the contradiction $g_e = 0$. So $f(2x) = 2g_e(x)f(x)/g_e(0)$. Using this derive that $f(2(x + y))f(2(x - y)) = f(2x)^2 - f(2y)^2$ for all $x, y \in \mathbb{R}$.

We recall from Theorem 15.3 that there are two possibilities for f:

(i) $f = c(\chi - \check{\chi})$, where $c \in \mathbb{C}$ is a constant, and χ is a character on $\mathbb{R}$.

(ii) f is additive.

(f) When $f = c(\chi - \check{\chi})$ prove that $f(u+v) - f(u-v) = [\chi(u) + \check{\chi}(u)]f(v)$, so that $g_e(u) = \frac{1}{2}g_e(0)[\chi(u) + \check{\chi}(u)]$.

(g) Finish the proof when $f = c(\chi - \check{\chi})$.

(h) Proceed similarly when f is additive. □

15.6 Notes and remarks

Most of the results on non-abelian groups are due to Corovei [58], and the present chapter builds upon [58]. We extend his work by replacing the assumption that the underlying group is 2-divisible by the assumption that it is generated by its squares. Aczél [2, Section 2.5.2] and Kannappan [132, Section 3.4] have references to earlier works about the sine functional equation on abelian groups.

Sinopoulos [175, 176, 177] studies functional equations of the form $f(x+y)f(x-y) = P[f(x), f(y)]$ and Pexiderized versions on abelian groups. P is a polynomial. He asks for which P will non-trivial solutions exist?

Theorem 15.3 extends Corovei [58, Theorem 3] from 2-divisible groups to groups generated by their squares. Corovei considers functions taking values in a quadratically closed field of characteristic $\neq 2$, whereas our functions are complex-valued. Our procedure extends from $\mathbb{C}$ to his case.

Corollary 15.5 can be found as Aczél and Dhombres [6, Ch. 13, Theorem 14] in which our range space $\mathbb{C}$ replaced by a quadratically closed field divisible by 2 (fields are by definition commutative).

Example 15.7: For the history of Example 15.7 see Aczél [2, Section 2.5.2].

The result of Example 15.9 is due to E. Thue Poulsen (private communication).

Corollary 15.13 is Kannappan [126, Theorem 2]. Another characterization of sine functions, due to van Vleck [206], is discussed in Exercise 9.18.

Exercise 15.11: Kannappan [132, Theorem 3.52] finds the real-valued solutions of (15.10).

Exercise 15.6: The result was derived by other methods for locally integrable, real-valued functions by Chung and Kim [48].

Exercise 15.7 is essentially Kurepa [136, Theorem 2].

Exercise 15.8 is modified from Kurepa [136].

Chapter 16

The Cocycle Equation

This chapter makes a very brief foray into the area of the cocycle equation, discussing essentially only the compact and abelian cases. We are not able to discuss the cocycle equation on groups and semigroups in full generality, because the subject is too large as is apparent from the literature, both in mathematics and physics. We strive to make our presentation of the cocycle equation self-contained and elementary. Consult monographs on the subject like [35] and [202] for other results and heavier algebraic machinery.

Historically, the cocycle equation plays a central role in the theory of projective representations (also known as ray representations) of groups in quantum mechanics, going back to the seminal paper of Bargmann [23] in 1954. Cocycles are known by various other names. For example, Bargmann [23] working with a Lie group G terms a continuous solution $F : G \times G \to \mathbb{R}$ of (16.1) an *exponent* if it also satisfies the normalizing condition $F(e, e) = 0$. Varadarajan [202] calls such a map a *multiplier* (or a K-*multiplier* if F takes values in a locally compact group K).

Throughout the chapter G denotes a group with neutral element e and $(H, +)$ an abelian group with neutral element 0.

16.1 Introduction

We have on various occasions met the Cauchy difference $Cf(x, y) = f(xy) - f(x) - f(y)$ of a map $f : G \to H$. A natural question is: Which maps $F : G \times G \to H$ can be written in the form $F = Cf$ for some $f : G \to H$? As is easy to check, a necessary condition is that F is a cocycle as defined below. The condition is sufficient for some groups G and H, but in general not.

Definition 16.1.

(a) A map $F : G \times G \to H$ satisfying *the cocycle equation*

$$F(x,y) + F(xy,z) = F(x,yz) + F(y,z) \qquad \text{for all } x,y,z \in G \quad (16.1)$$

is called a *cocycle* on G to H. The set of cocycles is an abelian group that we denote $Z^2(G,H)$.

(b) For each function $f : G \to H$ we define $df : G \times G \to H$ by

$$df(x,y) := f(x) + f(y) - f(xy), \qquad x,y \in G. \quad (16.2)$$

A map of the form df is a *coboundary* with generator f. It is easy to verify that every coboundary df is a cocycle. The set of coboundaries is a subgroup $B^2(G,H)$ of $Z^2(G,H)$.

(c) The quotient group $H^2(G,H) := Z^2(G,H)/B^2(G,H)$ is called the *second cohomology group*.

The formula $Cf = -df$ in which Cf denotes the Cauchy difference of f, points out a connection to the previous theory. That $df = 0$ means in the terminology of Chapter 2 that f is additive.

Note that two generators of the same coboundary differ by a solution of the additive Cauchy equation. A generator of a continuous coboundary need not be continuous even on $\mathbb{R}$: There are discontinuous solutions of $0 = f(x) - f(y) - f(x+y)$ (Proposition 2.8). A remedy for that calamity is Gajda's result (Corollary D.12).

Lemma 16.2. *Let F be a cocycle on G to H. Then*

$$F(x,e) = F(e,x) = F(e,e) \qquad \text{for all } x \in G.$$

If F is a coboundary, say $F = df$, then

$$F(x,e) = F(e,x) = f(e) \qquad \text{for all } x \in G.$$

Proof. Putting $y = e$ in (16.1) we get the first statement The next is immediate from $F = df$ and the definition of df. □

If $f : G \to H$ is a constant map, say $f(x) = \eta$, then $df = \eta$. Thus a cocycle F can be normalized so that $F(e,e) = 0$: Replace F by $F - df$, where $f(x) = F(e,e)$, $x \in G$.

The cocycle equation is important in several mathematical subjects. For example in the theory of group extensions: An extension of G by H is a short exact sequence $0 \to H \to W \to G \to \{e\}$ of groups. The following lemma gives examples of group extensions W of G by H of the form $W := G \times H$, constructed from a map $F : G \times G \to H$. In the sequence H is mapped into $G \times H$ by the injection map $h \mapsto (e, h + F(e,e))$, and $G \times H$ is mapped onto G by $\pi_1 : (x,h) \mapsto x$ (the projection onto the first variable). Both maps will be homomorphisms.

Lemma 16.3. *Let $F : G \times G \to H$. Equip the set $W := G \times H$ with the composition*

$$(x, h_1)(y, h_2) := (xy, h_1 + h_2 - F(x,y)) \qquad \text{for } (x,h_1),(y,h_2) \in G \times H.$$

(a) *The composition is associative if and only if F is a cocycle.*

(b) *If F is a cocycle, then W is a group. Its neutral element is $(e, F(e,e))$, and the inversion law is*

$$(x,h)^{-1} = (x^{-1}, -h + F(x, x^{-1}) + F(e,e)) \qquad \text{for } (x,h) \in W.$$

The group W is abelian if and only if G is abelian and F is symmetric.

(c) *Let $f : G \to H$. The graph $\{(x, f(x)) \mid x \in G\} \subseteq W$ of f is a subgroup of W if and only if $F(x,y) = f(x) + f(y) - f(xy)$ for all $x, y \in G$, i.e., if and only if F is a coboundary with f as generator.*

(d) *If there exists a subgroup S of W such that $\pi_1|S$ maps S bijectively onto G, then F is a coboundary.*

Proof. The proof is straightforward, so we skip it. □

16.2 Compact groups

Theorem 16.4. *On any compact group every continuous real-valued cocycle is a coboundary with a continuous generator.*

Proof. Let K be a compact group and $F : K \times K \to \mathbb{R}$ a continuous cocycle. Let λ be the normalized Haar measure on K. When we integrate the cocycle equation (16.2) over $z \in K$ with respect to λ we find by the left invariance of λ that

$$F(x,y) + \int_K F(xy,z)\,d\lambda(z) = \int_K F(y,z)\,d\lambda(z) + \int_K F(x,yz)\,d\lambda(z)$$
$$= \int_K F(y,z)\,d\lambda(z) + \int_K F(x,z)\,d\lambda(z),$$

which shows that F is a coboundary with generator

$$f(x) := \int_K F(x,z)\,d\lambda(z), \qquad x \in K.$$

It is easy to see from the formula defining f that f is continuous. □

The averaging argument above works equally well if F takes its values in a sequentially complete, locally convex vector space instead of in $\mathbb{R}$, because then integration can be performed. However, if F takes its values in an abelian group like $\mathbb{T}$, which is not a vector space, integration may not make sense. But we can still manage in some cases where the averaging becomes a finite sum: A finite group is compact, a Haar measure being the counting measure, and the above arguments carry through if the range group H is divisible and torsion free (Exercise 16.3). $\mathbb{R}$, $\mathbb{C}$ and any vector space are divisible and torsion free.

16.3 Abelian groups

Key results about cocycles on abelian groups are expressed below in Theorems 16.7 and 16.8. On an abelian group any coboundary $df(x,y) = f(x) + f(y) - f(xy)$ is clearly symmetric. In this section we derive converse statements.

Lemma 16.5 motivates us to consider skew-symmetric and symmetric cocycles separately. Of course, only the symmetric ones can be coboundaries.

Lemma 16.5. *Let G be abelian. If H is uniquely 2-divisible, then every cocycle on G to H may in exactly one way be written as a sum af a skew-symmetric and a symmetric cocycle on G to H.*

Proof. Let $F : G \times G \to H$ be a cocycle. Then $F'(x,y) := F(y,x)$, $x, y \in G$, is also a cocycle. Indeed, G is abelian, so when we in the computations below rewrite the term $F(z, xy)$ to $F(z, yx)$ and apply (16.1) to it we get

$$\begin{aligned}
&F'(x,y) + F'(xy,z) - F'(x,yz) - F'(y,z) \\
&\quad = F(y,x) + F(z,xy) - F(yz,x) - F(z,y) \\
&\quad = F(y,x) + F(z,yx) - F(zy,x) - F(z,y) \\
&\quad = F(y,x) + (F(z,y) + F(zy,x) - F(y,x)) - F(zy,x) - F(z,y) \\
&\quad = 0.
\end{aligned}$$

The desired decomposition of F is

$$F(x,y) = \frac{F(x,y) - F(y,x)}{2} + \frac{F(x,y) + F(y,x)}{2}, \qquad x, y \in G. \qquad \square$$

Theorem 16.6 presents a general result about skew-symmetric cocycles.

Theorem 16.6. *If $F : G \times G \to H$ is a skew-symmetric cocycle on an abelian group G, then $2F$ is bi-additive.*

Proof. Since F is skew-symmetric, it suffices to prove that $2F$ is additive in the first variable. This is achieved by the following calculation which uses the cocycle equation (16.1) three times.

$$\begin{aligned}
2F(xy,z) &= F(xy,z) - F(z,xy) \\
&= [F(x,yz) + F(y,z) - F(x,y)] - [F(zx,y) + F(z,x) - F(x,y)] \\
&= F(x,yz) - F(zx,y) + F(y,z) - F(z,x) \\
&= F(x,zy) - F(xz,y) + F(y,z) - F(z,x) \\
&= -F(z,y) + F(x,z) + F(y,z) - F(z,x) = 2F(x,z) + 2F(y,z). \quad \square
\end{aligned}$$

A partial inverse of Theorem 16.6: Any bi-additive map $F : G \times G \to H$ satisfies the cocycle equation (16.1). For that statement G need not be abelian.

The above motivates us to turn our attention from the skew-symmetric cocycles to a study of symmetric cocycles.

Theorem 16.7. *If G is abelian and H divisible, then all symmetric cocycles on G to H are coboundaries.*

Proof. We use the notation of Lemma 16.3, except that we write $+$ for the group composition in G, G being abelian. We note that W is abelian here, because G is abelian and F is symmetric. We let $+$ denote the group composition in W and 0_W its neutral element. According to Lemma 16.3(d) we shall produce a subgroup S of W such that $\pi_1|S$ maps S bijectively onto G.

The collection of subgroups S of W such that $\pi_1|S$ is injective is not empty, because it contains $S = \{0_W\}$. Ordering the collection by inclusion we get by Zorn's lemma a maximal subgroup S_1 of W such that $\pi_1|S_1$ is injective. If $\pi_1(S_1) = G$ we may use this S_1 as S, so it is left to derive a contradiction from the assumption that $\pi_1(S_1) \neq G$. Let $x_0 \in G \setminus \pi_1(S_1)$. For any $h \in H$ the subgroup $S_2 := S_1 + \mathbb{Z}(x_0, h)$ of W is strictly bigger than S_1, because $(x_0, h) \notin S_1$. To get a contradiction (to the maximality of S_1) it thus suffices to produce an $h \in H$ such that $\pi_1|S_2$ is injective. In other words, produce an $h \in H$ such that for all $(x_1, h_1) \in S_1$ and $n \in \mathbb{Z}$ the following implication holds:

$$\pi_1((x_1, h_1) + n(x_0, h)) = x_1 + nx_0 = 0 \quad \Rightarrow \quad (x_1, h_1) + n(x_0, h) = 0_W.$$

To obtain this implication it suffices to produce an $h \in H$ such that

$$\text{for all } n \in \mathbb{Z}: \; nx_0 \in \pi_1(S_1) \quad \Rightarrow \quad n(x_0, h) \in S_1. \tag{16.3}$$

Indeed, assume the implication (16.3) holds. If $\pi_1((x_1, h_1) + n(x_0, h)) = x_1 + nx_0 = 0$ then $nx_0 = -x_1 \in \pi_1(S_1)$, so we infer from (16.3) that $n(x_0, h) \in S_1$. Now $(x_1, h_1) + n(x_0, h) \in S_1$ and $\pi_1((x_1, h_1) + n(x_0, h)) = 0$. By the injectivity of π_1 on S_1 we conclude that $(x_1, h_1) + n(x_0, h) = 0_W$.

We proceed to produce an $h \in H$ such that (16.3) holds. The elements of the set $\{n \in \mathbb{Z} \,|\, nx_0 \in \pi_1(S_1)\}$ are integer multiples of a fixed $q \in \mathbb{N} \cup \{0\}$, so it suffices to produce an $h \in H$ such that $q(x_0, h) \in S_1$. Since $qx_0 \in \pi_1(S_1)$ there exists an $\eta_0 \in H$ such that $(qx_0, \eta_0) \in S_1$. If $q = 0$ we may choose any $h \in H$. For $q \geq 1$ the formula

$$q(x_0, h) = \Big(qx_0, qh - \sum_{l=1}^{q-1} F(x_0, lx_0)\Big),$$

shows that it suffices to choose $h \in H$ such that $qh - \sum_{l=1}^{q-1} F(x_0, lx_0) = \eta_0$. That is possible due to the divisibility of H. □

A new notion is in play in Theorem 16.8: A group is a *free abelian group*, if it is abelian and has a subset (called a basis) such that every element of the group can be written in one and only one way as a linear combination of elements of the basis with integer coefficients (linear combinations are always finite). $\mathbb{Z}$ and more generally $\mathbb{Z}^n$ for $n \in \mathbb{N}$ are examples of free abelian groups, whereas the groups $\mathbb{Q}$ and $\mathbb{R}$ are not free abelian (we do not prove these statements here). Theorem 16.8 and Proposition 16.9 have no restrictions on H in contrast to Theorem 16.7.

Theorem 16.8. *If G is a free abelian group, then all symmetric cocycles on G to H are coboundaries.*

Proof. We use the notation of Lemma 16.3, except that we write $+$ for the group composition in G, G being abelian. We note that W is abelian here, because G is abelian and F is symmetric. We let $+$ denote the group composition in W. According to Lemma 16.3(d) we shall produce a subgroup S of W such that $\pi_1|S$ maps S bijectively onto G. Being a free abelian group G has a basis $\{x_i \mid i \in I\}$. As S we take the subgroup of W consisting of elements of the form $\sum_{l=1}^{N} c_l(x_{i_l}, 0)$ where $c_1, c_2, \ldots, c_N \in \mathbb{Z}$ and N varies in $\mathbb{N}$. Since $\pi_1|S$ is a homomorphism we get that $\pi_1(\sum_{l=1}^{N} c_l(x_{i_l}, 0)) = \sum_{l=1}^{N} c_l x_{i_l}$. This formula shows immediately that $\pi_1|_S$ maps S onto G. To see that $\pi_1|_S$ is injective we shall show that its kernel is trivial. But $\pi_1(\sum_{l=1}^{N} c_l(x_{i_l}, 0)) = 0$ means that $\sum_{l=1}^{N} c_l x_{i_l} = 0$. And in that case $c_1 = c_2 = \cdots = c_N = 0$ so that $\sum_{l=1}^{N} c_l(x_{i_l}, 0) = 0$. □

The proofs of Theorems 16.7 and 16.8 are non-constructive, so regrettably the theorems do not provide formulas for the generators of the coboundaries.

Our next results are about cyclic groups. Such groups are abelian, so they fall within this section.

Proposition 16.9. *On cyclic groups all cocycles are symmetric.*

Proof. Let F be a cocycle on a cyclic group G. By induction up and down on $m, n \in \mathbb{N}$, and the formula $F(a, a^{-1}) = F(a^{-1}, a)$ that we get by putting $x = z = a$ and $y = a^{-1}$ in the cocycle equation, we find

$$F(a^n, a^m) = F(a^m, a^n) \qquad \text{for any } a \in G \text{ and any } m, n \in \mathbb{Z}.$$ □

Proposition 16.10. *Every cocycle on a cyclic group to a divisible group is a coboundary.*

Proof. Combine Theorem 16.7 and Proposition 16.9. □

Let us bring together some of the results about cocycles on abelian groups.

Theorem 16.11. *Let G be an abelian group, H a divisible group which is 2-torsion-free. Let $F : G \times G \to H$ be a cocycle on G to H.*

(a) *The decomposition*

$$F(x,y) = \frac{F(x,y) - F(y,x)}{2} + \frac{F(x,y) + F(y,x)}{2}$$

splits F into a skew-symmetric and a symmetric cocycle.

(b) *The skew-symmetric part is bi-additive.*

(c) *The symmetric part is a coboundary.*

Proof. (a) is Lemma 16.5, (b) is Theorem 16.6 and (c) is Theorem 16.7. □

Our considerations have been purely algebraic up to now. However some results about continuous cocycles are needed in concrete instances. The following corollary covers some cases of continuous cocycles.

Corollary 16.12. *Let G be a locally compact, abelian group and $H = X$ a Banach space. Let $F : G \times G \to X$ be a continuous cocycle on G to X.*

The decomposition

$$F(x,y) = \frac{F(x,y) - F(y,x)}{2} + \frac{F(x,y) + F(y,x)}{2}$$

splits F into a continuous, skew-symmetric, bi-additive cocycle of $G \times G$ into X and a coboundary with a continuous generator.

Proof. The proof is immediate from Theorem 16.11, except for the continuity of the generator. For the existence of a continuous generator we refer to Corollary D.12. □

16.4 Examples

Example 16.13. On the cyclic group $\mathbb{Z}$ we can omit the hypothesis of Proposition 16.10 that H is divisible (by Theorem 16.8 because $\mathbb{Z}$ is a free abelian group). But we cannot do so for cyclic groups in general as the following example shows. Let $G = \mathbb{Z}_2 = (\{\overline{0}, \overline{1}\}, +)$ and $H = \mathbb{Z}$. Using the notation $\overline{i} := i + 2\mathbb{Z}$ for $i \in \mathbb{Z}$ we define $F : \mathbb{Z}_2 \times \mathbb{Z}_2 \to \mathbb{Z}$ by

$$F(\overline{0},\overline{0}) = F(\overline{0},\overline{1}) = F(\overline{1},\overline{0}) = 0 \quad \text{and} \quad F(\overline{1},\overline{1}) = 1.$$

Going through the possibilities for the independent variables it is easy to check that F is a cocycle. However, no map $f : \mathbb{Z}_2 \to \mathbb{Z}$ exists such that $F = df$: If f were such a map then $f(\overline{0}) = 0$ (Lemma 16.2). Putting $i = j = 1$ in the formula $F(\overline{i}, \overline{j}) = f(\overline{i}) + f(\overline{j}) - f(\overline{i} + \overline{j})$ we get $2f(\overline{1}) = 1$, which contradicts that f is integer valued.

Example 16.14. Example 16.13 generalizes from $\mathbb{Z}_2$ to $\mathbb{Z}_N$ where $N \geq 2$. We tell how without going into details. Using the notation $\overline{i} := i + N\mathbb{Z}$ for $i \in \mathbb{Z}$ we define $F : \mathbb{Z}_N \times \mathbb{Z}_N \to \mathbb{Z}$ by

$$F(\overline{i}, \overline{j}) := \begin{cases} 0 & \text{if } i + j < N \text{ when } 0 \leq i, j \leq N - 1 \\ 1 & \text{if } i + j \geq N \text{ when } 0 \leq i, j \leq N - 1. \end{cases}$$

F is a cocycle on $\mathbb{Z}_N$ to $\mathbb{Z}$, but it is not a coboundary. It is for $N = 10$ the familiar "carry bit" one is taught in primary school.

Example 16.15. Let $G = H = \mathbb{R}$. Let $F \in C(\mathbb{R} \times \mathbb{R}, \mathbb{R})$ be a cocycle on $\mathbb{R}$ to $\mathbb{R}$. The skew-symmetric part F_1 of F is by Corollary 16.12 bi-additive. Being also continuous, F_1 is bilinear, and hence $F_1(x, y) = xyF_1(1, 1) = xy \cdot 0 = 0$. From Corollary 16.12 we conclude that F is a coboundary with a continuous generator.

Exercise 16.7 finds an explicit formula for a continuous generator, when $F \in C^1(\mathbb{R} \times \mathbb{R}, \mathbb{R})$.

Example 16.16. Let $G = \mathbb{R}^n$ and $H = \mathbb{R}$. Let $F \in C(\mathbb{R}^n \times \mathbb{R}^n, \mathbb{R})$ be a cocycle. F splits according to Corollary 16.12 into a skew-symmetric, bi-additive map B and a coboundary with a continuous generator. B has the form

$$B(x, y) = \langle Ax, y \rangle, \qquad x, y \in \mathbb{R}^n,$$

where $A \in M(n \times n, \mathbb{R})$ is a skew-symmetric matrix (Lemma 2.14).

Example 16.17. Let $G = \mathbb{T}$ and $H = \mathbb{R}$. Here any continuous cocycle is a coboundary with a continuous generator (Theorem 16.4).

16.5 The cocycle equation on semidirect products

We will in this section very shortly describe the cocycle equation on semidirect products of groups. The philosophy is that a cocycle on a semidirect product $G = N \times_s H$ should be determined by its values on the factors N and on H plus a term involving their interaction.

When $G = N \times_s Q$ is a semidirect product we write $x = n(x)q(x)$ for any $x \in G$, where $n(x) \in N$ and $q(x) \in Q$. We also write $n^x = xnx^{-1}$ for $n \in N$ and $x \in G$.

We cite the following result from [71, Theorem 2.8] without proof:

Theorem 16.18. *Let $G = N \times_s Q$ be the topological semidirect product of a normal subgroup N of G by a subgroup Q of G. Let H be an abelian topological group. Let finally $F \in C(G \times G, H)$.*

F is a cocycle on G if and only if there exist cocycles $F_N \in C(N \times N, H)$ and $F_Q \in C(Q \times Q, H)$ and maps $\Phi \in C(Q \times N, H)$ and $f \in C(G, H)$ such that for all $x, y \in G$, $q, q_1, q_2 \in Q$, $n, n_1, n_2 \in N$:

$$F(x,y) = F_N(n(x), n(y)^{q(x)}) + \Phi(q(x), n(y)) + F_Q(q(x), q(y)) + df(x,y),$$
$$\Phi(q_1 q_2, n) = \Phi(q_1, n^{q_2}) + \Phi(q_2, n),$$
$$\Phi(q, n_1 n_2) - \Phi(q, n_1) - \Phi(q, n_2) = F_N(n_1^q, n_2^q) - F_N(n_1, n_2).$$

Simplifications occur, when the subgroups N and Q are compact or abelian, which allows us to compute explicit formulas for the continuous cocycles on, e.g., the $(ax+b)$-group, the Heisenberg groups H_n and the Euclidean motion group. The methods are elementary, but not quite simple. [70] and [71] have the details.

16.6 Exercises

Exercise 16.1. Let $f : G \to H$. Show for any $a \in G$ that

$$f(a^n) = nf(a) - \sum_{l=1}^{n-1} df(a, a^l) \qquad \text{for } n \geq 2.$$

Exercise 16.2. Define $F : \mathbb{R} \times \mathbb{R} \to \mathbb{R}$ by $F(x,y) := xy$, $x, y \in \mathbb{R}$. Show that F a symmetric cocycle on $\mathbb{R}$ to $\mathbb{R}$. Is it a coboundary? If so, then find a generator or preferably all continuous generators.

Same questions for the function $F : \mathbb{R}^n \times \mathbb{R}^n \to \mathbb{R}$ defined by $F(x,y) := \langle x, y \rangle$, $x, y \in \mathbb{R}^n$.

Hint: $\|x+y\|^2 - \|x\|^2 - \|y\|^2 = 2\langle x, y \rangle$.

Exercise 16.3. Let F be a cocycle on a finite group G to an abelian group H which is divisible and torsion free. Show that F is a coboundary.

Hint: You may adapt the procedure of Section 16.2.

Exercise 16.4. Define $F : \mathbb{Z} \times \mathbb{Z} \to \mathbb{Z}$ by $F(m,n) := -mn$, $m, n \in \mathbb{Z}$. Show that F a symmetric cocycle on $\mathbb{Z}$ to $\mathbb{Z}$. Infer by Theorem 16.8 that F is a coboundary. Show that $f(n) := n(n+1)/2$, $n \in \mathbb{Z}$, is a generator of the coboundary F.

Exercise 16.5. Let X be a topological vector space. Show that all cocycles $F \in C(\mathbb{R} \times \mathbb{R}, X)$ on $\mathbb{R}$ to X are coboundaries.

Exercise 16.6. Verify the statements of Example 16.14.

Exercise 16.7. Let $F \in C^1(\mathbb{R} \times \mathbb{R}, \mathbb{R})$ be a cocycle. According to Example 16.15 F is a coboundary with a continuous generator. We shall in this exercise find an explicit formula for a continuously differentiable generator.

(a) Show that the cocycle functional equation for F reads

$$F(x,y)+F(x+y,z) = F(x,y+z)+F(y,z) \quad \text{for all } x,y,z \in \mathbb{R}. \tag{16.4}$$

(b) Show that we may assume that $F(0,0) = 0$.

(c) Differentiate (16.4) with respect to z at $z = 0$ and derive that

$$\partial_2 F(x,y) = \partial_2 F(x+y,0) - \partial_2 F(y,0) \qquad \text{for all } x,y \in \mathbb{R},$$

where ∂_2 denotes partial derivation with respect to the second coordinate.

(d) Integrating this over $[0,y]$ with respect to the second variable obtain that

$$F(x,y) = g(x)+g(y)-g(x+y), \quad \text{where } g(x) := -\int_0^x \partial_2 F(u,0)\,du.$$

(e) Note that F is a coboundary with a generator $g \in C^1(\mathbb{R})$.

16.7 Notes and remarks

Ebanks and Ng [68] found formulas for the general solution of the following extension of the cocycle equation

$$F_1(x+y,z)+F_2(y+z,x)+F_3(z+x,y)+F_4(x,y)+F_5(y,z)+F_6(z,x)=0$$

with six unknown functions $F_i : G \times G \to X$, $i = 1, 2, \ldots, 6$, where G is an abelian group and X a vector space over $\mathbb{Q}$.

Davison and Ebanks [65] discuss the cocycle equation (16.1) in a more general set up than we do: They replace our group G by a cancellative, abelian monoid M. Their main result is that every symmetric cocycle on M to H is a coboundary when the abelian group H is divisible.

Theorems 16.7 and 16.8: Our derivations follow Jessen, Karpf and Thorup [120, Theorem 1].

Exercise 16.1 presents a formula which is valid for $n \geq 2$. The corresponding formula for $n < 0$ can be found in Prunesco [160, Remark 1.2].

Exercise 16.7 is inspired by Prunesco [160, Theorem 3.2].

Appendices

Appendix A

Basic Terminology and Results

This appendix contains background material like basic terminology, notation, definitions and simple consequences of the definitions that we shall need. Furthermore it contains a number of examples of groups and semigroups. Surely the reader knows most of the material, so the appendix is not meant to be read from one end to the other, but as a place where the basic terminology, notation, definitions etc. are fixed and can be looked up.

A.1 Numbers and matrices

$A := B$ means that the left hand side A is defined by or is a notation for the expression B on the right hand side.

We let $\mathbb{N} := \{1, 2, \ldots\}$, $\mathbb{Z}$, $\mathbb{Q}$, $\mathbb{R}$ and $\mathbb{C}$ denote the sets of the natural numbers, the integers, the rational numbers, the real numbers and the complex numbers. We let $\mathbb{N}_0 := \{0\} \cup \mathbb{N}$. For any $N \in \mathbb{N}$ we let $\mathbb{Z}_N := \mathbb{Z}/N\mathbb{Z}$ be the quotient group. We view the vectors in $\mathbb{R}^n$ and $\mathbb{C}^n$, $n \in \mathbb{N}$, as column vectors. We give $\mathbb{R}^+ := \{t \in \mathbb{R} \,|\, t > 0\}$, $\mathbb{R}^* := \{t \in \mathbb{R} \,|\, t \neq 0\}$, $\mathbb{C}^* := \{z \in \mathbb{C} \,|\, z \neq 0\}$ and *the circle group* $\mathbb{T} := \{z \in \mathbb{C} \,|\, |z| = 1\}$ the multiplication from $\mathbb{C}$, making them into groups. All the sets above are equipped with their usual algebraic and topological structures.

The inner product on $\mathbb{C}^n$ and by restriction also on $\mathbb{R}^n \subseteq \mathbb{C}^n$ is defined by

$$\langle w, z \rangle = \sum_{i=1}^{n} w_i \overline{z_i} \quad \text{for } w = (w_1, w_2, \ldots, w_n)^t, z = (z_1, z_2, \ldots, z_n)^t \in \mathbb{C}^n,$$

so the inner product is linear in the first variable and conjugate linear in the second.

If A is a matrix then A^t denotes its transpose and A^* its adjoint matrix. That a matrix A is *symmetric* means that $A^t = A$.

$M(n \times m, \mathbb{C})$ denotes the complex vector space of all $n \times m$ matrices with complex entries. As a finite-dimensional vector space it has exactly one topology that makes it a topological vector space (see Theorem E.4). We equip it with that topology. By the uniqueness it is the one induced by the inner product

$$\langle A, B\rangle_{HS} := \operatorname{tr}(AB^*) = \sum_{i=1}^{n}\sum_{j=1}^{m} a_{ij}\overline{b_{ij}}\,,$$
$$A = \{a_{ij}\}, B = \{b_{ij}\} \in M(n \times m, \mathbb{C}),$$

or rather from the corresponding norm (*the Hilbert-Schmidt norm*) given by

$$\|A\|_{HS}^2 := \operatorname{tr}(AA^*) = \sum_{i=1}^{n}\sum_{j=1}^{m} |a_{ij}|^2, \;\; A = \{a_{ij}\} \in M(n \times m, \mathbb{C}).$$

It is immediate that the entries $A \mapsto a_{ij}$, $i = 1, \ldots, n$, $j = 1, \ldots, m$, are continuous functions on $M(n \times m, \mathbb{C})$, and that convergence of a sequence in $M(n \times m, \mathbb{C})$ means convergence of each entry.

We let $M(n \times m, \mathbb{R}) \subseteq M(n \times m, \mathbb{C})$ denote the real vector space of all $n \times m$ matrices with real entries.

A.2 Functions

In this section X, Y and Z are sets, unless otherwise specified.

The set of maps from X to Y is denoted Y^X. The composition of $f : X \to Y$ with $g : Y \to Z$ is denoted $g \circ f$. The identity map is denoted id.

We say that a function $F : X \times X \to Y$ is *symmetric* if $F(x_1, x_2) = F(x_2, x_1)$ for all $x_1, x_2 \in X$. If Y is an abelian group, we say that $F : X \times X \to Y$ is *skew-symmetric* if $F(x_1, x_2) = -F(x_2, x_1)$ for all $x_1, x_2 \in X$.

A function $f : X \to \mathbb{C}$ is said to be *non-zero* if there exists an $x \in X$ such that $f(x) \neq 0$. In other words if $f \neq 0$, where 0 denotes the function that vanishes everywhere on X. A non-zero function may well vanish at some points, but it may not vanish at all points. The concept should not be confused with the notion of a nullfunction from measure theory.

If $f_i : X_i \to \mathbb{C}$ is a function on the set X_i for $i = 1, 2, \ldots, n$ then we let $f_1 \otimes f_2 \otimes \cdots \otimes f_n : X_1 \times X_2 \times \cdots \times X_n \to \mathbb{C}$ denote the function

$$f_1 \otimes f_2 \otimes \cdots \otimes f_n(x_1, x_2, \ldots, x_n) := f_1(x_1) f_2(x_2) \ldots f_n(x_n),$$
$$(x_1, x_2, \ldots, x_n) \in X_1 \times X_2 \times \cdots \times X_n.$$

The algebra of continuous, complex-valued functions on a topological space X is denoted $C(X)$. We let $C_b(X)$, resp. $C_c(X)$, denote the subspace of $C(X)$ consisting of bounded, resp. compactly supported, functions. If X is a C^∞ manifold we let $C^\infty(X)$ denote the set of complex-valued C^∞ functions on X, and put $C_c^\infty(X) := C^\infty(X) \cap C_c(X)$.

If $\tau : X \to X$ we say that $f : X \to Y$ is *τ-invariant* or *even with respect to τ* if $f \circ \tau = f$. If Y is an abelian group we say that f is *odd with respect to τ* if $f \circ \tau = -f$. The special case in which τ is the inversion on a group is particularly important: For any function f on a group G we let $\check{f}$ denote the function defined by $\check{f}(x) := f(x^{-1})$, $x \in G$. A map $f : G \to H$, where G is a group and H an abelian group, is said to be *even* if $\check{f} = f$, and *odd* if $\check{f} = -f$. Any function $f : G \to \mathbb{C}$ can in exactly one way be written in the form $f = f_e + f_o$, where $f_e : G \to \mathbb{C}$ is even and $f_o : G \to \mathbb{C}$ is odd. $\overline{f}$ is the complex conjugate of f.

The symbol log denotes the natural logarithm on $\mathbb{R}^+$.

The *sign*, denoted sgn, of the complex number $z = e^{i\theta}|z|$ is

$$\operatorname{sgn}(z) := \begin{cases} e^{i\theta} & \text{if } z \neq 0 \\ 0 & \text{if } z = 0. \end{cases}$$

The indicator function $\mathbf{1}_X$ of the set X is defined by

$$\mathbf{1}_X(x) = \begin{cases} 1 & \text{if } x \in X \\ 0 & \text{otherwise.} \end{cases}$$

A.3 Vector spaces

Our vector spaces are complex unless otherwise specified or obvious from the context. Let V be a vector space.

We let $\mathcal{L}(V)$ denote the algebra of all linear maps from V to V, and $\mathrm{GL}(V)$ the subset of $\mathcal{L}(V)$ consisting of the bijective linear maps. $\mathrm{GL}(V)$ is a group under composition. $\operatorname{tr}(A)$ denotes the trace of the operator A and $\det A$ its determinant.

V^* denotes the dual vector space of V, i.e., the vector space consisting af all linear maps $v^* : V \to \mathbb{C}$. The value that $v^* \in V^*$ assumes at $v \in V$, is denoted $\langle v^*, v \rangle$.

If $\{V_i\}_{i \in I}$ are subspaces of V that form a direct sum then we write

$$\operatorname{span}\{V_i \mid i \in I\} = \bigoplus_{i \in I} V_i.$$

The case of just one summand in a direct sum is allowed.

We incorporate the Hausdorff property in the definition of a topological vector space.

A.4 Algebras

We write down the definition of an algebra, because it is not always presented in elementary algebra courses. We follow the standard convention that fields are commutative.

Definition A.1. Let $\mathcal{A}$ be a vector space over a field F. Suppose that for each pair $x, y \in \mathcal{A}$ a product $xy \in \mathcal{A}$ is defined such that $\mathcal{A}$ under this operation is a semigroup, and that this product is related to the vector space structure of $\mathcal{A}$ as follows:

(i) $x(y + z) = xy + xz$

(ii) $(x + y)z = xz + yz$

(iii) $(\alpha x)y = x(\alpha y) = \alpha(xy)$

for all $x, y, z \in \mathcal{A}$ and all $\alpha \in F$. Then $\mathcal{A}$ is said to be an *algebra over F*. An *algebra with unit* is an algebra having a multiplicative identity $1 \neq 0$; a *commutative algebra* is one for which $xy = yx$ for all $x, y \in \mathcal{A}$. *Real* and *complex algebras* are algebras over $\mathbb{R}$ and $\mathbb{C}$, respectively.

It is not required that algebras have unit elements, because some interesting algebras like the convolution algebra $L^1(\mathbb{R})$ do not.

Let $\mathcal{A}$ be a complex algebra. We let $\widehat{\mathcal{A}}$ denote the set of multiplicative, linear functionals $g : \mathcal{A} \to \mathbb{C}$ such that $g \neq 0$. If $\mathcal{A}$ has a unit $1 \in \mathcal{A}$, then we have for any multiplicative linear functional $g : \mathcal{A} \to \mathbb{C}$ that $g \in \widehat{\mathcal{A}}$ if and only if $g(1) = 1$ (Lemma A.28).

A.5 Groups

A.5.1 *General theory and notation*

Throughout the book we let G denote a group with neutral element e. We write the group composition multiplicatively, unless the group is abelian. If the group is abelian, we usually let $+$ denote the group composition and 0 the identity element of G, but on some groups, for example the circle group $\mathbb{T}$, we keep the multiplicative way of writing the group composition.

The subgroup of G generated by a subset A of G is denoted $\langle A\rangle$. We use the abbreviation $\langle x\rangle := \langle\{x\}\rangle$ when $x \in G$. We let $G^2 := \{x^2 \mid x \in G\}$. The subgroup $\langle G^2\rangle$ of G generated by the squares of the elements of G is a normal subgroup of G.

The center of G will be denoted $Z(G)$.

The *commutator* between $x \in G$ and $y \in G$ is $[x, y] := xyx^{-1}y^{-1}$.

If A and B are subsets of G we let $[A, B]$ denote the smallest subgroup of G containing the set $\{[a, b] \mid a \in A,\ b \in B\}$. From Lemma A.2(a) we see that the *commutator subgroup* $[G, G]$ (also called the *derived subgroup*) consists of elements of the form $[x_1, y_1][x_2, y_2]\cdots[x_n, y_n]$, where $x_1, x_2, \ldots, x_n, y_1, y_2, \ldots, y_n \in G$ and $n = 1, 2, \ldots$

$[G, G]$ and its subgroup $[G, [G, G]]$ are both normal subgroups of G. The group $[G, [G, G]]$ can be characterized as the subgroup of G generated by the set $\{[x, [y, z]] \mid x, y, z \in G\}$ (Lemma A.7). As is easy to see, the quotient group $G/[G, G]$ is abelian.

If G is a connected Lie group, then $[G, G]$ is also a connected Lie subgroup of G (see [201, Theorem 3.18.8]).

Let us list a couple of formulas that are handy in dealing with commutators:

Lemma A.2. *Let G be a group and $x, y, z \in G$.*

(a) $[x, y]^{-1} = [y, x]$.

(b) $[x, y] = x^2(x^{-1}y)^2y^{-2}$, *so the commutator subgroup $[G, G]$ is contained in the subgroup $\langle G^2\rangle$ of G generated by the squares of the elements of G.*

(c) $[x, yz] = [x, y]y[x, z]y^{-1}$. *In particular* $[x, y^2] = [x, y]y[x, y]y^{-1}$

(d) $[yz, x] = y[z, x]y^{-1}[y, x]$.

Definition A.3. A group G is *step 2 nilpotent*, if $[G, G] \subseteq Z(G)$, or equivalently if $[G, [G, G]] = \{e\}$.

It follows immediately from Lemma A.2(b) that a group G is step 2 nilpotent if $x^2 \in Z(G)$ for each $x \in G$. This applies for example to the quaternion group Q_8 (defined in Example A.17(h)), so Q_8 is step 2 nilpotent.

Definition A.4. A group G is said to be a P_3 *group* if and only if $[G,G]$ has order 1 or 2 and $x^2 \in Z(G)$ for all $x \in G$.

P_3 groups are particularly simple examples of step 2 nilpotent groups. The quaternion group Q_8 from Example A.17(h) is a P_3 group, while the Heisenberg group H_3 from Example A.17(a) is not, although it is step 2 nilpotent.

An important class of groups consists of the nilpotent ones. We proceed to introduce them and derive some of their properties.

Definition A.5. If G is a group we define the *descending central series* $\mathcal{C}^0G, \mathcal{C}^1G, \dots, \mathcal{C}^nG, \dots$ for G inductively as follows:

$$\mathcal{C}^0G := G \quad \text{and} \quad \mathcal{C}^nG := [G, \mathcal{C}^{n-1}G] \qquad \text{for } n = 1, 2, \dots$$

G is said to be *nilpotent*, if $\mathcal{C}^nG = \{e\}$ for some $n \in \mathbb{N}_0$.

Any abelian group and any step 2 nilpotent group G is nilpotent. Indeed, $\mathcal{C}^2G = \{e\}$ for such groups.

We list some properties of the descending central series:

Lemma A.6. *Let G be a group. Then $\mathcal{C}^nG$ is a normal subgroup of G, $\mathcal{C}^0G \supseteq \mathcal{C}^1G \supseteq \cdots \supseteq \mathcal{C}^nG \supseteq \cdots$ and $[\mathcal{C}^nG, \mathcal{C}^mG] \subseteq \mathcal{C}^{n+m+1}G$ for all $n, m \in \mathbb{N}_0$.*

Proof. Induction on n and m. □

$[G,G]$ is a proper subset of G when $G \neq \{e\}$ is nilpotent. However, that is not so in general: $[G,G] = G$ for connected, semisimple Lie groups like SO(3), SU(n), SL($n,\mathbb{R}$) and SL($n,\mathbb{C}$) when $n \geq 2$. See [201, Corollary 3.18.10].

The following lemma about the group $[G,[G,G]]$ simplifies some computations in the chapter about Jensen's functional equation.

Lemma A.7. *$[G,[G,G]]$ is generated by the set $\{[x,[y,z]] \mid x,y,z \in G\}$.*

Proof. Let H be the subgroup of G generated by $\{[x,[y,z]] \mid x,y,z \in G\}$. We shall prove that $[G,[G,G]] = H$. Since $[x,[y,z]] \in [G,[G,G]]$ for all $x,y,z \in G$, we get immediately that $H \subseteq [G,[G,G]]$.

It is left to prove that $[G, [G, G]] \subseteq H$. The elements $[x, c]$, where $x \in G$ and $c \in [G, G]$, generate $[G, [G, G]]$, so it suffices to prove that $[x, c] \in H$ for all $x \in G$ and $c \in [G, G]$. Any $c \in [G, G]$ has the form $c = c_1 c_2 \cdots c_n$, where $c_i = [x_i, y_i]$ for some $x_i, y_i \in G$ for $i = 1, 2, \ldots, n$. In principle we should in the expression for c also include inverses $[x_i, y_i]^{-1}$ among the factors, but $[x, y]^{-1} = [y, x]$ for all $x, y \in G$, so the inverses have the same form as the other factors. We shall thus prove that $[x, [x_1, y_1][x_2, y_2] \cdots [x_n, y_n]] \in H$ for all $x, x_1, y_1, \ldots, x_n, y_n \in G$, which we do by induction on n. The induction hypothesis at step n is that $[x, [x_1, y_1][x_2, y_2] \cdots [x_n, y_n]] \in H$ for all $x, x_1, y_1, \ldots, x_n, y_n \in G$. The case $n = 1$ is trivial. Assuming step n true we shall prove that $[x, c\,[x_{n+1}, y_{n+1}]] \in H$, where we for brevity write $c := [x_1, y_1][x_2, y_2] \cdots [x_n, y_n]$. Using the formula $[x, yz] = [x, y][i(y)x, i(y)z]$ where $i(a)b := aba^{-1}$, we get

$$\begin{aligned}
[x, c\,[x_{n+1}, y_{n+1}]] &= [x, c][i(c)x, i(c)[x_{n+1}, y_{n+1}]] \\
&= [x, c][i(c)x, [i(c)x_{n+1}, i(c)y_{n+1}]].
\end{aligned}$$

The first factor on the right is an element of H by the induction hypothesis, and the second one is one of the generators of H, so the product is in H. □

Definition A.8. If G is a group we define $\mathcal{D}^1(G) := [G, G]$ and inductively $\mathcal{D}^{n+1}(G) := [\mathcal{D}^n(G), \mathcal{D}^n(G)]$ for $n = 1, 2, \ldots$

We say that G is *solvable* if there exists $n \in \mathbb{N}$ such that $\mathcal{D}^n(G) = \{e\}$.

Definition A.9. Let $G, G_1, G_2, \ldots, G_n$ and H be groups.

(a) A *homomorphism* from G to H is a map $\phi : G \to H$ such that $\phi(xy) = \phi(x)\phi(y)$ for all $x, y \in G$. The set of homomorphisms from G to H is denoted $\mathrm{Hom}(G, H)$. If H is abelian we sometimes use the terminology *additive map* (or *additive function*) instead of homomorphism. An *isomorphism* between G and H is a bijective homomorphism from G to H. If G and H are topological groups then a *topological isomorphism* between G and H is an isomorphism $\phi : G \to H$ such that both ϕ and ϕ^{-1} are continuous.

(b) A map $\Phi : G_1 \times G_2 \to H$ is a *bi-homomorphism* if the maps $x_2 \mapsto \Phi(x_1^0, x_2)$ and $x_1 \mapsto \Phi(x_1, x_2^0)$ are homomorphisms for each fixed $x_1^0 \in G_1$ and $x_2^0 \in G_2$. If H is abelian we usually use the terminology *bi-additive map* instead of bi-homomorphism.

(c) A map $\phi : G_1 \times G_2 \times \cdots \times G_n \to H$ is said to be *multi-homomorphism*, if it is a homomorphism in any of its variables when the remaining variables are fixed. If H is abelian we usually use the terminology *multi-additive map* instead of multi-homomorphism.

A.5.2 *Groups generated by their squares*

In this subsection we discuss groups generated the squares of their elements and related topics.

Definition A.10. Let S be a semigroup. S is said to be *divisible*, if for any given $y \in S$ and $n \in \mathbb{N}$ the equation $x^n = y$ has a solution $x \in S$. S is said to be *2-divisible*, if $S = \{x^2 \mid x \in S\}$. It is *uniquely 2-divisible* if for any $a \in S$ the equation $x^2 = a$ has exactly one solution $x \in S$.

Some groups are 2-divisible ($\mathbb{R}$ is an example), others are not ($\mathbb{Z}$ is an example). Compact, connected Lie groups are 2-divisible, because the exponential map is surjective on such groups ([33, Theorem IV.2.2]).

Any 2-divisible group is generated by its squares (trivially). The converse is true for abelian, but not for general groups. To take an example, $\mathrm{SL}(2,\mathbb{R})$ is not 2-divisible, but it is nevertheless generated by its squares: Direct computations reveal that no matrix $X \in \mathrm{SL}(2,\mathbb{R})$ satisfies

$$X^2 = \begin{pmatrix} -2 & 0 \\ 0 & -\frac{1}{2} \end{pmatrix},$$

so $\mathrm{SL}(2,\mathbb{R})$ is not 2-divisible. On the other hand side it is known that $\mathrm{SL}(2,\mathbb{R})$ is connected, so $\bigcup_{n=0}^{\infty} U^n = \mathrm{SL}(2,\mathbb{R})$ for any neighborhood U of the identity, where $U^n := \{u_1 u_2 \cdots u_n \mid u_i \in U \text{ for } i = 1, 2, \ldots, n\}$ (see [111, Theorem 7.4]). So $\mathrm{SL}(2,\mathbb{R})$ is generated by its squares if it has a neighborhood of the identity consisting of squares. The image of the Lie algebra of $\mathrm{SL}(2,\mathbb{R})$ under the exponential map is such a neighborhood. This argument works not just for $\mathrm{SL}(2,\mathbb{R})$ but for any connected Lie group, so we have proved the first part of the following general result (the second part follows from Lemma A.2(b)):

Lemma A.11. *Connected Lie groups and simple, non-abelian groups are generated by their squares.*

So each member of two large classes of important groups has the algebraic property that it is generated by its squares. Thus we want whenever possible to replace the condition of 2-divisibility which is often used in the literature, by the weaker condition of being generated by the squares.

Definition A.12. Let G be a group. It is *2-torsion-free*, if for all $x \in G$: $x^2 = e$ implies $x = e$, or in other words if its only element of order 2 is e. G is *torsion-free* if its only element of finite order is e. We shall only need these notions for G abelian. $(\mathbb{R}, +)$ is an example of a torsion-free group. Although 2-divisible $\mathbb{T}$ is not 2-torsion-free.

A.5.3 *Topological groups*

Definition A.13. A *topological group* is a group G equipped with a topology such that the group operations $(x, y) \mapsto xy$ from $G \times G$ (with the product topology) to G and $x \mapsto x^{-1}$ from G to G are continuous.

Definition A.14. A *locally compact group* is a topological group the topology of which is locally compact (i.e., each point has a compact neighborhood) and Hausdorff. A *compact group* is a locally compact group which is compact.

$\mathbb{R}^n$ is a locally compact group for any $n \in \mathbb{N}$. Any group becomes a locally compact group when we equip it with the discrete topology.

Definition A.15. A *matrix group* is a closed subgroup of $\mathrm{GL}(n, \mathbb{C})$. We always equip a matrix group G with the topology inherited from $\mathrm{GL}(n, \mathbb{C}) \subseteq M(n \times n, \mathbb{C})$ (for the topology on $M(n \times n, \mathbb{C})$ see Section A.1). This makes G a locally compact group, actually a Lie group (see [201, Theorem 2.12.6]).

We use the standard notation and terminology for the names of groups. For example we let $\mathrm{SL}(n, \mathbb{C})$ denote the special linear group, i.e., the matrix group of all $n \times n$ complex matrices with determinant equal to 1. The neutral element of any matrix subgroup of $\mathrm{GL}(n, \mathbb{C})$ is the identity matrix

$$I = I_n = \begin{pmatrix} 1 & 0 & \dots & 0 \\ 0 & 1 & \dots & 0 \\ \vdots & \vdots & \ddots & 0 \\ 0 & 0 & \dots & 1 \end{pmatrix}.$$

That a group G comes equipped with a topology making it a topological semigroup (Definition A.25) does not force it to be a topological group, because the inversion map $x \mapsto x^{-1}$ may not be continuous. However, under some conditions topological semigroups are topological groups:

Theorem A.16. *Let G be a locally compact Hausdorff space with a group structure such that the maps $x \mapsto xy$ and $x \mapsto yx$ are continuous of G into G for all $y \in G$. Then G is a topological group.*

A.5.4 *Examples of groups*

Below we list most of the specific examples of groups that the solutions of our functional equations have as domain of definition. Although the groups have quite different properties almost all of them are matrix groups.

Examples A.17.

(a) For the *Heisenberg group*

$$H_3 := \left\{ [x,y,z] := \begin{pmatrix} 1 & x & z \\ 0 & 1 & y \\ 0 & 0 & 1 \end{pmatrix} \,\middle|\, x,y,z \in \mathbb{R} \right\}$$

we find that

$$[x,y,z]^{-1} = [-x,-y,-z+xy]$$

and then

$$[H_3,H_3] = \{[0,0,z] \mid z \in \mathbb{R}\} = Z(H_3),$$

so that H_3 is a step 2 nilpotent group. It is not abelian. It is uniquely 2-divisible (Definition A.10).

The Heisenberg group is a topological semi-direct product of $\mathbb{R}^2$ with $\mathbb{R}$: See Example A.22.

There are many variations of this Heisenberg group: Instead of ranging over $\mathbb{R}$, the entries x, y and z could range over a commutative ring as for example $\mathbb{C}$, $\mathbb{Z}$ or a finite field. We denote them $H_3(\mathbb{C})$, $H_3(\mathbb{Z})$ etc. The formulas above for the inverse and the commutator group hold also for such variations.

(b) The $(2n+1)$-*dimensional Heisenberg group*

$$H_{2n+1} := \left\{ \begin{pmatrix} 1 & x_1 & x_2 & \dots & x_n & z \\ 0 & 1 & 0 & \dots & 0 & y_1 \\ 0 & 0 & 1 & \dots & 0 & y_2 \\ \vdots & \vdots & \vdots & \vdots & \vdots & \vdots \\ 0 & 0 & 0 & \dots & 1 & y_n \\ 0 & 0 & 0 & \dots & 0 & 1 \end{pmatrix} \,\middle|\, \begin{array}{l} x_i, y_i, z \in \mathbb{R}, \\ i = 1,2,\dots,n \end{array} \right\}$$

is a step 2 nilpotent group. It is not abelian.

(c) Let $n \in \mathbb{N}$. The *general linear group* over $\mathbb{C}$, resp. $\mathbb{R}$, is the group of all invertible complex, resp. real, $n \times n$ matrices. It is denoted $\mathrm{GL}(n, \mathbb{C})$, resp. $\mathrm{GL}(n, \mathbb{R})$.

(d) Let $n \in \mathbb{N}$, $n \geq 2$. The *special linear group* over $\mathbb{C}$ is

$$\mathrm{SL}(n, \mathbb{C}) := \{A \in \mathrm{GL}(n, \mathbb{C}) \mid \det A = 1\},$$

while the *special linear group* over $\mathbb{R}$ is

$$\mathrm{SL}(n, \mathbb{R}) := \{A \in \mathrm{GL}(n, \mathbb{R}) \mid \det A = 1\}.$$

They are connected semisimple Lie groups. In particular $[G, G] = G$ for $G = \mathrm{SL}(n, \mathbb{C})$ and $G = \mathrm{SL}(n, \mathbb{R})$ (see [201, Corollary 3.18.10]).

(e) The *orthogonal group* $O(n)$, $n \in \mathbb{N}$, consists of the $n \times n$ real matrices O such that $O^*O = I$. The *special orthogonal group* is $\mathrm{SO}(n) := O(n) \cap \mathrm{SL}(n, \mathbb{R})$.

(f) The *unitary group* $U(n)$, $n \in \mathbb{N}$, consists of the $n \times n$ complex matrices U such that $U^*U = I$.

(g) The *special unitary group* $\mathrm{SU}(2) := U(2) \cap \mathrm{SL}(2, \mathbb{C})$ can be computed to be

$$\mathrm{SU}(2) = \left\{ \begin{pmatrix} \alpha & \beta \\ -\overline{\beta} & \overline{\alpha} \end{pmatrix} \,\middle|\, \alpha, \beta \in \mathbb{C},\ |\alpha|^2 + |\beta|^2 = 1 \right\}.$$

It is an example of a semisimple, compact connected Lie group. From the theory of semisimple Lie groups (see for instance [201, Corollary 3.18.10]) we then get that $[\mathrm{SU}(2), \mathrm{SU}(2)] = \mathrm{SU}(2)$. As we shall see later, $\mathrm{SU}(2)$ plays a special role for the bounded solutions of d'Alembert's functional equation (Corollary 8.27).

(h) Let $\mathbb{H} = \{a + bi + cj + dk \mid a, b, c, d \in \mathbb{R}\}$ denote the division algebra of the quaternions. For easy reference we recall the multiplication table

$$i^2 = j^2 = k^2 = -1,\ ij = -ji = k,\ jk = -kj = i,\ ki = -ik = j.$$

The map

$$\Pi : a + bi + cj + dk \mapsto \begin{pmatrix} a + ib & c + id \\ -(c - id) & a - ib \end{pmatrix} \tag{A.1}$$

is an injective algebra-homomorphism of $\mathbb{H}$ into the algebra of 2×2 matrices with complex entries, both viewed as algebras over the field $\mathbb{R}$.

$\mathbb{H}$ equipped with multiplication is a monoid and Π is a representation of it on $\mathbb{C}^2$. Note that

$$\det\{\Pi(a+bi+cj+dk)\} = a^2+b^2+c^2+d^2.$$

The *group of unit quaternions* is $\{a+bi+cj+dk \in \mathbb{H} \mid a^2+b^2+c^2+d^2=1\}$. It is well known that the restriction of the map Π to the group of unit quaternions is a topological isomorphism of it onto SU(2). (That the inverse map is continuous follows without work from Theorem D.5, or from the standard fact from point set topology that the inverse of a continuous bijection from a compact space to a Hausdorff space is continuous.)

The *quaternion group* is the subgroup $Q_8 := \{\pm 1, \pm i, \pm j, \pm k\}$ of the group of unit quaternions. It is a step 2 nilpotent group, being a P_3 group. Note that the commutator subgroup $[Q_8, Q_8] = \{\pm 1\}$, because it is contained in $\langle Q_8^2\rangle = \{\pm 1\}$ (Lemma A.2(b)). Q_8 is not generated by its squares.

(i) The *$(ax+b)$-group*

$$G := \left\{\begin{pmatrix} a & b \\ 0 & 1\end{pmatrix} \;\middle|\; a>0,\ b\in\mathbb{R}\right\}$$

is an example of a solvable Lie group. It is not step 2 nilpotent. It is uniquely 2-divisible (Definition A.10).

The name is due to the fact that it becomes the group of affine motions of $\mathbb{R}$ when you view $\mathbb{R}$ as $\{(x,1)^t \in \mathbb{R}^2 \mid x\in\mathbb{R}\}$:

$$\begin{pmatrix} a & b \\ 0 & 1\end{pmatrix}\begin{pmatrix} x \\ 1\end{pmatrix} = \begin{pmatrix} ax+b \\ 1\end{pmatrix}.$$

Note the useful formulas

$$\begin{pmatrix} a & b \\ 0 & 1\end{pmatrix}^{-1} = \begin{pmatrix} a^{-1} & -a^{-1}b \\ 0 & 1\end{pmatrix},$$

and

$$[G,G] = \left\{\begin{pmatrix} 1 & x \\ 0 & 1\end{pmatrix} \;\middle|\; x\in\mathbb{R}\right\},$$

G is the topological semidirect product of $(\mathbb{R},+)$ and $(\mathbb{R}^+,\cdot)$, i.e., $G = \mathbb{R}\times_s \mathbb{R}^+$. See Example A.22.

A.5.5 *Semidirect products of groups*

Definition A.18. Let N be a normal subgroup and Q a subgroup of a group G. We say that G is the *semidirect product* of N and Q and write $G = N \times_s Q$, if $G = NQ$ and $N \cap Q = \{e\}$.

The conditions for a semidirect product mean that any $x \in G$ in exactly one way factorizes as $x = n(x)h(x)$, where $n(x) \in N$ and $h(x) \in Q$. The maps $x \mapsto n(x)$ and $x \mapsto h(x)$ of G onto N and Q are called the *projections* of G onto N and Q respectively.

Using that N is normal we may express the group operations in $G = N \times_s Q$ in terms of the projections as follows, where $x, y \in G$:

$$\begin{aligned} n(xy) &= n(x)h(x)n(y)h(x)^{-1} && \text{and} \quad n(x^{-1}) = h(x)^{-1}n(x)^{-1}h(x), \\ h(xy) &= h(x)h(y), && \text{so that} \quad h(x^{-1}) = h(x)^{-1}. \end{aligned}$$

The semidirect product $G = N \times_s Q$ is a direct product if and only if $nh = hn$ for all $n \in N$ and $h \in Q$.

So much for the algebraic aspect. But there is also a topological aspect that we proceed to discuss.

Definition A.19. A topological group G is said to be a *topological semidirect product* of the subgroups N and Q, if

(i) $G = N \times_s Q$,

(ii) N and Q are closed subgroups of G, and

(iii) the projections on N and Q are continuous.

Theorem A.20. *Consider a semidirect product $G = N \times_s Q$. If G is a σ-compact, locally compact group, and N and Q are closed in G, then G is a topological semidirect product of N and Q.*

Proof. Let us equip N and Q with the topology from G, and let G_0 denote the group G with the product topology instead of the given topology. G_0 is a topological group and the identity map $i : G_0 \to G$ is continuous. Since the projection maps in G_0 are continuous, it suffices to prove that i is a homeomorphism. For that it suffices by the open mapping theorem for groups (Theorem D.5) to prove that G_0 is a locally compact and σ-compact space. N and Q, being closed subsets of the σ-compact, locally compact space G, are σ-compact and locally compact as well. And then $G_0 = N \times Q$ is σ-compact and locally compact in the product topology. □

Example A.21. A word of caution: The projection maps in a semidirect product are not always continuous, not even when the underlying group is uncomplicated from the algebraical point of view:

Equip $G := \mathbb{R} \times \mathbb{R}_d$ with the product topology, where $\mathbb{R}$ is given its usual topology, but where $\mathbb{R}_d$ denotes the real line with the discrete topology. $N := \{(x, x) \in G \mid x \in \mathbb{R}\}$ is a normal subgroup of G, $Q := (0, \mathbb{R})$ is a subgroup of G and $G = N \times_s (0, \mathbb{R})$.

The sequence $x_n = (-1/n, 0)$, $n = 1, 2, \ldots$, in G converges to $(0, 0)$ with respect to the topology of G, because the $\mathbb{R}$ of the first component of $G = \mathbb{R} \times \mathbb{R}_d$ has its usual topology. But the projection of x_n on Q in the algebraic semidirect product $G = N \times_s Q$ is not continuous, i.e., $(0, 1/n)$ does not converge to $(0, 0)$ in $Q \subseteq G$, because $Q = \mathbb{R}_d$ has the discrete topology.

It might be mentioned that both N and Q are closed in G, so only the σ-compactness is lacking in the hypotheses of Theorem A.20.

Examples A.22.

(a) The Heisenberg group H_3 is in the notation of Example A.17(a)) the topological semidirect product $H_3 = N \times_s Q$, where

$$N = \{[x, 0, z] \mid x, z \in \mathbb{R}\} \quad \text{and} \quad Q = \{[0, y, 0] \mid y \in \mathbb{R}\}.$$

(b) The $(ax+b)$-group (see Examples A.17(i)) is the topological semidirect product of the subgroups

$$N = \left\{ \begin{pmatrix} 1 & b \\ 0 & 1 \end{pmatrix} \;\middle|\; b \in \mathbb{R} \right\} \quad \text{and} \quad Q = \left\{ \begin{pmatrix} a & 0 \\ 0 & 1 \end{pmatrix} \;\middle|\; a > 0 \right\}.$$

(c) $\mathbb{R}^n \times_s Q$, where Q is a matrix subgroup of $\mathrm{GL}(n, \mathbb{R})$, with the group composition

$$(b_1, A_1)(b_2, A_2) = (b_1 + A_1 b_2, A_1 A_2) \qquad \text{for } b_1, b_2 \in \mathbb{R}^n, \text{ and } A_1, A_2 \in Q,$$

is a topological semidirect product of $\mathbb{R}^n$ and Q.

You may view $\mathbb{R}^n \times_s Q$ as a group of transformations of $\mathbb{R}^n$, acting by $(b, A)x = Ax + b$ for $b, x \in \mathbb{R}^n$ and $A \in Q$. With $Q = O(n)$ you get the group of rigid motions of $\mathbb{R}^n$.

(d) $\mathrm{GL}(n,\mathbb{R})$ is the topological semidirect product of $\mathrm{SL}(n,\mathbb{R})$ and $\mathbb{R}^*$ according to the following decomposition of $A = \{a_{ij}\} \in \mathrm{GL}(n,\mathbb{R})$:

$$\begin{pmatrix} a_{11}(\det A)^{-1} & a_{12} & \dots & a_{1n} \\ a_{21}(\det A)^{-1} & a_{22} & \dots & a_{2n} \\ \vdots & \vdots & \ddots & \vdots \\ a_{n1}(\det A)^{-1} & a_{n2} & \dots & a_{nn} \end{pmatrix} \begin{pmatrix} \det A & 0 & \dots & 0 \\ 0 & 1 & \dots & 0 \\ \vdots & \vdots & \ddots & \vdots \\ 0 & 0 & \dots & 1 \end{pmatrix}.$$

A.5.6 *Functions on coset spaces*

Let H be a subgroup of a group G. A function f on G which on each left coset xH is constant, defines a function F on the coset space G/H by the prescription $F(xH) := f(x)$, $xH \in G/H$. This leads us to the following standard terminology:

Definition A.23. Let H be a subgroup of a group G, $\pi : G \to G/H$ the canonical quotient map and Z be a set. Let $f : G \to Z$. If $f = F \circ \pi$ for some $F : G/H \to Z$, or equivalently if $f(xh) = f(x)$ for all $x \in G$ and $h \in H$ (i.e., f is constant on each coset xH), then we say that f is *a function on the coset space* with values in Z.

The function F in the definition is uniquely determined from f, because $F(xH) = f(x)$ for all $xH \in G/H$, and the map $f \mapsto F$ is a bijection of the functions on the coset space with values in Z onto the maps from G/H to Z. We usually do not distinguish between the function f on G which is constant on each coset and the corresponding function F on G/H, even though they are different (one being a function on G and the other a function on G/H), because it will be clear from the context which one of the two interpretations we have in mind.

If H is generated by a set M, then f is a function on G/H if and only if $f(xm) = f(x)$ for all $x \in G$ and $m \in M$.

Example A.24. Let $G = \mathbb{R}^2 = \mathbb{R} \times \mathbb{R}$, $H = \{0\} \times \mathbb{R}$ and f a function on G. Then f is a function on G/H if and only if $f(x,y) = f(x,0)$ for all $(x,y) \in \mathbb{R}^2$.

We shall always equip G/H with the quotient topology whenever G is a topological group: $O \subseteq G/H$ is open if and only if $\pi^{-1}(O)$ is open in G. The quotient topology has the property that a map $F : G/H \to Z$ (where Z now is a topological space) is continuous if and only if $F \circ \pi : G \to Z$ is continuous.

A.6 Semigroups and involutions

Definition A.25. Let S be a *semigroup*, i.e., a non-empty set equipped with an associative operation $(x, y) \mapsto xy$ of $S \times S$ into S.

A *subsemigroup* of S is a non-empty subset S_1 of S with the property that $s, t \in S_1 \Rightarrow st \in S_1$.

Two elements $x \in S$ and $y \in S$ are said to *commute*, if and only if $xy = yx$. A subset A of S is said to be *abelian*, if and only if $xy = yx$ for all $x, y \in A$. The center of S is $Z(S) := \{z \in S \mid zx = xz \text{ for all } x \in S\}$.

A *neutral element* in S is an element $e \in S$ such that $ex = xe = x$ for all $x \in S$. A *monoid* (also known as a unital semigroup) is a semigroup with a neutral element. We usually call a monoid M.

S is said to be a *topological semigroup*, if S is a topological space and the product map $(x, y) \mapsto xy$ of $S \times S$ into S is continuous with respect to the product topology on $S \times S$.

We write the operation of a semigroup multiplicatively, unless it is abelian in which case we often use $+$ as a symbol for the operation. A semigroup is characterized by associativity of its operation, i.e., that $x(yz) = (xy)z$ for all $x, y, z \in S$. The associativity implies that we need not write parentheses in products of elements from S, the result being the same for any position of the parentheses.

A semigroup has at most one neutral element, but it need not possess one, and its elements are not required to have inverses even when it has a neutral element. A group is a special instance of a monoid, so what is defined and derived below about semigroups and monoids also applies to groups.

Definition A.26. Let S be a semigroup. A map $\mu : S \to \mathbb{C}$ such that $\mu(xy) = \mu(x)\mu(y)$ for all $x, y \in S$ is said to be a *multiplicative function* or just to be *multiplicative.*

A non-trivial example of a multiplicative function is the determinant on the multiplicative semigroup of $n \times n$ matrices.

Lemma A.27. *Let S be a semigroup. If $\mu : S \to \mathbb{C}$ is multiplicative and bounded, i.e.,* $\sup|\mu(S)| < \infty$, *then $|\mu(x)| \leq 1$ for all $x \in S$.*

Proof. With $C := \sup|\mu(S)|$ we find for all $n \in \mathbb{N}$ that $|\mu(x)|^n = |\mu(x)^n| = |\mu(x^n)| \leq C$, so $|\mu(x)| \leq C^{1/n}$. Letting $n \to \infty$ we get $|\mu(x)| \leq 1$. □

Lemma A.28. *A multiplicative function μ on a monoid is non-zero if and only if $\mu(e) = 1$, where e denotes the neutral element of the monoid.*

Proof. Choose x_0 in the monoid such that $\mu(x_0) \neq 0$. From $\mu(x_0) = \mu(x_0 e) = \mu(x_0)\mu(e)$ we infer that $\mu(e) = 1$. □

Definition A.29. A map $\mu : M \to M'$ between two monoids M and M' with neutral elements e and e' respectively is said to be a *monoid homomorphism* if $\mu(e) = e'$ and $\mu(xy) = \mu(x)\mu(y)$ for all $x, y \in M$. A special case of a monoid homomorphism is a *monoid character* on M, defined as a non-zero multiplicative function on M.

Definition A.30. Let S be a semigroup. A map $\tau : S \to S$ is said to be an *involution* in S, if $\tau(xy) = \tau(y)\tau(x)$ and $\tau(\tau(x)) = x$ for all $x, y \in S$.

Lemma A.31. *Let M be a monoid with neutral element e. If $\tau : M \to M$ is an involution, respectively a homomorphism $\sigma : M \to M$ such that $\sigma \circ \sigma = I$, then $\tau(e) = e$, respectively $\sigma(e) = e$.*

Proof. We do the involution τ. The homomorphism σ can be treated similarly. $\tau(e) = \tau(e)e = \tau(e)\tau(\tau(e)) = \tau(\tau(e)e) = \tau(\tau(e)) = e$. □

Definition A.32. Let S be a semigroup with an involution $\tau : S \to S$. A function F on S is said to be *τ-invariant*, if $F \circ \tau = F$.

Examples A.33. Concrete semigroups with involutions occur throughout mathematics. The examples below illustrate the versatility of the concept.

(a) On any group the inversion map $x \mapsto x^{-1}$ is an involution.

(b) Complex conjugation in $(\mathbb{C}, +)$ is an involution. More generally, a reflection in a hyperplane in $\mathbb{R}^n$ is an involution.

(c) The closed unit disc $D := \{z \in \mathbb{C} \mid |z| \leq 1\}$ is an abelian monoid with the semigroup operation being multiplication of complex numbers. An involution is $\tau(z) := \overline{z}$, $z \in D$.

(d) On any abelian semigroup $(S, +)$ the identical involution is the involution defined by $\tau(x) := x$ or all $x \in S$.

(e) Let X be a non-empty set. Under intersection the set $\mathcal{P}(X)$ of subsets of X is an abelian monoid. As an involution we take the identical involution. The property that $A + A = A$ for all $A \in \mathcal{P}(X)$ shows that this monoid is far from being a group. It is not even a subset of a group.

(f) If $\alpha \in \mathbb{R}$, then $\mathbb{R}^2$ equipped with the composition $\cdot_\alpha$, given by

$$(x_1, y_1) \cdot_\alpha (x_2, y_2) := (x_1x_2 + \alpha y_1y_2, x_1y_2 + y_1x_2), \qquad (x_1, y_1), (x_2, y_2) \in \mathbb{R}^2,$$

is an abelian monoid with neutral element $(1, 0)$. The map $i(x, y) := (x, -y)$ (complex conjugation!) is an involution.

(g) Transposition in the additive group of $n \times n$ matrices is an involution.

(h) The adjoint $A \mapsto A^*$ is an involution in the matrix group $\mathrm{GL}(n, \mathbb{C})$, different from the group inversion.

(i) On the Heisenberg group H_3 (see Examples A.17(a)) the map

$$\tau \begin{pmatrix} 1 & x & z \\ 0 & 1 & y \\ 0 & 0 & 1 \end{pmatrix} = \begin{pmatrix} 1 & y & z \\ 0 & 1 & x \\ 0 & 0 & 1 \end{pmatrix}$$

is an involution.

(j) Let W be a 2-dimensional complex or real vector space. To any linear map $A \in \mathcal{L}(W)$ we define its *adjugate* $\mathrm{adj}(A) \in \mathcal{L}(W)$ by

$$A + \mathrm{adj}(A) = (\mathrm{tr}\, A)I. \tag{A.2}$$

Clearly $\mathrm{adj} : \mathcal{L}(W) \to \mathcal{L}(W)$ is linear. Furthermore

$$A\, \mathrm{adj}(A) = \mathrm{adj}(A)A = (\det A)I.$$

To see this, multiply (A.2) by A and use $A^2 - (\mathrm{tr}\, A)A + (\det A)I = 0$ by Cayley-Hamilton's theorem.

Also $\mathrm{adj}(BAB^{-1}) = B\, \mathrm{adj}(A)B^{-1}$, when $B \in \mathcal{L}(W)$ is invertible. If A with respect to a basis of W has the matrix

$$\begin{pmatrix} a & b \\ c & d \end{pmatrix}$$

then $\mathrm{adj}(A)$ has the matrix

$$\begin{pmatrix} d & -b \\ -c & a \end{pmatrix}.$$

From this matrix representation of $\mathrm{adj}(A)$ it is easy to check that

$$\mathrm{adj}(AB) = \mathrm{adj}(B)\, \mathrm{adj}(A) \quad \text{and} \quad \mathrm{adj}(\mathrm{adj}\, A) = A \qquad \text{for all } A, B \in \mathcal{L}(W),$$

which shows adj is an involution of the monoid $\mathcal{L}(W)$ (monoid with respect to composition of linear maps).

(k) $S := \{x + iy \mid x > 0,\ y > 0\}$ with $\tau(x + iy) := y + ix$ for $x + iy \in S$ is a semigroup with involution τ.

A.7 Exercises

Exercise A.1. Let G be a group and let $f : G \to \mathbb{C}$.

(a) Let $g : G \to \mathbb{C}$ be both even and odd. Show that $g = 0$.

(b) Show that a possible decomposition $f = f_e + f_o$ of f into an even function $f_e : G \to \mathbb{C}$ and an odd function $f_o : G \to \mathbb{C}$ is unique.

(c) Show that there exists a decomposition of f as $f = f_e + f_o$ where $f_e : G \to \mathbb{C}$ is even and $f_o : G \to \mathbb{C}$ is odd.

Hint: Consider the two functions $(f + \check{f})/2$ and $(f - \check{f})/2$, where $\check{f}(x) := f(x^{-1})$, $x \in G$.

(d) Show that $f_o(e) = 0$.

(e) Let G be a topological group. Show that $f \in C(G)$ if and only if f_e and f_o are continuous.

Exercise A.2. Let N be a subgroup of a group G. Show that N is normal, if and only if for all $x, y \in G$: $xy \in N \Rightarrow yx \in N$.

The above characterization of normal subgroups is used in the proof of Lemma 15.2(c).

Exercise A.3. Prove that the circle group $\mathbb{T}$, the Heisenberg group H_3 and the $(ax + b)$-group are 2-divisible. Which ones are uniquely 2-divisible?

Exercise A.4. Let G be a group and let $x, y \in G$. Derive the formulas

$$[x, y^2] = [x, y]y[x, y]y^{-1}, \qquad [x^2, y] = x[x, y]x^{-1}[x, y],$$
$$[x^2, y^2] = x[x, y]x^{-1}[x, y]yx[x, y]x^{-1}[x, y]y^{-1}.$$

Exercise A.5. Let G be a group. Show that $G/\langle G^2\rangle$ is an abelian group.

Hint: Lemma A.2(b).

Exercise A.6. Consider the set $\mathcal{U} \subseteq M(n \times n, \mathbb{C})$ of upper triangular matrices with 1's in the diagonal. So $\{a_{ij}\} \in \mathcal{U}$ if and only if $a_{ij} = 0$ for $i > j$ and $a_{ii} = 1$.

Show that $\mathcal{U}$ is a nilpotent subgroup of $\mathrm{GL}(n, \mathbb{C})$.

Hint: What are the elements of the superdiagonal of $[u_1, u_2]$ for $u_1, u_2 \in \mathcal{U}$?

Exercise A.7. Let H be a normal subgroup of the topological group G. Let $f \in C(G)$ be a function on G/H such that $f(xy) + f(xy^{-1}) = 2f(x)$, $x, y \in G$.

(a) Note that the quotient map $\pi : G \to G/H$ is a surjective homomorphism.

(b) Show that $F : G/H \to \mathbb{C}$ is well defined by $F(\pi(x)) := f(x)$, $x \in G$.

(c) Show that $F \in C(G/H)$.

(d) Show that $F(\xi\eta) + F(\xi\eta^{-1}) = 2F(\xi)$ for all $\xi, \eta \in G/H$.

Exercise A.8. Show that the only constant, multiplicative functions on a semigroup are 0 and 1.

Exercise A.9. If $\tau : S \to S$ is an involution of a semigroup S, then $\tau(x\tau(y)) = y\tau(x)$ for all $x, y \in S$ (see Definition A.30). In this exercise you are asked to derive the following partial inverse:

Proposition A.34. *Let M be a monoid and let $f : M \to M$ satisfy the functional equation $f(xf(y)) = yf(x)$ for all $x, y \in M$. Then $f|_{f(M)}$ is an involution on $f(M)$, and $f(e)^2 = f(e)$. Furthermore $f(e) = e$, if and only if $f(M) = M$. If M is a group, then $f(e) = e$.*

(a) Show that $(f \circ f)(y) = yf(e)$ and $f(xf(e)) = f(x)$ for all $x, y \in M$.

(b) Show by help of (a) and the functional quation that $f(xy) = f(y)f(x)$ for all $x, y \in M$.

(c) Show that $(f \circ f)(f(x)) = f(x)f(e) = f(x)$, so that $f \circ f$ is the identity on $f(M)$.

(d) Show that $f(M) = f(e)Mf(e)$.

(e) Show that $f(e)^2 = f(e)$.

Find an example of a solution f of the functional equation for which $f(e) \neq e$.
Ideas: $M = (\mathbb{R}, \cdot)$ or Example A.33(e).

Exercise A.10. Let V be a real normed vector space of dimension at least 2. Let f be a mapping from V into V, bounded on the unit sphere $\{x \in V \mid \|x\| = 1\}$, such that whenever x and y are in V, $f(x + f(y)) = f(x) + y$. Prove that f is a continuous, linear involution on V.

Hint: Combine Proposition A.34 and Exercise 2.19.

Exercise A.11. Consider the functional equation

$$f(x_1x_2+y_1y_2, x_1y_2+y_1x_2) = f(x_1, y_1)f(x_2, y_2), \quad (x_1, y_1), (x_2, y_2) \in \mathbb{R}^2,$$

that is connected to Example A.33(f) with $\alpha = 1$. Show that $f : \mathbb{R}^2 \to \mathbb{C}$ has the form $f(x, y) = M_1(x + y)M_2(x - y)$, where $M_1, M_2 : \mathbb{R} \to \mathbb{C}$ satisfy $M_i(xy) = M_i(x)M_i(y)$ for all $x, y \in \mathbb{R}$ and $i = 1, 2$.

A.8 Notes and remarks

Definition A.4: Corovei [56, 55] studies various functional equations on P_3 groups.

Theorem A.16 is from Ellis [72, Theorem 2].

Theorem A.20: For a generalization to more than 2 factors see Ebanks and Stetkær [70].

Example A.33(f): For motivation and more information about this monoid see Berrone and Dieulefait [27, Introduction]. The case of $\alpha = 1$ occurs in another connection in Chung and Sahoo [47].

Proposition A.34 and Exercise A.10: Exercise A.10 is a problem posed by Cuculière in [59]. Ger [99] generalized its set up to maps f having the homomorphism property, $f(xf(y)) = f(x)y$. Inspired by Ger's procedure we exploit in Proposition A.34 the corresponding anti-homomorphism property.

Appendix B

Substitutes for Commutativity

Throughout this chapter we let S be a semigroup, X a set and $f : S \to X$ a map from S to X.

Functional equations are usually much easier to handle on abelian than on non-abelian semigroups. But luckily, quite often a solution of a functional equation or a component of it has properties that compensate to some degree for the lacking commutativity of the underlying semigroup. The present section discusses three such properties of a function: Being central, satisfying Kannappan's condition and being abelian. These related notions are important for all of our functional equations.

B.1 Kannappan's condition etc

Definition B.1. f is *central*, if and only if $f(xy) = f(yx)$ for all $x, y \in S$. If S is a group, then f is central if and only if $f(xyx^{-1}) = f(y)$ for all $x, y \in S$.

Definition B.2. f satisfies *Kannappan's condition*, if and only if $f(xyz) = f(xzy)$ for all $x, y, z \in S$. Equivalently if and only if $Z(f) = S$, where

$$Z(f) := \{z \in S \mid f(xyz) = f(xzy) \text{ for all } x, y \in S\}. \tag{B.1}$$

It is obvious that $Z(S) \subseteq Z(f)$, where $Z(S)$ denotes the center of S.

Definition B.3. We say that f is *abelian* if and only if $f(x_1 x_2 \cdots x_n) = f(x_{\pi(1)} x_{\pi(2)} \cdots x_{\pi(n)})$ for all $x_1, x_2, \ldots, x_n \in S$, all permutations π of n elements and all $n = 1, 2, 3, \ldots$. Equivalently, f is abelian if and only if it is a central function satisfying Kannappan's condition.

We say that f is *non-abelian* or a *non-abelian function*, if it is not abelian.

On an abelian semigroup any function is abelian. A example of a non-abelian, but nevertheless central, function is the trace on $M(n \times n, \mathbb{C})$ for $n \geq 2$.

A functional equation may be of such a nature that it compels its solutions have one or more of the above properties. To take an example, all solutions $f : S \to \mathbb{C}$ of $f(xy) = f(x)f(y)$, $x, y \in S$, are abelian.

Lemma B.4 points out some algebraic properties of the set $Z(f)$ when the semigroup S is a monoid. For example: In the definition (B.1) of $Z(f)$ only the two letters y and z standing to the right are permuted. However, on a monoid it might as well be the two letters to the left, as shown by (b).

Lemma B.4. *Let $S = M$ be a monoid.*

(a) *f satisfies Kannappan's condition, if and only if f is abelian.*

(b) $Z(f) = \{z \in M \mid f(xzy) = f(zxy)$ *for all* $x, y \in M\}$.

(c) *If $z \in Z(f)$ and $x_1, x_2, \ldots, x_n \in M$, where $n \geq 1$, then for $1 \leq i < n$:*

$$f(zx_1x_2\cdots x_n) = f(x_1x_2\cdots x_i z x_{i+1}\cdots x_n) = f(x_1x_2\cdots x_n z).$$

Proof. (a) Take $x = e$ in the definition of Kannappan's condition.

(b) and (c) are left to the reader. Note in (c) that we may assume that there are at least 3 factors, because we can insert the neutral element e wherever we want. □

Let $S = M$ be a monoid. If $z \in Z(f)$ occurs as a factor inside the argument $(-)$ of $f(-)$ then the value of f does not change as z moves around inside the argument (Lemma B.4(c)). In this respect z behaves like an element of the center $Z(M)$ of M, which is why we use the notation $Z(f)$. It is also why we use the terminology abelian function.

In Lemma B.5 we specialize further by progressing from a monoid to a group.

Lemma B.5. *Let f be a function on a group G.*

(a) *$Z(f)$ is a normal subgroup of G.*

(b) *Let A be a subset of G. Then f is a function on $G/[G, A]$ if and only if $A \subseteq Z(f)$.*

(c) *f satisfies Kannappan's condition if and only if it is a function on $G/[G, G]$.*

(d) *f satisfies Kannappan's condition if it is a function on $G/\langle G^2\rangle$, where $\langle G^2\rangle$ denotes the subgroup of G generated by the squares.*

Proof. (a), (c) and the if part of (b) are left to the reader.

The only if part of (b): For that we note that if $a \in A$ then $f(xya^{-1}) = f(xa^{-1}y[y^{-1}, a]) = f(xa^{-1}y)$ for all $x, y \in G$, so that $a^{-1} \in Z(f)$ and hence $a \in Z(f)$.

(d) follows from Lemma A.2(b). □

On a group G Kannappan's condition on a function f means that f is a function on the abelian group $G/[G, G]$ (Lemma B.5(c)), so working with such a function we may often assume the underlying group to be abelian: If it is not abelian we replace it by $G/[G, G]$. However, there are certain advantages in working with Kannappan's condition instead of passing to the quotient group:

(1) It applies to semigroups, not just to groups, because inverses do not appear in the definition of Kannappan's condition.

(2) It is used in the literature. Kannappan [125] introduced it in 1968 in connection with his study of d'Alembert's functional equation, but it occurs in studies of other functional equations as well: See for example [41] or our discussion of the quadratic functional equation (Section 13).

(3) It lends itself to generalizations as found in papers by the Moroccan school (Akkouchi, Elqorachi et al.).

(4) In trying to establish that a given function is a function on the abelian group $G/[G, G]$ it is often technically expedient to show instead that it satisfies Kannappan's condition.

(5) The crucial point of many considerations is that the solution is abelian, not that the group is.

(6) It is not so that a function is abelian if and only if the underlying group is abelian. (There are even classes of non-abelian groups on which all solutions of a given functional equation are abelian. For pre-d'Alembert functions some examples of this phenomenon can be found at Theorem 8.16.) So it is necessary to distinguish between the two notions of the group being abelian and the solution being abelian.

B.2 Four derived functions

We will in this subsection consider four functions constructed from a given function $g : S \to \mathbb{C}$. They play crucial roles for several of the trigonometric functional equations that we study. The reader can anticipate prospective results by observing that the new functions (g_x, d, $\widetilde{\Delta}$ and Δ) are very simple, when g is the solution $g(x) = \cos x$ of d'Alembert's functional equation on the real line.

Definition B.6.

(a) $g_x : S \to \mathbb{C}$ is for any $x \in S$ defined by $g_x(y) := g(xy) - g(x)g(y)$, $y \in S$.

(b) $d = d_g : S \to \mathbb{C}$ is given by $d(x) := 2g(x)^2 - g(x^2)$ for $x \in S$.

(c) $\widetilde{\Delta} : S \times S \to \mathbb{C}$ is for any $x, y \in S$ defined by

$$\widetilde{\Delta}(x, y) := 2d(x)d(y) + g(xy^2x) + g(yx^2y) - 4g(xy)g(yx). \quad \text{(B.2)}$$

(d) $\Delta : S \times S \to \mathbb{C}$ is defined by

$$\Delta(x, y) := \tfrac{1}{2}[g(x^2y^2) - g((xy)^2)], \qquad x, y \in S. \quad \text{(B.3)}$$

Note that g is multiplicative if and only if $g_x = 0$ for all $x \in S$. $d : S \to \mathbb{C}$ is important, not just in the discussion of the symmetrized sine addition formula (6.1) and earlier (Theorem 4.1(f)), but also later in the theory of d'Alembert's functional equation. $\widetilde{\Delta}$ is closely related to the function $\Delta : S \times S \to \mathbb{C}$, that we need in Chapters 8 and 9. Indeed, if g is a pre-d'Alembert function, then $\widetilde{\Delta} = 4\Delta$ (Lemma 8.9). For reference we list in Lemma B.7 a couple of relations between d, $\widetilde{\Delta}$ and Δ.

Lemma B.7. *Let $g : S \to \mathbb{C}$ be any function.*

(a) *If g is central, then we get for all $x, y \in S$:*

$$\begin{aligned} \widetilde{\Delta}(x, y) &= 2[d(x)d(y) - d(xy)] + 2[g(x^2y^2) - g((xy)^2)] \\ &= 2[d(x)d(y) - d(xy)] + 4\Delta(x, y). \end{aligned}$$

(b) *If g is abelian, then $\widetilde{\Delta}(x, y) = 2(d(x)d(y) - d(xy))$ for all $x, y \in S$.*

(c) *If g is central and d is multiplicative, then $\widetilde{\Delta} = 4\Delta$.*

(d) *If $x \in S$ and $y \in S$ commute, then $\widetilde{\Delta}(x, y) = 2(d(x)d(y) - d(xy))$.*

(e) *If $\widetilde{\Delta} = 0$ and S is a group, then $d(x)d(y) = d(x^{-1})d(xyx)$ for all $x, y \in S$.*

Proof. (a) Since g is central, (B.2) reduces as follows:

$$\begin{aligned}\widetilde{\Delta}(x,y) &= 2d(x)d(y) + 2g(x^2y^2) - 4g(xy)^2 \\ &= 2d(x)d(y) + 2g(x^2y^2) - 2[2g(xy)^2 - g((xy)^2)] - 2g((xy)^2) \\ &= 2d(x)d(y) + 2g(x^2y^2) - 2d(xy) - 2g((xy)^2),\end{aligned}$$

which is the desired formula by the definition of Δ.

(b) and (c) follow immediately from (a).

(d) follows from (b), when we restrict our considerations to the abelian subsemigroup of S generated by x and y.

(e) $\widetilde{\Delta} = 0$ is equivalent to

$$2d(x)d(y) = 4g(xy)g(yx) - g(xy^2x) - g(yx^2y) \qquad \text{for all } x, y \in S. \quad \text{(B.4)}$$

Replacing x by x^{-1} and y by xyx the right hand side of (B.4) is unchanged while the left hand side becomes by $d(x^{-1})d(xyx)$. □

B.3 Exercises

Exercise B.1. Let $g = \cos$ on $\mathbb{R}$. Fix $x \in \mathbb{R}$. Show that the function g_x from Definition B.6 is a constant times sin.

Let $g = \cosh$ on $\mathbb{R}$. Fix $x \in \mathbb{R}$. Show that the function g_x from Definition B.6 is a constant times sinh.

Generalize to $g = (\chi + \check{\chi})/2$, where $\chi : G \to \mathbb{C}$ is a character on a group G.

Exercise B.2. Let S be a semigroup. Let $g = (\chi + \sigma)/2$, where $\chi, \sigma : S \to \mathbb{C}$ are multiplicative.

(a) Show for any $x \in S$ that

$$g_x = \frac{\chi(x) - \sigma(x)}{2} \frac{\chi - \sigma}{2}.$$

(b) Let $x \in S$. Show that $g_x(x) = 0 \Rightarrow g_x = 0$.

Exercise B.3. Let $\chi : S \to \mathbb{C}$ be a multiplicative function on a semigroup S. Show that any polynomial in χ is abelian.

Exercise B.4. Find a group and a central function g on it such that g does not satisfy Kannappan's condition.

Hint: Which functions on $\mathrm{GL}(2,\mathbb{R})$ or $\mathrm{SL}(2,\mathbb{R})$ occur in the characteristic polynomial?

Exercise B.5. Let $f : G \to \mathbb{C}$, where G is a group, and let $z \in Z(f)$. Show that any commutator of the form $[z,x]$ or $[x,z]$ where $x \in G$, can be deleted inside an argument of f, i.e., $f(a[z,x]b) = f(a[x,z]b) = f(ab)$ for all $a, b \in G$.

Exercise B.6. Let M and M' be monoids. Let $\tau : M' \to M$ be a surjective map which is either a homomorphism or an anti-homomorphism. Let $f : M \to \mathbb{C}$. Show that

(a) $Z(f \circ \tau) = \tau^{-1}(Z(f))$ and $\tau(Z(f \circ \tau)) = Z(f)$.

(b) f is abelian if and only if $f \circ \tau$ is abelian.

Exercise B.7. Let $\tau : S \to S$ be an involution of a semigroup S and let $f : S \to \mathbb{C}$ be τ-invariant. Show that $z \in Z(f) \Rightarrow \tau(z) \in Z(f)$.

Exercise B.8. Let G be a group. Let $f : G \to \mathbb{C}$ be abelian. Show that f_x is abelian for all $x \in G$.

Exercise B.9.

(a) Let $g(x) = \cos x$ for $x \in \mathbb{R}$. What are the corresponding functions g_x, d, $\widetilde{\Delta}$ and Δ?

(b) Let $g : \mathbb{R} \to \mathbb{C}$ be a non-zero solution of d'Alembert's functional equation

$$g(x+y) + g(x-y) = 2g(x)g(y), \qquad x, y \in \mathbb{R}.$$

What are the corresponding functions d, $\widetilde{\Delta}$ and Δ?

Exercise B.10. Let $g(x) = \frac{1}{2}\operatorname{tr}\rho(x)$, $x \in S$, where ρ is a 2-dimensional representation of a semigroup S. Show that $d(x) = \det \rho(x)$ for all $x \in S$, and that d is multiplicative.

B.4 Notes and remarks

Definition B.1: In the literature a central function is also called a class function.

Definition B.6: The function d is in the literature. See Parnami, Singh and Vasudeva [154] and their references. They define d on Banach spaces, not on groups. Davison [64, Definition 2.9] introduced it on monoids. Yang used "the magic function" (her phrase about d) in her paper [216] on the symmetrized Sine addition formula on groups. The function Δ stems from Davison [64]. Davison encountered it so often in his computations that he found it convenient to name it. $\widetilde{\Delta}$ occurs in Yang [216].

Lemma B.7(e) is due to Dilian Yang [216, Proposition 2.5(i)].

Appendix C

The Casorati Determinant

We shall in this appendix state and generalize a classical criterion that determines when a set of functions is linearly dependent (their Casorati determinant should vanish everywhere). We use it to discuss functions F of the form $F(x,y) = \sum_{l=1}^{N} g_l(x)h_l(y)$ and to prove that continuous solutions of certain trigonometric functional equations are smooth (Proposition D.7).

C.1 Around the Casorati determinant

Definition C.1. Let X be a set, $N \in \mathbb{N}$ and $f_1, f_2, \ldots, f_N : X \to \mathbb{C}$. The *Casorati matrix* of $(f_1, f_2, \ldots, f_N)$ at $(x_1, x_2, \ldots, x_N) \in X \times X \times \cdots \times X$ is

$$\begin{pmatrix} f_1(x_1) & f_2(x_1) & \ldots & f_N(x_1) \\ f_1(x_2) & f_2(x_2) & \ldots & f_N(x_2) \\ \vdots & \vdots & \ddots & \vdots \\ f_1(x_N) & f_2(x_N) & \ldots & f_N(x_N) \end{pmatrix}.$$

The *Casorati determinant* of $(f_1, f_2, \ldots, f_N)$ at $(x_1, x_2, \ldots, x_N) \in X \times X \times \cdots \times X$ is the determinant of the Casorati matrix.

The classical criterion about linear dependence in a space of complex valued functions is

Lemma C.2. *Let X be a set, $N \in \mathbb{N}$ and $f := (f_1, f_2, \ldots, f_N)^t : X \to \mathbb{C}^N$.*

(a) *$f_1, f_2, \ldots, f_N$ are linearly dependent if and only if their Casorati determinant vanishes at all points $(x_1, x_2, \ldots, x_N) \in X \times X \times \cdots \times X$.*

(b) *$f_1, f_2, \ldots, f_N$ are linearly independent if and only if there exists a point $(x_1, x_2, \ldots, x_N) \in X \times X \times \cdots \times X$ where their Casorati determinant does not vanish.*

(c) $f_1, f_2, \ldots, f_N$ *are linearly independent if and only if* $\text{span}\{f(x) \mid x \in X\} = \mathbb{C}^N$.

Proof. (a) is an easy consequence of Proposition C.3 below: Take Φ as the set of point evaluations (the point evaluation at $x \in X$ is the map $f \mapsto f(x)$, $f \in \mathbb{C}^X$). (b) is immediate from (a), and (c) is a consequence of (b). □

We need a modification of Lemma C.2 in the proof of Proposition D.7. The following proposition encompasses both versions. It also reveals that the key property of the point evaluations is that they separate functions.

Proposition C.3. *Let $\mathcal{F}$ be a real or complex vector space and let $\mathcal{F}^*$ be its dual space. The value of $\phi \in \mathcal{F}^*$ at $f \in \mathcal{F}$ is denoted $\langle \phi, f \rangle$. Let Φ be a subset of $\mathcal{F}^*$ that separates the points of $\mathcal{F}$, i.e., has the property*

$$f \in \mathcal{F} \quad \textit{and} \quad \langle \phi, f \rangle = 0 \quad \textit{for all} \quad \phi \in \Phi \quad \Rightarrow \quad f = 0.$$

Let finally $f_1, f_2, \ldots, f_N \in \mathcal{F}$, where $N \in \mathbb{N}$.

Then $f_1, f_2, \ldots, f_N$ are linearly dependent if and only if

$$\det \begin{pmatrix} \langle \phi_1, f_1 \rangle & \langle \phi_2, f_1 \rangle & \ldots & \langle \phi_N, f_1 \rangle \\ \langle \phi_1, f_2 \rangle & \langle \phi_2, f_2 \rangle & \ldots & \langle \phi_N, f_2 \rangle \\ \vdots & \vdots & \ldots & \vdots \\ \langle \phi_1, f_N \rangle & \langle \phi_2, f_N \rangle & \ldots & \langle \phi_N, f_N \rangle \end{pmatrix} = 0 \qquad \text{(C.1)}$$

for all $\phi_1, \ldots, \phi_N \in \Phi$.

Proof.

Only if. Assume $f_1, f_2, \ldots, f_N$ linearly dependent. If $N = 1$ then $f_1 = 0$, so the matrix in (C.1) vanishes. If $N > 1$ then one of $f_1, f_2, \ldots, f_N$ is a linear combination of the others, say $f_N = c_1 f_1 + c_2 f_2 + \cdots + c_{N-1} f_{N-1}$. But then the last column of the matrix in (C.1) is a linear combination of the first ones. And so its determinant vanishes.

If. First some considerations involving Φ, but not yet $f_1, f_2, \ldots, f_N$. For any finite-dimensional subspace V of $\mathcal{F}$ we look at the restriction map $r : \text{span}\,\Phi \to V^*$, defined by $r(\phi)(v) = \langle \phi, v \rangle$ for $v \in V$ and $\phi \in \text{span}\,\Phi$. We prove by contradiction that $r(\text{span}\,\Phi) = V^*$: If not, then $r(\text{span}\,\Phi)$ is a proper subspace of V^*. By linear algebra there then exists a $v_0 \in V$, $v_0 \neq 0$, such that $r(\text{span}\,\Phi)(v_0) = \{0\}$. In particular $\langle \phi, v_0 \rangle = r(\phi)(v_0) = 0$

for all $\phi \in \Phi$. By our assumption on Φ we arrive at the contradiction $v_0 = 0$. The left hand side of $r(\text{span } \Phi) = V^*$ is generated by $\{r(\phi) \mid \phi \in \Phi\}$, so we may choose $\phi_1, \phi_2, \ldots, \phi_N \in \Phi$, where $N = \dim V$, such that $r(\phi_1), r(\phi_2), \ldots, r(\phi_N)$ is a basis of V^*.

If $f_1, f_2, \ldots, f_N$ are linearly independent, then they form a basis of $V := \text{span}(f_1, f_2, \ldots, f_N)$. A well known fact from linear algebra says: If $f_1, f_2, \ldots, f_N$ is a basis of V and $v_1^*, v_2^*, \ldots, v_n^*$ is a basis of V^*, then the matrix $\{v_i^*(f_j)\}$ has maximal rank and so $\det\{v_i^*(f_j)\} \neq 0$. Apply this to the basis $v_i^* := r(\phi_i)$, $i = 1, 2, \ldots, N$. □

We give two examples of sets Φ for which the separation property of Proposition C.3 is not by point evaluations.

Example C.4. Let G be a locally compact group with a left Haar measure dy. Let $\mathcal{F} := C(G)$. We view $C_c(G)$ as a subspace of $\mathcal{F}^*$ by

$$\langle \phi, f \rangle := \int_G f(y)\phi(y)\, dy \qquad \text{for } \phi \in C_c(G) \text{ and } f \in C(G).$$

$\Phi := C_c(G)$ separates the points of $C(G)$. This is easy to derive from Urysohn's lemma and the fact that any non-empty, open subset of G has strictly positive measure.

The next example follows from Example C.4, because on a Lie group G any function in $C_c(G)$ can be approximated uniformly by functions from $C_c^\infty(G)$ having their supports in a joint compact set.

Example C.5. Let G be a Lie group with a left Haar measure dy. Let $\mathcal{F} := C(G)$. We view $C_c^\infty(G)$ as a subspace of $\mathcal{F}^*$ by

$$\langle \phi, f \rangle := \int_G f(y)\phi(y)\, dy \qquad \text{for } \phi \in C_c^\infty(G) \text{ and } f \in C(G).$$

$\Phi := C_c^\infty(G)$ separates the points of $C(G)$.

We apply Lemma C.2 to get a useful corollary about functions of two variables:

Corollary C.6. *Let X and Y be non-empty sets and let $N \in \mathbb{N}$. Let $F : X \times Y \to \mathbb{C}$ and $g_1, \ldots, g_N : X \to \mathbb{C}$, $h_1, \ldots, h_N : Y \to \mathbb{C}$ be related by*

$$F(x, y) = \sum_{l=1}^{N} g_l(x) h_l(y) \qquad \textit{for all } (x, y) \in X \times Y. \tag{C.2}$$

(a) *If* $\{g_1, g_2, \ldots, g_N\}$ *is a linearly independent set of functions on* X, *then there exist* $x_1, x_2, \ldots, x_N \in X$ *and* $\beta_{kl} \in \mathbb{C}$ *for* $k, l = 1, 2, \ldots, N$ *such that*

$$h_l(y) = \sum_{k=1}^{N} \beta_{kl} F(x_k, y) \qquad \textit{for all } y \in Y \textit{ and } l = 1, 2, \ldots, N. \quad \text{(C.3)}$$

(b) *If* $\{h_1, h_2, \ldots, h_N\}$ *is a linearly independent set of functions on* Y, *then there exist* $y_1, y_2, \ldots, y_N \in Y$ *and* $\alpha_{kl} \in \mathbb{C}$ *for* $k, l = 1, 2, \ldots, N$ *such that*

$$g_l(x) = \sum_{k=1}^{N} \alpha_{kl} F(x, y_k) \qquad \textit{for all } x \in X \textit{ and } l = 1, 2, \ldots, N. \quad \text{(C.4)}$$

Loosely speaking, Corollary C.6 says: If $\{g_1, g_2, \ldots, g_N\}$ is a linearly independent set then we may divide by it in (C.2) and thus obtain $h_1, h_2, \ldots, h_N$ with the formula (C.3) giving a more precise description. And similarly, if $\{h_1, h_2, \ldots, h_N\}$ is a linearly independent then we may divide by it in (C.2) and thus obtain (C.4).

Proof. $g_1, g_2, \ldots, g_N$ being linearly independent, there exist by Lemma C.2 points $x_1, x_2, \ldots, x_N \in X$ such that the Casorati matrix

$$\begin{pmatrix} g_1(x_1) & g_2(x_1) & \ldots & g_N(x_1) \\ g_1(x_2) & g_2(x_2) & \ldots & g_N(x_2) \\ \vdots & \vdots & \ldots & \vdots \\ g_1(x_N) & g_2(x_N) & \ldots & g_N(x_N) \end{pmatrix}$$

is invertible. Expressing $F(x_i, y) = \sum_{l=1}^{N} g_l(x_i) h_l(y)$, $i = 1, 2, \ldots, N$, in matrix form gives us that

$$\begin{pmatrix} F(x_1, y) \\ F(x_2, y) \\ \vdots \\ F(x_N, y) \end{pmatrix} = \begin{pmatrix} g_1(x_1) & g_2(x_1) & \ldots & g_N(x_1) \\ g_1(x_2) & g_2(x_2) & \ldots & g_N(x_2) \\ \vdots & \vdots & \ldots & \vdots \\ g_1(x_N) & g_2(x_N) & \ldots & g_N(x_N) \end{pmatrix} \begin{pmatrix} h_1(y) \\ h_2(y) \\ \vdots \\ h_N(y) \end{pmatrix},$$

which implies part (a) of the corollary. (b) can be proved in a similar way, so we skip the proof. □

C.2 Exercises

Exercise C.1. Let S be a semigroup.

(a) Let $\chi_0, \chi_1, \ldots, \chi_N : S \to \mathbb{C}$ be multiplicative functions such that $\chi_0 \neq \chi_i$ for $i = 1, 2, \ldots, N$. Let $g = \chi_0 + \sum_{i=1}^{N} \alpha_i \chi_i$, where $\alpha_1, \alpha_2, \ldots, \alpha_N \in \mathbb{C}$. Show that $\chi_0 \in \operatorname{span}\{R(x)g \mid x \in S\}$.

Hint: You may combine Theorem 3.18(b) and Corollary C.6(b) with $F(x, y) = g(xy)$.

(b) Let $\tau : S \to S$ be an involution, $\chi : S \to \mathbb{C}$ be multiplicative and $g := (\chi + \chi \circ \tau)/2$. Show that $\chi \in \operatorname{span}\{R(x)g \mid x \in S\}$.

C.3 Notes and remarks

For more about finite sum decompositions of functions with the variables separated like in (C.2) consult the monograph Rassias and Šimša [163].

For historical references and applications of Lemma C.2 see Aczél and Dhombres [6, Lemma 14.1, pp. 228–230]. Vincze [203, 204, 205] used the lemma in his studies of complex-valued solutions of trigonometric functional equations on abelian semigroups.

Appendix D

Regularity

In this appendix we derive regularity properties of maps between locally compact groups, in particular of homomorphisms including additive maps, characters and representations. Many solutions of functional equations can be expressed in terms of additive maps and characters, or more generally representations, so it is worth to know results about regularity of homomorphisms.

In Section D.2 we collect useful information about continuity of homomorphisms between locally compact groups. In Section D.3 we discuss differentiability of continuous solutions of various trigonometric functional equations. Finally Section D.4 presents a result about continuity of a function f, given that its Cauchy difference Cf is continuous.

D.1 Introduction

To impose a *regularity condition* on a function is to require it to possess a local property like being continuous, differentiable, C^∞, measurable or locally integrable. A regularity property like differentiability is useful, because it allows us to apply the techniques and results from the well developed theory of differential equations. A prime example of this is Example 3.7(a) in which we compute the continuous characters of $\mathbb{R}$. A regularity assumption involves a topology, a differentiable structure or a measure on the space on which the function is defined. Algebraic or set theoretical conditions are not counted as regularity conditions.

Experience shows that a weak regularity assumption on a solution of a functional equation often implies a stronger regularity property that may be more convenient to work with. As the proverb says: Give the devil a little finger and he will take the whole hand. This phenomenon is not limited to solutions of functional equations. The ability of algebraic structures to upgrade weak regularity to strong regularity is a recurring theme in mathematics. Let us as an illustration consider an example from the theory of functional equations: If a character on a locally compact group G is assumed measurable then it is actually continuous (Proposition D.4(b)), and if G is a Lie group then it is even C^∞ (Proposition D.6). In particular any measurable character on $\mathbb{R}$ has the form $x \mapsto \exp(\lambda x)$ for some $\lambda \in \mathbb{C}$ (Example 3.7), and so it is even real-analytic (meaning it has a power series expansion).

D.2 Continuity of homomorphisms

In this section we collect some useful results about continuity of homomorphisms between locally compact groups. We recall that a topological space is *σ-compact*, if it is a countable union of compact subsets.

Lemma D.1. *A homomorphism $\pi : G \to H$ between two topological groups G and H is continuous if and only if it is continuous at the neutral element $e \in G$. In particular, a character on a topological group is continuous if and only if it is continuous at the neutral element.*

Proof. We leave the proof to the reader. □

Steinhaus' theorem is a technical tool that we need.

Theorem D.2 (Steinhaus' theorem). *Let G be a locally compact group with left Haar measure λ. If A is a λ-measurable subset of G such that $0 < \lambda(A) < \infty$ then AA^{-1} contains a neighborhood of e.*

Proof. We refer to [196] for a short, elementary proof. □

Definition D.3. Let G be a locally compact group with a left (or right) Haar measure λ. Let Z be a topological space. We say that a map $F : G \to Z$ is *measurable* if $F^{-1}(U)$ is λ-measurable for all open subsets U of Z.

If G is a topological group we shall usually not work with all of its characters, but only the continuous ones. You might think that it could pay to relax the assumption of continuity to, say measurability. But at least on locally compact groups that will not give you any more characters. More precisely:

Proposition D.4. *Let G be a locally compact group.*

(a) *Let H be a σ-compact topological group, and let $\pi : G \to H$ be a homomorphism which is measurable. Then π is continuous.*

(b) *Let $\chi : G \to \mathbb{C}^*$ be a measurable character. Then χ is continuous.*

(c) *Let $a : G \to \mathbb{C}$ be an measurable, additive function. Then a is continuous.*

(d) *Let $\pi : G \to \mathrm{GL}(V)$ be a finite-dimensional representation of G such that its matrix-coefficients are measurable. Then π is continuous.*

Proof. We need only prove (a), since (b), (c) and (d) are immediate consequences of (a).

It suffices by Lemma D.1 to verify that π is continuous at $e \in G$. This means that given any neighborhood U_H of the neutral element $e_H \in H$ we shall produce a neighborhood V of $e \in G$ such that $\pi(v) \in U_H$ for all $v \in V$. The main technical tool for doing so is Steinhaus' theorem:

Choose first an open neighborhood W of $e_H \in H$ such that $WW^{-1} \subseteq U_H$. By the σ-compactness of H there exists a sequence $z_1, z_2, \ldots$ in H such that $\bigcup_{n=1}^{\infty}(Wz_n) = H$. Consequently $G = \bigcup_{n=1}^{\infty} \pi^{-1}(Wz_n)$.

Next take a compact subset $K \subseteq G$ such that $0 < \lambda(K)$, where λ is a left Haar measure on G. (You may for instance choose K as a compact neighborhood of $e \in G$.) Of course $\lambda(K) < \infty$, because K is compact. From $G = \bigcup_{n=1}^{\infty} \pi^{-1}(Wz_n)$ we find that $K = \bigcup_{n=1}^{\infty}(K \cap \pi^{-1}(Wz_n))$. Since K is not a nullset, not all of the terms on the right can be nullsets, so $0 < \lambda(K \cap \pi^{-1}(Wz_n)) < \infty$ for some $n \in \mathbb{N}$. Taking $A := K \cap \pi^{-1}(Wz_n)$ in Steinhaus' theorem we get a neighborhood V of $e \in G$ such that

$$V \subseteq (K \cap \pi^{-1}(Wz_n))\,(K \cap \pi^{-1}(Wz_n))^{-1} \subseteq \pi^{-1}(Wz_n)\,(\pi^{-1}(Wz_n))^{-1}.$$

Any $v \in V$ can thus be written in the form $v = ab^{-1}$, where $a, b \in \pi^{-1}(Wz_n)$, so there exist $w_a, w_b \in W$ such that $\pi(a) = w_a z_n$ and $\pi(b) = w_b z_n$. Now we finally get $\pi(v) = \pi(ab^{-1}) = \pi(a)\pi(b)^{-1} = w_a z_n (w_b z_n)^{-1} = w_a w_b^{-1} \in WW^{-1} \subseteq U_H$. □

A comment to Proposition D.4(c): On monoids, in contrast to groups, continuity does not follow from measurability. On the monoid $M = ([0, \infty[, +)$ the additive function $\mathbf{1}_{\{0\}}$ is discontinuous, although it is measurable.

The following result is sometimes helpful in topological considerations.

Theorem D.5 (Open mapping theorem for groups). *Let G and H be locally compact groups such that G is σ-compact. Let $\phi : G \to H$ be a continuous homomorphism of G onto H.*

(a) *ϕ is an open mapping.*

(b) *The induced map of the quotient group $G/\ker\phi$ onto H is a topological isomorphism.*

(c) *If ϕ is injective, then $\phi^{-1} : H \to G$ is a continuous mapping, so ϕ is a topological isomorphism.*

Proof. See [111, (5.29) Theorem]. □

D.3 Smoothness of solutions

We shall derive smoothness of continuous, complex valued solutions of trigonometric functional equations on a Lie group. Smooth means here C^∞.

Throughout this section G denotes a Lie group, dx a left Haar measure on it and $\tau : G \to G$ a continuous involution. $\mathbb{R}^n$ and matrix groups are examples of Lie groups. A possibility for τ is the group inversion $\tau(x) = x^{-1}$, $x \in G$. Proposition D.6 shows that τ is not just continuous, but actually smooth.

Proposition D.6. *Let G and H be Lie groups and $\pi : G \to H$ a continuous homomorphism or anti-homomorphism of G into H. Then π is C^∞. In particular all continuous characters on G and all continuous, finite-dimensional representations of G are C^∞.*

Proof. For the case of a homomorphism see for example [107, Theorem I.4.14] or [108, Theorem II.2.6]. For the case of an anti-homomorphism compose π with the group inversion $i_H : h \mapsto h^{-1}$ in H, and note that $i_H \circ \pi$ is a continuous homomorphism. □

The reader can by direct inspection of the examples of Section 3.2 confirm that continuous characters on Lie groups are C^∞.

Smoothness can sometimes be inferred if the structure of the solutions is known. If, say, g is a continuous pre-d'Alembert function on G, then there exists a continuous, 2-dimensional representation ρ of G such that $g(x) = \frac{1}{2}\operatorname{tr}\rho(x)$, $x \in G$ (Davison's structure theorem = Theorem 8.26). And a continuous representation is smooth (Proposition D.6), so g is smooth.

Proposition D.7 does in contrast to this not rely on knowledge of the structure of the solutions. Its proof exploits that some trigonometric functional equations have a special form so that the technique of convolution by smooth functions works. The proposition covers among other equations Levi-Civita's equation, The Sine Addition Formula, d'Alembert's equations, Wilson's equation and the quadratic equation, but not the sine functional equation. We use it to compute the continuous solutions of d'Alembert's long equation and of Wilson's equation on the $(ax+b)$-group (Examples 10.8 and 11.14), because we do not know the structure of the solutions on this group: The proposition tells us that the continuous solutions are smooth, which allows us to differentiate the functional equation as many times as needed and to bring differential equations into play.

Proposition D.7. *Assume that*

$$w_1(xy) + w_2(x\tau(y)) + w_3(yx) + w_4(\tau(y)x) = \sum_{i=1}^{N} f_i(x)g_i(y) \quad \textit{for all } x, y \in G, \quad \text{(D.1)}$$

where $N \geq 1$ *and* $w_1, w_2, w_3, w_4, f_1, \ldots, f_N, g_1, \ldots, g_N \in C(G)$.

If $g_1, g_2, \ldots, g_N$ *are linearly independent in* $C(G)$, *then* $f_1, \ldots, f_N \in C^\infty(G)$.

If $f_1, f_2, \ldots, f_N$ *are linearly independent in* $C(G)$, *then* $g_1, \ldots, g_N \in C^\infty(G)$.

Proof. Let us multiply (D.1) by $\phi(y)$, where $\phi \in C_c^\infty(G)$ is arbitrary, and integrate the result over G. We get, with $\langle g, \phi \rangle := \int_G g(y)\phi(y)\,dy$, that

$$\begin{aligned}\int_G w_1(xy)\phi(y)\,dy &+ \int_G w_2(x\tau(y))\phi(y)\,dy \\ &+ \int_G w_3(yx)\phi(y)\,dy + \int_G w_4(\tau(y)x)\phi(y)\,dy \\ &= \sum_{i=1}^{N} f_i(x)\int_G g_i(y)\phi(y)\,dy = \sum_{i=1}^{N} f_i(x)\langle g_i, \phi\rangle \qquad \text{for all } x \in G.\end{aligned}$$

We will prove that each of the four terms on left hand side is a smooth function of $x \in G$. It suffices to consider the second term $\int_G w_2(x\tau(y))\phi(y)\,dy$, because the other terms can be treated similarly. From the theory of Lie groups it is known that the convolutions and similarly looking maps like

$$x \mapsto \int_G w(y)\psi(yx)\,dy, \qquad x \in G,$$

is smooth for any $w \in C(G)$ and $\psi \in C_c^\infty(G)$, so we want to reformulate the second term to such a map. To do so we apply the following well known formula for changing variables:

$$\int_G w(ya)\,dy = \Delta(a^{-1}) \int_G w(y)\,dy \qquad \text{for all } \psi \in C_c(G) \text{ and } a \in G,$$

where $\Delta \in C(G)$ denotes G's modular function. Now,

$$\begin{aligned} x \mapsto \int_G w_2(x\tau(y))\phi(y)\,dy &= \int_G (w_2 \circ \tau)(y\tau(x))\phi(y)\,dy \\ &= \Delta(\tau(x^{-1})) \int_G (w_2 \circ \tau)(y)\phi(y\tau(x^{-1}))\,dy \end{aligned}$$

is smooth as compositions of smooth maps. Indeed $\tau : G \to G$ and the inversion map are smooth (Proposition D.6) and so is the modular function $\Delta : G \to]0, \infty[$, being a continuous homomorphism (Proposition D.6).

So $\sum_{i=1}^N \langle g_i, \phi\rangle f_i \in C^\infty(G)$. Replacing ϕ by $\phi_1, \phi_2, \ldots, \phi_N \in C_c^\infty(G)$ we find that

$$\begin{pmatrix} \langle g_1, \phi_1\rangle & \langle g_2, \phi_1\rangle & \ldots & \langle g_N, \phi_1\rangle \\ \langle g_1, \phi_2\rangle & \langle g_2, \phi_2\rangle & \ldots & \langle g_N, \phi_2\rangle \\ \vdots & \vdots & \ldots & \vdots \\ \langle g_1, \phi_N\rangle & \langle g_2, \phi_N\rangle & \ldots & \langle g_N, \phi_N\rangle \end{pmatrix} \begin{pmatrix} f_1 \\ f_2 \\ \vdots \\ f_N \end{pmatrix} \in C^\infty(G, \mathbb{C}^N).$$

Since $g_1, g_2, \ldots, g_N$ are linearly independent in $C(G)$ we see by help of Proposition C.3 and Example C.5 that there exist $\phi_1, \phi_2, \ldots, \phi_N \in C_c^\infty(G)$ such that the matrix on the left hand side is invertible. It follows that $f_1, f_2, \ldots, f_N \in C^\infty(G)$. Quite analogous arguments show that $g_1, g_2, \ldots, g_N \in C^\infty(G)$ if $f_1, f_2, \ldots, f_N$ are linearly independent. □

Example D.8. Any continuous solution g of d'Alembert's functional equation (9.6) on $\mathbb{R}^n$ is in $C^\infty(\mathbb{R}^n)$.

Example D.9. If $g \in C(G)$ satisfies d'Alembert's long functional equation (10.1) for a continuous involution $\tau : G \to G$, i.e., if

$$g(xy) + g(yx) + g(x\tau(y)) + g(\tau(y)x) = 4g(x)g(y) \qquad \text{for all } x, y \in G,$$

then $g \in C^\infty(G)$.

Example D.10. If the pair $(f, g) \in (C(G) \setminus \{0\}) \times C(G)$ satisfies Wilson's functional equation (11.3) for a continuous involution $\tau : G \to G$, i.e., if

$$f(xy) + f(x\tau(y)) = 2f(x)g(y) \qquad \text{for all } x, y \in G,$$

then $f, g \in C^\infty(G)$.

D.4 Gajda's result

We recall that the Cauchy difference of a map $f : S \to H$ from a semigroup S to an abelian group H is $(Cf)(x, y) = f(xy) - f(x) - f(y)$ for $x, y \in S$.

We work often inside the framework of continuous functions on a topological group G, even though our formulas make sense for any function. In this connection a naive question about a function $f : G \to \mathbb{C}$ is whether $Cf \in C(G \times G)$ implies $f \in C(G)$, so that we can count on staying inside the framework of continuous functions once we have got there. This is too naive a question because in general there are discontinuous additive functions on G (Proposition 2.8): And if $a : G \to \mathbb{C}$ is such a function then $Ca = 0 \in C(G \times G)$. So the answer is no. A better question to ask is whether given $Cf \in C(G \times G)$ there exists an $f_0 \in C(G)$ such that $Cf = Cf_0$. Or equivalently whether f can be written in the form $f = f_0 + a$, where $f_0 \in C(G)$ and $a : G \to \mathbb{C}$ is additive. In other words, can the singularities of f be relegated out to an additive map? This not so naive question has a positive, satisfactory answer for locally compact groups. The answer (Corollary D.12) is immediate from a technical result due to Gajda [92, Theorem 1]. We feel that Gajda's result has a certain beauty and charm, stemming from its mixture of non-measurable and continuous functions. Gajda works with a complex Banach space, but his result is equally true for a real Banach space (complexify it!).

Theorem D.11 (Gajda). *Let G be a locally compact group, and let X be a Banach space. If $f : G \to X$ is such that for each $y \in G$ the function $x \mapsto f(xy) - f(x)$ is continuous on G and the function $x \mapsto f(yx) - f(x)$ is Borel measurable on G, then $f = f_0 + a$, where $f_0 \in C(G, X)$ and where $a : G \to X$ is additive.*

Proof. We refer to [92, Theorem 1]. □

Corollary D.12. *Let G be a locally compact group, and let X be a Banach space. If the Cauchy difference $Cf(x,y) = f(xy) - f(x) - f(y)$ of a function $f : G \to X$ is separately continuous in $x \in G$ and in $y \in G$, then f can be written in the form $f = f_0 + a$, where $f_0 \in C(G, X)$, and where $a : G \to X$ is additive.*

We use Corollary D.12 to pass from an algebraic result to the corresponding topological one in our discussion of Jensen's functional equation (see the proof of Theorem 12.20). We shall also need it in our discussion of the cocycle equation (Chapter 16).

D.5 Exercises

Exercise D.1. Let G be the Heisenberg group from Example A.17(a). Show that $G/[G,G]$ is topologically isomorphic to $\mathbb{R}^2$.

Exercise D.2. Let G be the $(ax+b)$-group from Example A.17(i). Show that $G/[G,G]$ is topologically isomorphic to $(\mathbb{R}^+, \cdot)$.

D.6 Notes and remarks

Our interest in the present book lies in continuous solutions of functional equations, but sporadically we mention measurable solutions. For a systematic study of measurable solutions consult Járai's monograph [116]. How measurability implies regularity can be found in the survey papers Ger [97, 98].

Proposition D.4: The proof in the text is adapted from Hewitt and Ross [111, Theorem 22.18].

Proposition D.4: Unitary characters need not be continuous: Let us view $(\mathbb{R}, +)$ and $(\mathbb{T}, \cdot)$ as $\mathbb{Z}$-modules (by $n \cdot x = nx$ for $n \in \mathbb{Z}$ and $x \in \mathbb{R}$, and $n \cdot z = z^n$ for $n \in \mathbb{Z}$ and $z \in \mathbb{T}$). Take a basis $\mathcal{B}$ of the $\mathbb{Z}$-module $\mathbb{R}$ such that $\mathcal{B}$ contains 1 and π. Define a homomorphism $\phi : \mathbb{R} \to \mathbb{T}$ by defining it on the basis elements as follows: $\phi(1) = e^i$ and $\phi(b) = 1$ for all $b \in \mathcal{B} \setminus \{1\}$. Now ϕ is a unitary character and we show by contradiction that it is not continuous. If ϕ were continuous it would according to Example 3.7(a) have the form $\phi(x) = \exp(i\lambda x)$, where λ is a real constant. Now, $e^i = \phi(1) = e^{i\lambda}$, so $\lambda - 1 = 2\pi n$ for some $n \in \mathbb{Z}$. Similarly $1 = \phi(\pi) = e^{i\lambda\pi}$, so $\lambda = 2m$ for some $m \in \mathbb{Z}$. Thus $2m - 1 = 2\pi n$ which contradicts the fact that π is irrational.

Appendix E

Matrix-Coefficients of Representations

In this appendix we discuss matrix-coefficients of operator-valued functions, where the operators act on finite-dimensional vector spaces. We may apply the set up and the results to representations, because they are operator-valued functions of a special form.

E.1 Matrix-coefficients

Throughout this section we let W denote an n-dimensional, complex vector space where $1 \leq n < \infty$. Its dual vector space is denoted W^*, and the value of $w^* \in W^*$ on $w \in W$ is denoted $\langle w^*, w \rangle$. The expression $\langle v^*, v \rangle$ is bilinear. For $w \in W$ and $w^* \in W^*$ we define $w \otimes w^* \in \mathcal{L}(W)$ by $w \otimes w^*(v) := \langle w^*, v \rangle w$ for $v \in W$.

If $w_1, w_2, \ldots, w_n$ is a basis of W and $w_1^*, w_2^*, \ldots, w_n^*$ is the dual basis of W^*, then $\{w_i \otimes w_j^* \,|\, i, j = 1, 2, \ldots, n\}$ is a basis of $\mathcal{L}(W)$, so any $L \in \mathcal{L}(W)$ can be decomposed into $L = \sum_{i,j=1}^n L_{ij} w_i \otimes w_j^*$, where $L_{ij} \in \mathbb{C}$. $\{L_{ij}\}$ can also be characterized as the matrix of the linear operator L with respect to the basis $w_1, w_2, \ldots, w_n$ of W. Note the formulas $L_{ij} = \langle w_i^*, L w_j \rangle$ and $\operatorname{tr}((w_j \otimes w_i^*) L) = L_{ij}$.

Definition E.1. Let X be a set and $F : X \to \mathcal{L}(W)$. A *matrix-coefficient* of F is a function on X of the form $x \mapsto \langle w^*, F(x) w \rangle$, $x \in X$, where $w^* \in W^*$ and $w \in W$. Matrix-coefficients are sometimes called matrix-elements. The vector space of functions on X spanned by the matrix-coefficients of F is called the *space of matrix-coefficients* of F and denoted $C(F)$.

Proposition E.2. *Let X be a set and $F : X \to \mathcal{L}(W)$. Letting $w_1, w_2, \ldots, w_n$ be a basis of W we write $F(x)w_j = \sum_{i=1}^n F_{ij}(x)w_i$, where $F_{ij} : X \to \mathbb{C}$. Then*

(a) *each F_{ij}, $i, j = 1, \ldots, n$, is a matrix-coefficient of F.*

(b) $C(F) = \operatorname{span}\{F_{ij} \mid i, j = 1, 2, \ldots, n\}$, *so* $\dim C(F) \leq n^2$.

(c) *$C(F)$ consists of the functions $x \mapsto \operatorname{tr}(AF(x))$, $x \in X$, where A ranges over $\mathcal{L}(W)$.*

(d) *In particular $x \mapsto \operatorname{tr} F(x)$ belongs to $C(F)$.*

Proof. We leave (a) and (b) to the reader.

(c) Let $\{w_1, w_2, \ldots, w_n\}$ be a basis of W and let $\{w_1^*, w_2^*, \ldots, w_n^*\}$ be its dual basis. Now, $F_{ij}(x) = \operatorname{tr}((w_j \otimes w_i^*)F(x))$, so

$$\begin{aligned} C(F) &= \operatorname{span}\{F_{ij}(-) \mid i, j = 1, 2, \ldots, n\} \\ &= \operatorname{span}\{\operatorname{tr}((w_j \otimes w_i^*)F(-)) \mid i, j = 1, 2, \ldots, n\} \\ &= \operatorname{span}\{\operatorname{tr}(AF(-)) \mid A \in \mathcal{L}(W)\}. \end{aligned}$$ □

Proposition E.2(a) motivates the terminology matrix-coefficient: Indeed, $\{F_{ij}(x)\}$ is the matrix of the linear operator $F(x) : W \to W$ with respect to the basis $w_1, w_2, \ldots, w_n$ of W. The individual matrix-coefficients F_{ij} depend on the choice of the basis $w_1, w_2, \ldots, w_n$, but the vector space $C(F)$ does not.

The following simple result on invariant subspaces is useful:

Lemma E.3. *Let X be a set and $F : X \to \mathcal{L}(W)$. Let furthermore W' be a subspace of W such that $F(x)(W') \subseteq W'$ for all $x \in X$. Define $F' : X \to \mathcal{L}(W')$ by $F'(x)w' := F(x)w'$ for $x \in X$ and $w' \in W'$.*

Then any matrix-coefficient of F' is a matrix-coefficient of F, and so $C(F') \subseteq C(F)$.

Proof. Let $f(x) = \langle \phi, F'(x)w' \rangle$, $x \in X$, be a matrix-coefficient of F'. So $\phi \in (W')^*$ and $w' \in W'$. Extending ϕ from the subspace W' of W to a linear functional $w^* \in W^*$ on W we have $f(x) = \langle w^*, F(x)w' \rangle$, which shows that f is a matrix-coefficient of F. □

There is no need to worry about the topology of a finite-dimensional vector space, because it is unique. More precisely formulated ([166, Theorem 1.21(a)]):

Theorem E.4. *There is exactly one topology on W that makes it a Hausdorff topological vector space. Any isomorphism of $\mathbb{C}^n$ onto W is a homeomorphism, when W is equipped with this topology.*

We shall always equip a finite-dimensional vector space with this topology. The algebraically dual space of W then equals the topologically dual space W^* of continuous linear functionals on W, because this is the case for $\mathbb{C}^n$.

Lemma E.5. *Let X be a topological space and $F : X \to \mathcal{L}(W)$. Then*

(a) *F is continuous iff each matrix-coefficient of F is continuous.*

(b) *F is bounded iff each matrix-coefficient of F is bounded.*

Proof. From Proposition E.2, because F is continuous, resp. bounded, iff the special matrix-elements F_{ij} are continuous, resp. bounded. After all they are the coordinates of F with respect to the basis $\{w_i \otimes w_j^* \mid i, j = 1, 2, \ldots, n\}$ of the vector space $\mathcal{L}(W)$ (we use the notation from the proof of Proposition E.2(c)). □

Let $F : X \to \mathcal{L}(W)$. The expression $\langle w^*, F(-)w\rangle$ is bilinear in $(w^*, w) \in W^* \times W$, so to produce the space of matrix-coefficients $C(F)$ it suffices to take the span of $\langle w^*, F(-)w\rangle$ where w^* and w range over generating subsets of the vector spaces W^* and W respectively. Proposition E.2(c) is an example with the basis $w_1, w_2, \ldots, w_n$ of W as the generating subset of W and the dual basis as generating subset of W^*. This construction of $C(F)$ via a basis and its dual applies to any finite-dimensional vector space W, but sometimes specific generating subsets of W^* and/or W are more pertinent for concrete vector spaces W. For instance if W is a space of functions which is a case of interest for us.

Definition E.6. Let W be a subspace of the complex-valued functions on a set Y. For any $y \in Y$ we define the *point evaluation* $\mathrm{ev}_y \in W^*$ by $\langle \mathrm{ev}_y, f\rangle := f(y)$, $f \in W$.

Lemma E.7. *If W is a finite-dimensional subspace of the complex-valued functions on a set Y, then $W^* = \mathrm{span}\{\mathrm{ev}_y \mid y \in Y\}$.*

For any $F : X \to \mathcal{L}(W)$ its space of matrix-coefficients $C(F)$ is spanned by the functions $x \mapsto \langle \mathrm{ev}_y, F(x)f\rangle$, $x \in X$, where f ranges over W and y over Y.

Proof. If $\text{span}\{\text{ev}_y \mid y \in Y\}$ were a proper subspace of the finite-dimensional vector space W^* then there would exist an element $w^{**} \in W^{**} \setminus \{0\}$ such that $w^{**}(\text{ev}_y) = 0$ for all $y \in Y$. The natural embedding $i : W \to W^{**}$ is surjective, because $\dim W < \infty$, so $w^{**} = i(f)$ for some $f \in W$. From $w^{**}(\text{ev}_y) = 0$ for all $y \in Y$ we get that $f(y) = 0$ for all $y \in Y$, so that $f = 0$. But $f \neq 0$ because $i(f) = w^{**} \neq 0$, so we have arrived at a contradiction. □

E.2 On representations

We collect in this section certain standard notions, definitions and results from the theory of representations.

In many parts of mathematics you meet a group or semigroup S acting on a set X. For instance the group $\mathbb{R}$ acts on $X = \mathbb{R}$ by translations and the orthogonal group $O(n)$ acts on $\mathbb{R}^n$ by rotations. The corresponding action of S on the vector space of functions on X is called the regular representation of S. Harmonic analysis is the study of the interplay between the functions on X and the semigroup action (the regular representation). Harmonic analysis is more generally viewed as the study of functions by means of representations. It contains as a subtopic abelian harmonic analysis which is the study of functions on abelian groups by help of unitary characters, i.e., 1-dimensional unitary representations. On the real line and on the circle group the continuous unitary characters are the exponential functions of modulus 1, so abelian harmonic analysis comprises the Fourier analysis. In other words, harmonic analysis is Fourier analysis on groups which need not be abelian. In our studies of functional equations we shall meet both abelian and non-abelian harmonic analysis. It is of course a welcome surprise that harmonic analysis enters in the study of functional equations, because representation theory is well developed. That abelian harmonic analysis enters (via characters) was already pointed out in Section 1.2 (our basic philosophy). But we need also the non-abelian harmonic analysis: d'Alembert functions on a monoid S can be described by 2-dimensional representations of S (according to Davison's structure theorem Theorem 8.26).

Definition E.8. Let S be a semigroup and V a vector space (as always complex).

(a) A *representation* of S on V is a map $\pi : S \to \mathcal{L}(V)$ such that

$$\pi(xy) = \pi(x)\pi(y) \qquad \text{for all } x, y \in S.$$

If S is a monoid with neutral element e we furthermore demand that $\pi(e) = I$ (the identity map). V is called the *representation space* of π. The *dimension of the representation* π is $\dim \pi := \dim V$.

(b) Let π_1 and π_2 be representations of S on vector spaces V_1 and V_2, respectively. A linear map $A : V_1 \to V_2$ is an *intertwining operator* between π_1 and π_2 if $A\pi_1(x) = \pi_2(x)A$ for all $x \in S$. If there exists an intertwining isomorphism between π_1 and π_2 then we say that π_1 and π_2 are *algebraically equivalent*, and write $\pi_1 \simeq \pi_2$.

(c) Let π be a representation of S on V. A subspace W of V is said to be *invariant* under π, if $\pi(x)W \subseteq W$ for all $x \in S$. In that case we let $\pi|_W$ denote the representation $x \mapsto \pi(x) : W \to W$ of S on W. $\pi|_W$ is said to be a *subrepresentation* of π.

(d) Let π be a representation π of S on a finite-dimensional vector space W. A vector $w \in W$ is said to be *cyclic* if $\operatorname{span}\{w, \pi(x)w \mid x \in S\} = W$.

(e) A representation π of S on V is *algebraically irreducible*, if $\{0\}$ and V are the only invariant subspaces of V. More generally, a collection of linear transformations of V is said to be algebraically irreducible, if it has only trivial invariant subspaces.

(f) Let π be a representation of S on a vector space V. The functions on S of the form $x \mapsto \langle v^*, \pi(x)v \rangle$, $x \in S$, where v ranges over V and v^* over the dual vector space V^*, are called *matrix-coefficients* of π (See Appendix E.1). The vector space $C(\pi)$ spanned by is the matrix-coefficients is called the *space of matrix-coefficients* of π.

(g) Let π be a representation of S on a finite-dimensional vector space. Its *character* is the function $\chi_\pi : S \to \mathbb{C}$ defined by $\chi_\pi(x) := \operatorname{tr}(\pi(x))$, $x \in S$. Note that $\chi_\pi \in C(\pi)$.

(h) A representation π of S on a finite dimensional vector space W is said to be *bounded* if each matrix-coefficient is bounded, or equivalently if $\|\pi(-)\|$ is bounded for any norm $\|-\|$ on W.

(i) A representation π of S on a complex Hilbert space is said to be *unitary* if $\pi(x)$ is a unitary operator for each $x \in S$.

If S is a group, and π a representation of S on V then $\pi(S) \subseteq \mathrm{GL}(V)$, because $\pi(e) = I$. Definition E.8(a) does not require $\pi(S) \subseteq \mathrm{GL}(V)$, because the right regular representation R from Definition E.15 below need not consist of invertible operators.

Lemma E.9. *Let π be a finite-dimensional representation of a semigroup S. Then*

(a) *If π' is algebraically equivalent to π, then π' and π have the same matrix-coefficients.*

(b) *If π' is algebraically equivalent to π, then $\chi_{\pi'} = \chi_\pi$.*

(c) *χ_π is a central function on S.*

(d) *χ_π is a finite sum of characters of irreducible representations. The sum of the dimensions of these irreducible representations is* $\dim \pi$.

Proof. (a) and (b) are left to the reader.

(c) Let W be the representation space of π. Since $\mathrm{tr}(AB) = tr(BA)$ for any $A, B \in \mathcal{L}(W)$ we find that $\chi_\pi(xy) = \mathrm{tr}\,\pi(xy) = \mathrm{tr}(\pi(x)\pi(y)) = \mathrm{tr}(\pi(y)\pi(x)) = \mathrm{tr}\,\pi(yx) = \chi_\pi(yx)$ for all $x, y \in S$.

(d) Let $A \in \mathcal{L}(W)$. If V is a subspace of W which is invariant under A, then A induces a map $\widetilde{A} \in \mathcal{L}(W/V)$ by $\widetilde{A}(w + V) := Aw + V$ for $w \in W$. And $\mathrm{tr}(A) = \mathrm{tr}(A|_V) + \mathrm{tr}\,\widetilde{A}$. Observe that $\dim(W/V) < \dim W$ if V is a proper subspace of W. Apply this to the operators $\pi(x)$, $x \in S$. □

Schur's lemma and Burnside's theorem are fundamental results in representation theory.

Lemma E.10 (Schur's lemma). *Let π be an algebraically irreducible representation of a semigroup S on a finite dimensional, complex vector space W. If $A \in \mathcal{L}(W)$ and $A\pi(x) = \pi(x)A$ for all $x \in S$, then $A = \lambda I$ for some $\lambda \in \mathbb{C}$.*

Proof. We may assume that $\dim W \geq 2$. A has an eigenvalue $\lambda \in \mathbb{C}$, because W is a finite dimensional, complex vector space. By the irreducibility the corresponding eigenspace is W. □

Theorem E.11 (Burnside 1905). *Let W be a finite-dimensional vector space of dimension strictly larger than 1 over an algebraically closed field. If $\mathcal{A}$ is an algebraically irreducible algebra of linear transformations of W containing I, then $\mathcal{A} = \mathcal{L}(W)$.*

Proof. For two short proofs of it that involve only elementary linear algebra, see [140] or [142]. □

Burnside's theorem allows us to sharpen Proposition E.2(b) and (c), when F is an irreducible representation. Under the circumstances of Corollary E.12 we get equality and not just an inequality in Proposition E.2(b):

Corollary E.12. *Let ρ be an algebraically irreducible representation of a monoid M on a finite-dimensional vector space W such that* $\dim W \geq 2$.

Let $c_A(x) := \frac{1}{2}\operatorname{tr}(A\rho(x))$, $x \in M$. The map $A \mapsto c_A$ is an isomorphism of $\mathcal{L}(W)$ onto the space $C(\rho)$ of matrix-coefficients of ρ. In particular, $\dim C(\rho) = (\dim \rho)^2$.

Proof. Due to Proposition E.2(c) it suffices to prove that $A \mapsto c_A$ is injective. So assuming $\operatorname{tr}(A\rho(x)) = 0$ for all $x \in M$ we shall infer that $A = 0$. $\operatorname{span}\{\rho(x) \,|\, x \in M\}$ is an algebra containing I, because ρ is a representation of a monoid, and we may apply Burnside's theorem to it, because ρ is algebraically irreducible. Burnside's theorem tells us that $\operatorname{span}\{\rho(x) \,|\, x \in M\}$ is all of $\mathcal{L}(W)$, which implies that $\operatorname{tr}(AX) = 0$ for all $X \in \mathcal{L}(W)$. And then $A = 0$ as desired. □

Theorem E.13. *Let $\pi_1, \pi_2, \dots, \pi_n$ be algebraically irreducible representations of a semigroup S such that no two of them are algebraically equivalent.*

(a) *$C(\pi_1) + C(\pi_2) + \cdots + C(\pi_n)$ form a direct sum in $\mathbb{C}^S$.*

(b) *In particular, if $\pi_1, \pi_2, \dots, \pi_n$ are finite-dimensional, then their characters $\chi_{\pi_1}, \chi_{\pi_2}, \dots, \chi_{\pi_n}$ span a direct sum:*

$$\mathbb{C}\chi_{\pi_1} \oplus \mathbb{C}\chi_{\pi_2} \oplus \cdots \oplus \mathbb{C}\chi_{\pi_n} \subseteq C(\pi_1) \oplus C(\pi_2) \oplus \cdots \oplus C(\pi_n).$$

Proof. (a) can be found in Bourbaki [32, Proposition 2 of Chap. VIII, §13, no. 3].

(b) is an immediate consequence of (a). □

If π is a 1-dimensional representation, say $\pi(x) = \chi(x)I$, $x \in S$, where $\chi : S \to \mathbb{C}$ is a multiplicative function, then $C(\pi) = \mathbb{C}\chi$. We know from Theorem 3.18(a) that spaces spanned by different multiplicative functions form direct sum.

Definition E.14. A *continuous representation* of a topological semigroup S on a topological vector space V is a representation π such that the map $(x, v) \mapsto \pi(x)v$ from $S \times V$ to V is continuous.

The 1-dimensional representations of a semigroup S are the ones of the form $\pi(x)w = \chi(x)w$, $x \in S$, $w \in W$, where $\chi : S \to \mathbb{C}$ is multiplicative, and where furthermore $\chi(e) = 1$ if S is a monoid with neutral element $e \in S$. Here W is a 1-dimensional vector space. So the action of π is multiplication by χ. π is continuous if and only if χ is. This gives a connection to the earlier definition of a group character (Definition 3.1).

E.3 On the regular representations

In this section we connect the theory of matrix-coefficients to the right and left regular representations. These representations play a crucial role in our discussions of Levi-Civita's functional equation (Chapter 5), the pre-d'Alembert functional equation and d'Alembert's functional equation (Chapters 9 and 11) and Wilson functions (Chapter 11).

Definition E.15. Let S be a semigroup. For any $x \in S$ we define $R(x) : \mathbb{C}^S \to \mathbb{C}^S$ by $[R(x)f](y) := f(yx)$, $y \in S$. Similarly we define $L'(x) : \mathbb{C}^S \to \mathbb{C}^S$ by $[L'(x)f](y) := f(xy)$, $y \in S$.

R is a representation of S on $\mathbb{C}^S$, called the *right regular representation* of S.

L' has the property that $L'(xy) = L'(y)L'(x)$ for all $x, y \in S$, i.e., it reverses the order of the factors, so it is not a representation. We call L' *the left regular antirepresentation* of S. If S is a group then $L(x) := L'(x^{-1})$ is a representation of S on $\mathbb{C}^S$ called *the left regular representation* of S.

A map of the form $R(x)$, resp. $L'(x)$, is called a *right translation*, resp. a *left translation*.

The right and left regular representations stem from the fundamental operations of right and left translations on a semigroup S. S is not assumed abelian, so translation by an element in S from the right and from the left need not give the same result. A convenient fact is that right and left translations commute: $L'(x)R(y) = R(y)L'(x)$ for all $x, y \in S$.

Proposition E.16. *Let M be a topological monoid. Let W be an R-invariant, finite-dimensional subspace of $C(M)$.*

(a) *$R|_W$ is a continuous representation of M on W, and $W \subseteq C(R|_W)$.*

(b) *If W is also L'-invariant, then $W = C(R|_W)$.*

Proof. (a) $R|_W$ is clearly a representation of M on W. To get its continuity we consider the special matrix-coefficients $x \mapsto \langle \mathrm{ev}_y, R(x)f\rangle = f(yx)$ where $y \in M$ and $f \in W$. They are continuous, because M is a topological monoid and $f \in C(G)$. Since they span the space of all matrix-coefficients (Lemma E.7), we may refer to Lemma E.5(a) to infer the continuity of $R|_W$.

From $\langle \mathrm{ev}_e, R(x)f\rangle = f(ex) = f(x)$ we see that any $f \in W$ is a matrix-coefficient, so $W \subseteq C(R|_W)$.

(b) According to (a) it is left to prove that $W \supseteq C(R|_W)$. Now $C(R|_W)$ is spanned by the matrix-coefficients $x \mapsto [R(x)f](y)$, where $f \in W$ and $y \in G$ (by Lemma E.7). Thus it suffices to prove that they belong to W. Now, $[R(x)f](y) = f(yx) = [L'(y)f](x)$, i.e., $[x \mapsto [R(x)f](y)] = L'(y)f$. Finally, $L'(y)f \in W$ by the assumption of W being L'-invariant. □

E.4 On compact groups

We saw in Corollary 3.20 that characters are linearly independent. On compact groups even more is true for the continuous characters. A well known generalization of the following elementary fact from Fourier series

$$\int_{-\pi}^{\pi} e^{inx}\, \overline{e^{imx}}\, \frac{dx}{2\pi} = \delta_{nm} \qquad \text{for } n, m \in \mathbb{Z},$$

is that they form an orthonormal system in L^2:

Lemma E.17 (The orthogonality relations). *Let K be a compact group with normalized Haar measure dk. If χ_1 and χ_2 are continuous characters on K, then*

$$\int_K \chi_1(k)\, \overline{\chi_2(k)}\, dk = \begin{cases} 0 & \text{if } \chi_1 \neq \chi_2 \\ 1 & \text{if } \chi_1 = \chi_2 \end{cases}.$$

Proof. [111, Lemma 23.19] has a short and easy proof. □

Lemma E.18. *Let K be a compact group with normalized Haar measure dk. Let $g \in C(K)$. If $\widehat{g}(\pi) := \int_K g(k)\pi(x^{-1})\, dk = 0$ for all irreducible, continuous, unitary representations π of K, then $g = 0$.*

Proof. For any irreducible, continuous, unitary representation π of K we get that $\int_K g(k)\langle \xi, \pi(k)\eta\rangle\, dk = 0$ for all ξ, η in the representation space of π.

According to Peter-Weyl's theorem the matrix-elements $k \mapsto \langle \xi, \pi(k)\eta \rangle$ span a dense subspace of $L^2(K)$ for varying π, ξ and η. So $g = 0$ almost everywhere. Due to its continuity $g = 0$. □

Appendix F

The Small Dimension Lemma

In this appendix we state and prove a result due to Yang [214, 215], which is useful in applications of the Fourier transform to functional equations on compact groups. Yang calls it the *Small dimension lemma* because it implies that an irreducible, unitary representation of a group under certain assumptions must have dimension ≤ 2.

Lemma F.1. *Let K be an algebraically irreducible subgroup of $U(n)$ where $n \geq 2$. Let $w_0 \in \mathbb{C}^n$, $w_0 \neq 0$, and suppose that for each $k \in K$ there exists a constant $c(k) \in \mathbb{C}$ such that $(k + k^{-1})w_0 = c(k)w_0$.*

Then $n = 2$, $K \subseteq SU(2)$ and $k + k^{-1} = c(k)I$ for all $k \in K$.

Proof. We proceed in four steps.

In the first step we show that c is central, i.e., that $c(k_1^{-1}kk_1) = c(k)$ for all $k, k_1 \in K$.

(α) For any $k, k_1 \in K$ we get from the computation

$$k_1^{-1}(k + k^{-1})k_1 w_0 = (k_1^{-1}kk_1 + (k_1^{-1}kk_1)^{-1})w_0 = c(k_1^{-1}kk_1)w_0,$$

that $(k+k^{-1})k_1 w_0 = c(k_1^{-1}kk_1)k_1 w_0$. Thus $k_1 w_0$, resp. w_0, is an eigenvector for the selfadjoint matrix $k + k^{-1} = k + k^*$ corresponding to the eigenvalue $c(k_1^{-1}kk_1)$, resp. $c(k)$. Eigenvectors corresponding to different eigenvalues are orthogonal to one another, so $c(k_1^{-1}kk_1) = c(k)$ except possibly when $(w_0, k_1 w_0) = 0$.

In the exceptional case of $(w_0, k_1 w_0) = 0$ we find that

$$\begin{aligned} 0 = 0 + 0 &= (k_1 w_0, w_0) + (w_0, k_1 w_0) = ((k_1 + k_1^*)w_0, w_0) \\ &= ((k_1 + k_1^{-1})w_0, w_0) = c(k_1)(w_0, w_0), \end{aligned}$$

so $c(k_1) = 0$, and hence $k_1 w_0 = -k_1^{-1} w_0$. Now, for any $k \in K$ we get using the formula $(k + k^{-1})k_1 w_0 = c(k_1^{-1} k k_1) k_1 w_0$ from above near the end

$$\begin{aligned}[c(k_1^{-1}k) + c(k_1^{-1}k^{-1})]w_0 &= k_1^{-1}kw_0 + k^{-1}k_1w_0 + k_1^{-1}k^{-1}w_0 + kk_1w_0 \\ &= k_1^{-1}(k + k^{-1})w_0 + (k + k^{-1})k_1w_0 = k_1^{-1}c(k)w_0 + c(k_1^{-1}kk_1)k_1w_0 \\ &= (-c(k) + c(k_1^{-1}kk_1))k_1w_0.\end{aligned}$$

The two sides are orthogonal to one another, so both sides are 0. In particular $c(k) - c(k_1^{-1}kk_1) = 0$. We conclude that c is central as desired.

(β) We obtain that $(k + k^{-1})k_1w_0 = c(k_1^{-1}kk_1)k_1w_0 = c(k)k_1w_0$ for all $k, k_1 \in K$, when we use that c is central. It follows that $\{w \in \mathbb{C}^n \mid \text{for all } k \in K : (k + k^{-1})w = c(k)w\}$ is a K-invariant subspace of $\mathbb{C}^n$, so from the irreducibility $k + k^{-1} = c(k)I$ for all $k \in K$.

(γ) Let us analyze the situation for a fixed $k \in K$. From $c(k)I = k + k^{-1}$ we see that $-2 \le c(k) \le 2$, so that the two roots of the polynomial $\lambda^2 - c(k)\lambda + 1$ are complex conjugate to one another. Let $\alpha \in \mathbb{C}$ be one of the roots. The other root is $\overline{\alpha}$. If $\alpha = \overline{\alpha}$ then $2\alpha = c(k)$.

Being unitary $k \in K$ is a normal matrix, so it can be diagonalized by a unitary matrix $U \in U(n)$. Let us write

$$k = U \begin{pmatrix} \kappa_1 & 0 & \cdots & 0 \\ 0 & \kappa_2 & \cdots & 0 \\ \vdots & \vdots & \ddots & \vdots \\ 0 & 0 & \cdots & \kappa_n \end{pmatrix} U^{-1},$$

where $\kappa_1, \kappa_2, \ldots, \kappa_n \in \mathbb{C}$ are the eigenvalues of k. Since k is unitary $|\kappa_i| = 1$ for $i = 1, 2, \ldots, n$. From $k + k^{-1} = c(k)I$ we get that $\kappa_i + \kappa_i^{-1} = c(k)$ for $i = 1, 2, \ldots, n$. Thus κ_i is a root of the polynomial $\lambda^2 - c(k)\lambda + 1$. So $\kappa_i = \alpha$ or $\kappa_i = \overline{\alpha}$ for $i = 1, 2, \ldots, n$. If the roots coincide, then $k = \alpha I$.

(δ) Since K is irreducible, there exists a $k_0 \in K$ with two different eigenvalues, say α and $\overline{\alpha}$ with $\overline{\alpha} - \alpha \ne 0$. We fix such an element $k_0 \in K$. As above we diagonalize k_0 by a unitary matrix U. Possibly replacing K by $U^{-1}KU$, which preserves unitarity of the elements of K, we may assume that $U = I$, which means k_0 is diagonal. We may furthermore assume that $k_0 = \operatorname{diag}(\alpha I_r, \overline{\alpha} I_s)$, where $r \ge s \ge 1$. It now suffices to prove that $r = 1$. We do this by contradiction, so we assume $r \ge 2$. For any $k = \{k_{ij}\} \in K$ we reformulate $k + k^{-1} = c(k)I$ to $k + k^* = c(k)I$ and from this we observe that $k_{21} + \overline{k_{12}} = 0$. Applied to $kk_0 \in K$ instead of to $k \in K$ the observation says that $\alpha k_{21} + \overline{\alpha k_{12}} = 0$. However, the system of $k_{21} + \overline{k_{12}} = 0$ and $\alpha k_{21} + \overline{\alpha}\overline{k_{12}} = 0$ is only satisfied by $k_{21} = \overline{k_{12}} = 0$, because its determinant $\overline{\alpha} - \alpha \neq 0$. So $k_{12} = k_{21} = 0$ for all $k \in K$. This contradicts Burnside's theorem (Theorem E.11). □

We leave it to the reader to extend Lemma F.1 by incorporating a character (Lemma F.2.) Yang's result is the case in which the character is 1.

Lemma F.2. *Let K be an irreducible subgroup of $U(n)$ where $n \geq 2$. Let $\chi : K \to \mathbb{C}^*$ be a character on K. Let $w_0 \in \mathbb{C}^n$, $w_0 \neq 0$, and suppose that for each $k \in K$ there exists a constant $c(k) \in \mathbb{C}$ such that $(k + \chi(k)k^{-1})w_0 = c(k)w_0$.*

Then $n = 2$, $k + \chi(k)k^{-1} = c(k)I$ and $\det k = \chi(k)$ for all $k \in K$.

Appendix G

Group Cohomology

Throughout this appendix G is a group and H an abelian group.

We have in Appendix E.2 applied group representations to study groups. A generalization of the notion of a representation is the notion of a G-module: A G-module is an abelian group H together with a group action of G on H (as defined in Definition 14.2 with the discrete topologies) such that every element of G acts as an endomorphism of H. The action may be the trivial one, meaning that $g \cdot h = h$ for all $g \in G$ and $h \in H$.

The algebraic concept of a complex has found its way into several different branches of mathematics. To mention some examples: Differential topology has the de Rham complex, the theory of holomorphic functions of several variables has the Cousin complex and Lie algebra cohomology has the Koszul complex. We shall in this appendix introduce the concept in connection with group cohomology and view some of our functional equations from this perspective.

Definition G.1. A *complex* is a sequence $\{C_n\}_{n=0}^{\infty}$ of abelian groups and homomorphisms $d^{(n)} : C_n \to C_{n+1}$ between them:

$$C_0 \xrightarrow{d^{(0)}} C_1 \xrightarrow{d^{(1)}} C_2 \xrightarrow{d^{(2)}} C_3 \xrightarrow{d^{(3)}} \cdots$$

such that $d^{(n+1)} \circ d^{(n)} = 0$ for all $n = 0, 1, 2, \ldots$ (the crucial property of a complex).

The *n-cocycles* are the elements of the normal subgroup $Z^n := \{\phi \in C_n \mid d^{(n)}\phi = 0\}$ of C_n, i.e., the kernel of $d^{(n)} : C_n \to C_{n+1}$, while the *n-coboundaries* are the elements of the subgroup $B^n := \{d^{(n-1)}(\phi) \mid \phi \in C_{n-1}\}$ of C_n, i.e., the image of $d^{(n-1)} : C_{n-1} \to C_n$.

The condition $d^{(n)} \circ d^{(n-1)} = 0$ ensures that $B^n \subseteq Z^n$, so that the quotient group $H^n := Z^n/B^n$ (called the *nth cohomology group*) makes sense for $n = 1, 2, \ldots$

The mathematical literature uses often the abbreviation d for $d^{(n)}$. We will do as well.

Definition G.2. Let H be a G-module. For $n \in \mathbb{N} \cup \{0\}$ let $C^n(G,H)$ be the abelian group of all functions from $G^n = G \times G \times \cdots \times G$ to H. $C^0(G,H)$ is interpreted as all maps from the one-element group $G^0 = \{e\}$ to H, so $C^0(G,H) = H$. The *coboundary homomorphisms* $d^{(n)} : C^n(G,H) \to C^{n+1}(G,H)$ are for $\phi \in C^n(G,H)$ defined by

$$\begin{aligned}(d^{(n)}\phi)(x_1, x_2, \dots, x_{n+1}) &:= x_1 \cdot (\phi(x_2, \dots, x_{n+1})) \\ &+ \sum_{l=1}^{n} (-1)^l \phi(x_1, \dots, x_{l-1}, x_l x_{l+1}, x_{l+2}, \cdots, x_{n+1}) \\ &+ (-1)^{n+1} \phi(x_1, x_2, \dots, x_n) \qquad \text{for } x_1, x_2, \dots, x_n, x_{n+1} \in G.\end{aligned}$$

The sequence of pairs $(\{C^n(G,H)\}, \{d^{(n)}\})$ is called the *cochain complex*.

Let $n \in \mathbb{N} \cup \{0\}$. We define the group of *n-cocycles* as $Z^n(G,H) := \ker d^{(n)}$, and the group of *n-coboundaries* as

$$B^n(G,H) := \begin{cases} \{0\} & \text{for } n = 0 \\ d^{(n-1)}(C^{n-1}(G,H)) & \text{for } n \geq 1. \end{cases}$$

The crucial property $d^{(n+1)} \circ d^{(n)} = 0$ allows us to define the nth *cohomology group* as the quotient group $Z^n(G,H)/B^n(G,H)$.

Let us see what d amounts to for $n = 0, 1, 2$. For any $x, y, z \in G$ we find:
If $\eta \in C^0(G,H) = H$, then $(dh)(x) = x \cdot \eta - \eta$.
If $f \in C^1(G,H)$, then

$$(df)(x,y) = x \cdot f(y) - f(xy) + f(x) = f(x) + x \cdot f(y) - f(xy).$$

If $F \in C^2(G,H)$, then

$$(dF)(x,y,z) = x \cdot F(y,z) - F(xy,z) + F(x,yz) - F(x,y).$$

From the formulas for d we find that $B^1(G,H)$ consists of the maps $G \to H$ of the form $x \mapsto x \cdot \eta - \eta$, where $\eta \in H$, while $Z^1(G,H)$ consists of the maps $\phi \in C^1(G,H)$ such that $\phi(x_1 x_2) = \phi(x_1) + x_1 \cdot (\phi(x_2))$.

We find that $B^2(G,H)$ consists of the maps in $C^2(G,H)$ of the form $(d\phi)(x_1,x_2) = \phi(x_1) + x_1 \cdot (\phi(x_2)) - \phi(x_1 x_2)$ with $\phi \in$

$C^1(G,H)$, while $Z^2(G,H)$ consists of the maps $F \in C^2(G,H)$ such that $x_1 \cdot (F(x_2,x_3)) - F(x_1x_2,x_3) + F(x_1,x_2x_3) - F(x_1,x_2) = 0$, i.e., such that

$$F(x_1,x_2) + F(x_1x_2,x_3) = F(x_1,x_2x_3) + x_1 \cdot (F(x_2,x_3)).$$

If G acts trivially on H the formulas reduce:

$B^1(G,H) = \{0\}$, while $Z^1(G,H)$ consists of the maps $\phi : G \to H$ such that $\phi(x_1x_2) = \phi(x_1) + \phi(x_2)$, i.e., the additive maps from G to H.

$B^2(G,H)$ consists of the maps from $G \times G$ to H of the form $(df)(x_1,x_2) = f(x_1) + f(x_2) - f(x_1x_2)$ with $f : G \to H$. In our earlier terminology $df = -Cf$, where C denotes the Cauchy difference. Next, $Z^2(G,H)$ consists of the maps $F : G \times G \to H$ such that $F(x_2,x_3) - F(x_1x_2,x_3) + F(x_1,x_2x_3) - F(x_1,x_2) = 0$, i.e., such that

$$F(x_1,x_2) + F(x_1x_2,x_3) = F(x_1,x_2x_3) + F(x_2,x_3). \tag{G.1}$$

(G.1) is called the cocycle equation. So $H^2(G,H) = \{0\}$ means in other words that every solution $F : G \times G \to H$ of the cocycle equation (G.1) has the form $F(x,y) = f(x) + f(y) - f(xy)$ for some $f : G \to H$.

Bibliography

[1] Aczél, János, *The general solution of two functional equations by reduction to functions additive in two variables and with aid of Hamel-bases.* Glasnik Mat.-Fiz. Astronom. Društvo Mat. Fiz. Hrvatske (2), **20** (1965), 65–73. 39.30. MR0198023.

[2] Aczél, János: *Lectures on functional equations and their applications.* Mathematics in Science and Engineering, Vol. 19. Academic Press, New York-London 1966. xx+510 pp. 39.00. MR0208210 (34 #8020).

[3] Aczél, János: *On applications and theory of functional equations.* Elemente der Mathematik vom höheren Standpunkt, Band V. Birkhäuser Verlag, Basel-Stuttgart 1969. 64 pp. 39.30. MR0240491 (39 #1838).

[4] Aczél, János; Baker, John A.; Djoković, D. Ž.; Kannappan, Palaniappan and Radó, Ferenc, *Extensions of certain homomorphisms of semigroups to homomorphisms of groups.* Aequationes Math. **6** (1971), 263–271. MR:45 #12004.

[5] Aczél, János; Chung, Jukang K.; Ng, Che Tat, *Symmetric second differences in product form on groups.* In: Th. M. Rassias (ed.), *Topics in mathematical analysis,* 1–22, Ser. Pure Math. 11, World Sci. Publ., Teaneck, NJ, 1989. 39B52. MR1116572 (92g:39007).

[6] Aczél, János; Dhombres, Jean: *"Functional equations in several variables. With applications to mathematics, information theory and to the natural and social sciences."* Encyclopedia of Mathematics and its Applications, 31. Cambridge University Press, Cambridge, 1989. (39B40). MR1004465 (90h:39001).

[7] Aczél, János; Haruki, H.; McKiernan, M. A.; Sakovič, G. N., *General and regular solutions of functional equations characterizing harmonic polynomials.* Aequationes Math. **1** (1968), 37–53. 39.30. MR0279471 (43 #5193).

[8] Aït Sibaha, M.; Akkouchi, M.; Bakali, A.; Bouikhalene, B., *Caractérisation des fonctions μ-sphériques par des Équations intégro-différentielles.* (French) [Characterization of μ-spherical functions by integrodifferential equations]

Math-Rech. Appl. **3** (2001), 29–38. 43A90 (22D20 33C80 43A85). MR1917184 (2003k:43015).

[9] Akkouchi, Mohamed; Bakali, Allal; Khalil, Idriss, *A class of functional equations on a locally compact group.* (English summary) J. London Math. Soc. (2) **57** (1998), no. 3, 694–705. 43A99 (39B22). MR1659865 (99k:43015).

[10] Akkouchi, Mohamed; Bouikhalene, Belaid; Elqorachi, Elhoucien, *Functional equations and μ-spherical functions.* Georgian Math. J. **15** (2008), no. 1, 1–20. 39B52 (43A90 47G10). MR2418872 (2009g:39049).

[11] Akkouchi, Mohamed; Bakali, Allal; El Qorachi, El Houcien, *D'Alembert's functional equation on compact Gel'fand pairs.* Maghreb Math. Rev. **8** (1999), no. 1-2, 1–9. 43A90 (22D99 39B52 43A05). MR1871524 (2003b:43008).

[12] Akkouchi, M.; Bakali, A.; Bouikhalene, B.; Elqorachi, E., *On generalized d'Alembert functional equation.* Extracta Math. **21** (2006), no. 1, 67–82, 39B32 (39B42 47B99). MR2258342 (2007e:39041).

[13] d'Alembert, Jean Le Rond, *Addition au Mémoire sur la courbe que forme une corde tendue mise en vibration.* Hist. Acad. Berlin, 1750, 355–360.

[14] Alsina, Claudi; Sikorska, Justyna; Tomás, M. Santos: *"Norm derivatives and characterizations of inner product spaces."* World Scientific Publishing Co. Pte. Ltd., Hackensack, NJ, 2010. x+188 pp. ISBN: 978-981-4287-26-5; 981-4287-26-1. 46Cxx. MR2590240.

[15] Amir, Dan: *"Characterizations of Inner Product Spaces."* Operator Theory: Advances and Applications, 20. Birkhäuser Verlag, Basel, 1986. (46B20 46C05). MR:88m #46001.

[16] An, Jinpeng; Yang, Dilian, *Nonabelian harmonic analysis and functional equations on compact groups.* J. Lie Theory **21** (2011), no. 2, 427–456. 39B52 (22C05 43A30). MR2828724.

[17] An, Jinpeng; Yang, Dilian, *A Levi-Civitá equation on compact groups and nonabelian Fourier analysis.* Integral Equations Operator Theory **66** (2010), no. 2, 183–195. 39B52. MR2595652.

[18] Artin, Emil: *"Galois theory."* Notre Dame Math. Lectures, No. 2. 1942. Eighth printing 1985.

[19] Badora, Roman, *On a joint generalization of Cauchy's and d'Alembert's functional equations.* Aequationes Math. **43** (1992), no. 1, 72–89. 39B52 (43A99). MR1144590 (93a:39022).

[20] Badora, Roman, *On a generalized Wilson functional equation.* Georgian Math. J. **12** (2005), no. 4, 595–606. 39B32 (39B52 43A25 43A60 43A90). MR2197183 (2006j:39038).

[21] Bakali, Allal; Bouikhalene, Belaid, *On the generalized d'Alembert functional equation.* Aequationes Math. **71** (2006), no. 3, 209–227. 39B32 (22D05 39B42 43A99). MR2236399 (2007b:39050).

[22] Baker, John A., *The stability of the cosine equation.* Proc. Amer. Math. Soc. **80** (1980), no. 3, 411–416. 39B70 (39B20). MR0580995 (81m:39015).

[23] Bargmann, V., *On unitary ray representations of continuoys groups.* Ann. of Math. (2) **59** (1954), 1–46. 20.0X. MR0058601 (15,397b).

[24] Bass, Hyman: *"Introduction to some methods of algebraic K-theory."* Expository Lectures from the CBMS Regional Conference held at Colorado State University, Ft. Collins, Colo., August 24–28, 1973. Conference Board of the Mathematical Sciences Regional Conference Series in Mathematics, No. 20. American Mathematical Society, Providence, R.I., 1974. 18F25 (18-02). MR0347942.

[25] Benson, Chal; Jenkins, Joe; Ratcliff, Gail, *On Gel'fand pairs associated with solvable Lie groups.* Trans. Amer. Math. Soc. **321** (1990), no. 1, 85–116. 22E25 (22D25 43A20). MR1000329 (90m:22022).

[26] Benson, Chal; Jenkins, Joe; Ratcliff, Gail, *Bounded K-spherical functions on Heisenberg groups.* J. Funct. Anal. **105** (1992), no. 2, 409–443. 22E30 (22E25 33C80). MR1160083 (93e:22017).

[27] Berrone, Lucio R.; Dieulefait, Luis V., *A functional equation related to the product in a quadratic number field.* Aequationes Math. **81** (2011), no. 1–2, 167–175. 39B22. MR2773097.

[28] Boswell, R. D., Jr., *Classroom Notes: On Two Functional Equations.* Amer. Math. Monthly **66** (1959), no. 8, 716.

[29] Bouikhalene, Belaid, *On the generalized d'Alembert's and Wilson's functional equations on a compact group.* Canad. Math. Bull. **48** (2005), no. 4, 505–522. 39B32 (22C05 39B42). MR2176149 (2006g:39069).

[30] Bouikhalene, Belaid; Kabbaj, Samir, *Gelfand pairs and generalized D'Alembert's and Cauchy's functional equations.* Georgian Math. J. **12** (2005), no. 2, 207–216. 39B52 (22D45 39B42). MR2174177 (2006g:39070).

[31] Bouikhalene, Belaid; Elqorachi, Elhoucien, *On Stetkaer type functional equations and Hyers-Ulam stability.* (English summary) Publ. Math. Debrecen **69** (2006), no. 1-2, 95–120. 39B42 (22D12 39B32 39B82). MR2228478 (2007b:39051).

[32] Bourbaki, Nicholas: *"Éleménts de Mathématique. Livre II. Algèbre."* Hermann, Paris 1958.

[33] Bröcker, Theodor; Dieck, Tammo tom: *"Representations of Compact Lie Groups."* Graduate Texts in Mathematics, 98. Springer-Verlag, New York, 1985. 22E45 (57-01). MR0781344 (86i:22023).

[34] Browder, Andrew: *"Introduction To function Algebras."* W. A. Benjamin, Inc. New York, Amsterdam. 1969. 46.55. MR0246125 (39 #7431).

[35] Brown, Kenneth S.: *"Cohomology of Groups."* Graduate Texts in Mathematics, 87. Springer-Verlag, New York-Berlin, 1982. (18-01 20F32 20J05 55-01). MR0672956 (83k:20002).

[36] Butler, Steven. Problem no. 11030. Amer. Math. Monthly **110** (2003), 637.

[37] Chojnacki, Wojciech, *On some functional equation generalizing Cauchy's and d'Alembert's functional equations.* Colloq. Math. **55** (1988), no. 1, 169–178. 39B20 (39B50 43A99). MR0964333.

[38] Chojnacki, Wojciech, *On group decompositions of bounded cosine sequences.* Studia Math. **181** no.1, (2007), 61–85. 47D09 (39A20 47B40). MR2317854.

[39] Choquet, Gustave Alfred Arthur: *"L'enseignement de la géométrie."* Enseign. des sciences. Paris: Hermann & Cie. 1964. Zbl 0124.00110.

[40] Christensen, Ole: *"Frames and Bases. An Introductory Course."* Applied and Numerical Harmonic Analysis. Birkhäuser Boston, Inc., Boston, MA, 2008. 42-01 (42C10 46B15 46C05 46E30 94A12). MR2428338 (2010i:42001).

[41] Chung, Jukang K.; Ebanks, Bruce R.; Ng, Che Tat; Sahoo, Prasanna K., *On a quadratic-trigonometric functional equation and some applications.* Trans. Amer. Math. Soc. **347** (1995), 1131–1161. 39B52 (39B22, 39B32). MR1290715 (95j:39041).

[42] Chung, Jukang K.; Ebanks, Bruce R.; Ng, Che Tat; Sahoo, Prasanna K., Erratum to *On a quadratic-trigonometric functional equation and some applications.* [Trans. Amer. Math. Soc. **347** (1995), no. 4, 1131–1161]. Trans. Amer. Math. Soc. **349** (1997), no. 11, 4691. 39B52 (39B22 39B32) MR1422599 (98b:39026).

[43] Chung, Jukang K.; Ebanks, Bruce R.; Ng, Che Tat; Sahoo, Prasanna K.; Zeng, Wei-Bin, *On generalized rectangular and rhombic functional equations.* Publ. Math. Debrecen **47** (1995), no. 3-4, 249–270. 39B52. MR1362288 (97b:39048).

[44] Chung, Jukang K.; Ebanks, Bruce R.; Sahoo, Prasanna K., *On a functional equation of Świątek on groups.* Aequationes Math. **45** (1993), no. 2-3, 246–266. 39B52. MR1212390 (94b:39041).

[45] Chung, Jukang K.; Kannappan, Palaniappan; Ng, Che Tat, *A generalization of the cosine-sine functional equation on groups.* Linear Algebra Appl. **66** (1985), 259–277. 39B50. MR0781306.

[46] Chung, Jukang K.; Kannappan, Palaniappan; Ng, Che Tat, *On two trigonometric functional equations.* Math. Rep. Toyama Univ. **11** (1988), 153–165. 39B30 (39B50). MR0974257 (89j:39010).

[47] Chung, Jukang K.; Sahoo, Prasanna K., *General solution of some functional equations related to the determinant of some symmetric matrices.* Demonstratio Math. **35** (2002), no. 3, 539–544. 39B22. MR1917096.

[48] Chung, Jaeyoung; Kim, Dohan, *Sine functional equation in several variables.* Arch. Math. **86** no. 5, (2006), 425–429. 39B22. MR2229358.

[49] Corovei, Ilie, *The functional equation* $f(xy)+f(yx)+f(xy^{-1})+f(y^{-1}x)=4f(x)f(y)$ *for nilpotent groups.* (Romanian) Bul. Ştiinţ. Instit. Politehn. Cluj-Napoca Ser. Mat.-Fiz.-Mec. Apl. **20** (1978), 25–28. 39B50 (20F18). MR0630129 (82j:39005).

[50] Corovei, Ilie, *The cosine functional equation for nilpotent groups.* Aequationes Math. **15** nr. 1, (1977), 99–106. 39A15 (39A40). MR0435645.

[51] Corovei, Ilie, *The functional equation* $f(xy)+f(xy^{-1})=2f(x)g(y)$ *for nilpotent groups.* Mathematica (Cluj) **22** (45) no.1, (1980), 33–41. 39B50 (20F18 39B20). MR0618025 (82j:39007).

[52] Corovei, Ilie, *The functional equation* $f(xy)+f(yx)=f(x)f(y)+f(y)f(x)$ *on a semigroup into a ring.* Mathematica (Cluj) **27** (52) (1985), no. 2, 97–99. 39B50 (20M15). MR0853495 (87i:39019).

[53] Corovei, Ilie, *On semi-homomorphisms and Jensen's equation.* Mathematica **37** (60) no. 1–2, (1995), 59–64. 20F99. MR1607839 (99d:20066).

[54] Corovei, Ilie, *The d'Alembert functional equation on metabelian groups.* Aequationes Math. **57** no. 2–3, (1999), 201–205. 39B52 (20K99). MR1689193 (2000g:39027).

[55] Corovei, Ilie, *Symmetric second differences in product form on* P_3*-groups.* Mathematica **43 (66)** no. 2, (2001), 167–175. 39B52. MR2002647.

[56] Corovei, Ilie, *Wilson's functional equation on* P_3*-groups.* Aequationes Math. **61** nr. 3, (2001), 212–220. 39B52. MR1833141.

[57] Corovei, Ilie, *Wilson's functional equation on metabelian groups.* Mathematica **44 (67)** no. 2, (2002), 137–146. 39B52 (20F99). MR2032428.

[58] Corovei, Ilie, *The sine functional equation on 2-divisible groups.* Mathematica **47**(70) (2005), no. 1, 49–52. MR2165077 (2006d:39044). 39B52.

[59] Cuculière, Roger, *Problem 11578.* Amer. Math. Monthly **118** (2011), p. 464.

[60] Czerwik, Stefan: *"Functional equations and inequalities in several variables."* World Scientific Publishing Co., Inc., River Edge, NJ, 2002. x+410 pp. 39Bxx (26-01 39-02 47H99 47N20). MR1904790 (2003b:39027).

[61] Dacić, R., *The cosine functional equation for groups.* Mat. Vesnik **6** (21) (1969) 339–342. 39.38. MR0271570 (42 #6453).

[62] Davison, Thomas M. K., *D'Alembert's functional equation and Chebyshev polynomials.* Ann. Acad. Paedagog. Crac., Stud. Math. **4**, No. 1, 31–38 (2001). ISSN 1643-6555. MSC2000: 39B52.

[63] Davison, Thomas M. K., *D'Alembert's functional equation on topological groups.* Aequationes Math. **76** no. 1–2, (2008), 33–53. 39B52 (22A10). MR2443461 (2009k:39028).

[64] Davison, Thomas M. K., *D'Alembert's functional equation on topological monoids.* Publ. Math. Debrecen **75** 1/2 (2009), 41–66. 39B52. MR2560023 (2010h:39040).

[65] Davison, Thomas M. K.; Ebanks, Bruce R., *Cocycles on cancellative semigroups.* Publ. Math. Debrecen **46** 1/2 (1995), 137–147. 20M50. MR1316656 (96c:20120).

[66] Drygas, Hilmar, *Quasi-inner products and their applications.* Advances in multivariate statistical analysis, 13–30, Theory Decis. Lib. Ser. B: Math. Statist. Methods, Reidel, Dordrecht, 1987. 46C99 (15-01 15A63 62J05 90C25). MR0920424.

[67] Dym, Harry; McKean, Henry P.: *"Fourier series and integrals."* Probability and Mathematical Statistics, No. 14. Academic Press, New York-London, 1972. 42-02. MR0442564.

[68] Ebanks, Bruce R.; Ng, Che Tat, *On generalized cocycle equations.* Aequationes Math. **46** (1993), no. 1-2, 76–90. 39B52 (39A10). MR1220723 (95e:39005).

[69] Ebanks, Bruce R., *A Functional Equation.* Solution of Problem 11030. Amer. Math. Monthly **112** (2005), 371.

[70] Ebanks, Bruce R.; Stetkær, Henrik, *Continuous cocycles on locally compact groups.* Aequationes Math. **78** no. 1–2, (2009), 123–145. 39B52 (22E41). MR2552528.

[71] Ebanks, Bruce R.; Stetkær, Henrik, *Continuous cocycles on locally compact groups II.* Aequationes Math. **80** no. 1–2, (2010), 57–83. 39B52 (22E41 57T10). MR2736940.

[72] Ellis, Robert, *Locally compact transformation groups.* Duke Math. J. **24** (1957), 119–125. 22.0X. MR0088674 (19,561b).

[73] Elqorachi, Elhoucien; Akkouchi, Mohamed, *On generalized d'Alembert and Wilson functional equations.* Aequationes Math. **66** no. 3, (2003), 241–256. 39B32 (39B42). MR2028561.

[74] Elqorachi, Elhoucien; Akkouchi, Mohamed, *Functional equations and Gelfand measures.* Publ. Math. Debrecen **63** no. 4, (2003), 643–666. 39B42 (43A90 47G10). MR2020778.

[75] Elqorachi, Elhoucien; Akkouchi, Mohamed, *On Cauchy-type functional equations.* Int. J. Math. Math. Sci. 2004, no. **13-16**, 847–859. 39B52 (39B42). MR2054189 (2005e:39064).

[76] Elqorachi, Elhoucien; Akkouchi, Mohamed; Bakali, Allal; Bouikhalene, Belaid, *Badora's Equation on non-Abelian Locally Compact Groups.* Georgian Math. J. **11** (2004), no. 3, 449–466. 39B05 (43A90). MR2081737 (2005f:39060).

[77] Elqorachi, Elhoucien; Redouani, Ahmed, *Trigonometric formulas and μ-spherical functions.* Aequationes Math. **72** (2006), no. 1-2, 60–77. 39B32 (43A90 45N05). MR2258807 (2008d:39028).

[78] Faraut, Jacques; Harzallah, Khélifa, *Formes quadratiques généralisées et fonctions définies négatives.* C. R. Acad. Sci. Paris Sér. A-B **273** (1971), A1034–A1036. 43A85 (53C35). MR0293020 (45 #2100).

[79] Faĭziev, Valeriĭ A.; Sahoo, Prasanna K., *On Drygas functional equation on groups.* Int. J. Appl. Math. Stat. **7** (2007), No. Fe07, 59–69. 39B52 (39B72). MR2339509 (2008f:39034).

[80] Fechner, Żywilla, *Systems of trigonometric functional equations in Banach algebras.* Aequationes Math. **77** no. 1-2 (2009), 187–195.39B72. MR2495726 (2010b:39038).

[81] Fechner, Żywilla, *A generalization of Gajda's equation.* J. Math. Anal. Appl. **354** (2009), no. 2, 584–593.39B52. R2515239 (2010e:39025).

[82] Fechner, Żywilla, *Wilson's functional equation in Banach algebras.* Acta Sci. Math. (Szeged) **75** (2009), no. 1-2, 131–142. 39B52 (46H99). MR2533406 (2010d:39040).

[83] Fechner, Żywilla, *Wilson's functional equation in algebras.* Accepted for publication by Studia Scientiarum Mathematicarum Hungarica on April 16, 2009. 13 pp.

[84] Fechner, Żywilla, *Trigonometric identities in *-algebras.* Personal communication. Manuscript received June 26, 2010. 9 pages.

[85] Fechner, Żywilla, *A note on a modification of Gajda's equation.* Aequationes Math. **82** no. 1-2 (2011), 135–141.

[86] Feldman, Gennadiy: *Functional equations and characterization problems on locally compact abelian groups.* EMS Tracts in Mathematics, 5. European Mathematical Society (EMS), Zürich, 2008. xii+256 pp. ISBN: 978-3-03719-045-6. 43A25 (39B52 43A35 60B15 62E10). MR2412878 (2011e:43006).

[87] Förg-Rob, Wolfgang; Schwaiger, Jens, *A generalization of the cosine equation to n summands.* Grazer Math. Ber. **316** (1992), 219-226.

[88] Fréchet, Maurice, *Sur la définition axiomatique d'une classe d'espaces vectoriels distanciés applicables vectoriellement sur l'espace de Hilbert.* Ann. of Math. (2) **36** (1935), no. 3, 705–718. MR1503246.

[89] Friis, Peter de Place; Stetkær, Henrik, *On the cosine-sine functional equation on groups.* Aequationes Math. **64** (2002), no. 1-2, 145–164. 39B52. MR1929257 (2003h:39022).

[90] Friis, Peter de Place, *d'Alembert's and Wilson's functional equations on Lie groups.* Aequationes Math. **67** no. 1–2, (2004), 12–25. 39B52. MR2049602.

[91] Friis, Peter de Place; Stetkær, Henrik, *On the quadratic functional equation on groups.* Publ. Math. Debrecen **69** no. 1–2, (2006), 65–93. 39B52 (20D40). MR2228477 (2007b:39058).

[92] Gajda, Zbigniew, *Difference properties of higher orders for continuity and Riemann integrability.* Colloq. Math. **53** no. 2, (1987), 275–288. 22B05 (22C05 26A15 26A42). MR0924073.

[93] Gajda, Zbigniew, *A generalization of d'Alembert's functional equation.* Funkcial. Ekvac. **33** (1990), no. 1, 69–77. 39B52 (43A99). MR1065469 (92a:39014).

[94] Gajda, Zbigniew, *Unitary Representations of Topological Groups and Functional Equations.* J. Math. Anal. Appl. **152** no. 1, (1990), 6–19. 39B52 (22D10). MR1072924.

[95] Gajda, Zbigniew, *A remark to the talk of W. Förg-Rob.* Grazer Math. Ber. **316** (1992), 234–237.

[96] Gaur, A. K.; Nandakumar, N. R., *Derivation pairs of operators on algebras of analytic functions.* Functional analysis, 104–110, Narosa, New Delhi, 1998. 46J15 (16W25). MR1668797 (2000b:46089).

[97] Ger, Roman, *Christensen measurability and functional equations.* Selected topics in functional equations (Graz, 1986), Ber. No. 289, 17 pp., Ber. Math.-Statist. Sekt. Forschungsgesellsch. Joanneum, 285–296, Forschungszentrum Graz, Graz, 1988. 39B70 (28E10 39C05). MR1012745 (90h:39023).

[98] Ger, Roman, *Joint continuity of Christensen measurable multiadditive maps.* Tatra Mt. Math. Publ. **2** (1993), 93–103. 28B20. MR1251043 (95e:28008).

[99] Ger, Roman, *On a problem of Roger Cuculière.* J. Inform. Math. Sci. 1 (2009), no. 2-3, 157–163. 39B22. MR2780329.

[100] Ger, Roman, *Additivity and exponentiality are alien to each other.* Aequationes Math. **80**, No. 1–2 (2010), 111–118. 39B52. MR2736943.

[101] Ger, Roman; Kominek, Zygfryd, *An interplay between Jensen's and Pexider's functional equations on semigroups.* Ann. Univ. Sci. Budapest. Sect. Comput. 35 (2011), 107–124. 39B52. MR2894556.

[102] Glaeser, Georges, *Dérivation des algèbres commutatives.* C. R. Acad. Sci. Paris **233**, (1951), 1550–1552. MR0045104 (13,527e).

[103] Gronau, Detlef, *A remark on Sincov's functional equation.* Notices of the South African Mathematical Society **31** (2000) No. 1, 1–8.

[104] Halmos, Paul R.: *"Finite-dimensional vector spaces."* 2nd ed. The University Series in Undergraduate Mathematics. D. Van Nostrand Co., Inc., Princeton-Toronto-New York-London, 1958. viii+200 pp. 15.0X. MR0089819 (19,725b).

[105] Halter-Koch, Franz; Reich, Ludwig; Schwaiger, Jens, *On products of additive functions.* Aequationes Math. **45** (1993), no. 1, 83–88. 39B52 (13F20). MR1201399.

[106] Hamel, Georg, *Eine Basis aller Zahlen und die unstetigen Lösungen der Funktionalgleichung* $f(x+y) = f(x) + f(y)$. Math. Ann. **60** (1905), 459–462. Fortschr.: **36**, 446 (1905).

[107] Hausner, Melvin; Schwartz, Jacob T.: *"Lie groups; Lie algebras."* Gordon and Breach Science Publishers, New York-London-Paris. 1968. x+229 pp. 22.50. MR0235065 (38 #3377).

[108] Helgason, Sigurdur: *"Differential geometry, Lie groups, and symmetric spaces."* Pure and Applied Mathematics, 80. Academic Press, Inc. [Harcourt Brace Jovanovich, Publishers], New York-London, 1978. 53C35 (22E10 22E46 22E60 53C21). MR0514561 (80k:53081).

[109] Helgason, Sigurdur: *"Groups and Geometric Analysis. Integral geometry, invariant differential operators, and spherical functions."* Pure and Applied Mathematics, 113. Academic Press, Inc., Orlando, FL, 1984. 22E30 (22-02 43A85 53-02 53C65 58G35). MR0754767 (86c:22017).

[110] Heuvers, Konrad; Ng, Che Tat, *Multiplicative quadratic forms on the quaternions and related functional equations.* Sci. Math. Jpn. **70** (2009), no. 1, 29–41. 39B52 (15A33 16K20 39B72). MR2542004.

[111] Hewitt, Edwin; Ross, Kenneth A.: *"Abstract harmonic analysis. Vol. I: Structure of topological groups. Integration theory, group representations."* Second edition. Grundlehren der Mathematischen Wissenschaften [Fundamental Principles of Mathematical Sciences], 115. Springer-Verlag, Berlin-New York, 1979. 43-02 (22-01). MR0551496.

[112] Hewitt, Edwin; Ross, Kenneth A.: *"Abstract harmonic analysis. Vol. II: Structure and analysis for compact groups. Analysis on locally compact Abelian groups."* Die Grundlehren der mathematischen Wissenschaften, Band 152 Springer-Verlag, New York-Berlin 1970. 42.50 (22.00). MR0262773.

[113] Hosszú, Miklós, *A remark on scalar valued multiplicative functions of matrices.* Publ. Math. Debrecen **6** (1959), 288–289. 15.00 (39.00). MR0112898 (22 #3744).

[114] Hosszú, Miklós, *A remark on the square norm.* Aequationes Math. **2** (1969), 190–193. 39.38. MR0246011.

[115] Jacobson, Nathan: *"Lectures in Abstract Algebra. Volume II – Linear Algebra."* Reprint of the 1953 edition [Van Nostrand, Toronto, Ont.]. Graduate Texts in Mathematics, No. 31. Springer-Verlag, New York-Berlin, 1975. 15-02. MR0369381.

[116] Járai, Antal: *"Regularity Properties of Functional Equations in Several Variables."* Advances in Mathematics (Springer), 8. Springer, New York, 2005. 39-02 (39-00 39B05 39B52 39B72). MR2130441.

[117] Jensen, Johan Ludvig William Valdemar, *Om Fundamentalligningers "Opløsning" ved elementære Midler.* [In Danish. The title translates to "On the

solution of fundamental equations by elementary means."] Tidsskr. Math. (4) **2** (1878), 149–155. Fortschr. **10** (1878), 289.

[118] Jensen, Johan Ludvig William Valdemar, *Om konvekse Funktioner og Uligheder mellem Middelværdier.* [In Danish. The title translates to "On convex functions and inequalities between mean values."] Nyt Tidsskrift for Mathematik B. **16**, (1905), 49–68.

[119] Jensen, Johan Ludvig William Valdemar, *Sur les fonctions convexes et des inégalités entre les valeurs moyennes.* Acta Math. **30** (1906), 175–93. Fortschr. **37**, 422 (1906).

[120] Jessen, Børge; Karpf, J.; Thorup, Anders, *Some functional equations in groups and rings.* Math. Scand. **22** (1968), 257–265. 39.38 (13.00). MR0284739.

[121] Jones, Floyd Burton, *Connected and disconnected plane sets and the functional equation* $f(x) + f(y) = f(x + y)$. Bull. Amer. Math. Soc. **48** (1942), 115–120. MR0005906 (3,229e).

[122] Jordan, Pascual; Neumann, John von, *On inner products in linear, metric spaces.* Ann. of Math. (2) **36** (1935), no. 3, 719–723. MR1503247.

[123] Jung, Soon-Mo, *On the quadratic functional equation modulo a subgroup.* Indian J. Pure Appl. Math. **36** (2005), no. 8, 441–450. 39B55. MR2199219 (2006i:39041).

[124] Kaczmarz, Stefan, *Sur l'équation fonctionelle* $f(x)+f(x+y) = \phi(y)f(x+\frac{y}{2})$. Fund. Math. **6** (1924), 122–129. Fortschr.: **50**, 183 (1924).

[125] Kannappan, Palaniappan, *The functional equation* $f(xy) + f(xy^{-1}) = 2f(x)f(y)$ *for groups.* Proc. Amer. Math. Soc. **19** (1968), 69–74. 39.38. MR0219936.

[126] Kannappan, Palaniappan, *On sine functional equation.* Studia Sci. Math. Hungar. **4** (1969), 331–333. 39.31. MR0249859 (40 #3100).

[127] Kannappan, Palaniappan *A note on cosine functional equation for groups.* Mat. Vesnik **8**(23) (1971), 317–319. 39A40. MR0296557 (45 #5616)

[128] Kannappan, Palaniappan, *P177S2.* Aequationes Math. **19** (1979), 114–115.

[129] Kannappan, Palaniappan, *Cauchy equations and some of their applications.* Topics in mathematical analysis, 518–538, Ser. Pure Math., 11, World Sci. Publ., Teaneck, NJ, 1989. 39B52 (39B32). MR1116594 (92k:39014)

[130] Kannappan, Palaniappan, *On quadratic functional equation.* Int. J. Math. Stat. Sci. **9** (2000), no. 1, 35–60. 39B52 (39-02). MR1770925.

[131] Kannappan, Palaniappan, *Klee's trigonometry problem.* Amer. Math. Monthly **110** (2003), no. 10, 940–944. 39B22 (39B32 39B52). MR2025002 (2005b:39031).

[132] Kannappan, Palaniappan: *"Functional Equations and Inequalities with Applications."* Springer Monographs in Mathematics. Springer, New York, 2009. 39-02 (39Bxx). MR2524097.

[133] Kannappan, Palaniappan; Nandakumar, Nagaiah Rangaswamy, *On a cosine functional equation for operators on the algebra of analytic functions in a domain.* Aequationes Math. **61** (2001), no. 3, 233–238. MR1833143 (2002g:39016). 39B32 (30H05).

[134] Kobal, Damjan, *Inner product space and circle power.* Publ. Math. Debrecen **81**/1–2 (2012), 1–9. 39B22 (46C15). MR2957497.

[135] Kuczma, Marek: *"An Introduction to the Theory of Functional Equations and Inequalities: Cauchy's Equation and Jensen's Inequality."* Second edition. Edited and with a preface by Attila Gilányi. Birkhäuser Verlag, Basel, 2009. 39-02 (39Bxx). MR2467621.

[136] Kurepa, Svetozar, *On the functional equation* $f(x+y)f(x-y) = f^2(x) - f^2(y)$. Ann. Polon. Math. **10** (1961) 1–5. 39.35. MR0126644 (23 #A3940).

[137] Kurepa, Svetozar, *The Cauchy functional equation and scalar product in vector spaces.* Glasnik Mat.-Fiz. Astronom. Ser. II Društvo Mat. Fiz. Hrvatske **19** (1964), 23–36. 39.31. MR0171100.

[138] Kurepa, Svetozar, *Quadratic and sesqui-linear functionals.* Glasnik Mat.-Fiz. Astronom. Ser. II Društvo Mat. Fiz. Hrvatske **20** (1965) 79–92. 39.30 (15.99). MR0193390.

[139] Kurepa, Svetozar, *On bimorphisms and quadratic forms on groups.* Aequationes Math. **9** (1973), 30–45. 39A40 (20F99). MR0326221.

[140] Lam, T. Y., *A theorem of Burnside on matrix rings.* Amer. Math. Monthly **105** (1998), no. 7, 651–653. 20C15 (16S50). MR1633085 (99e:20011).

[141] Lê, Công-Trình; Thái, Trung-Hiêu, *Jensen's functional equation on the symmetric group* S_n. Aequationes Math. **82** nr. 3, (2011), 269–276. 39B52 (20B30). MR2855066.

[142] Lomonosov, Victor; Rosenthal, Peter, *The simplest proof of Burnside's theorem on matrix algebras.* Linear Algebra Appl. **383** (2004), 45–47. 16S50 (47A15). MR #2073890 (2005c:16046).

[143] Łukasik, Radosław, *Some generalization of Cauchy's and the quadratic functional equations.* Aequationes Math. **83** no. 1-2, (2012), 75–86. 39B52. MR2885500.

[144] Maksa, Gyula, *A remark on symmetric biadditive functions having nonnegative diagonalization.* Glasnik Mat. Ser. III. **15** (35) (1980), no. 2, 279–282. 39B70 (39B40 46C99). MR0608877.

[145] McKiernan, M. A., *The matrix equation* $a(x \circ y) = a(x) + a(x)a(y) + a(y)$. Aequationes Math. **15** (1977), no. 2-3, 213–223. 39A40. MR0481702.

[146] Nagy, Béla Szökefalvi-, *A sine functional equation in Banach algebras.* Publ. Math. Debrecen **24** (1977), no. 1–2, 77–90. MR0460954 (57 #943). 39A40 (47D99).

[147] Neder, Ludwig, *Lösung zu 401.* Arch. Math. Phys. (3) **22** (1914), 359–360.

[148] Ng, Che Tat, *Jensen's functional equation on groups.* Aequationes Math. **39** no. 1, (1990), 85–99. 39B52 (15A36 20E05). MR1044167.

[149] Ng, Che Tat, *Jensen's functional equation on groups, II.* Aequationes Math. **58** no. 3, (1999), 311–320. 39B52 (20E05 20G20). MR1715402.

[150] Ng, Che Tat, *Jensen's functional equation on groups, III.* Aequationes Math. **62** no. 1-2, (2001), 143–159. 39B52 (20E05 20G20 39A70 39B42). MR1849146.

[151] Ng, Che Tat, *A Pexider-Jensen functional equation on groups.* Aequationes Math. **70** (2005), no. 1-2, 131–153. 39B42 (39B52). MR2167991 (2006c:39042).

[152] O'Connor, T. A., *A solution of the functional equation* $\phi(x-y) = \sum_1^n a_j(x)\overline{a_j(y)}$ *on a locally compact Abelian group.* J. Math. Anal. Appl. **60** no. 1, (1977), 120–122. 39A40. MR0454444.

[153] Orosz, Ágota, *Sine and cosine equation on discrete polynomial hypergroups.* Aequationes Math. **72** (2006), no. 3, 225–233. 43A62 (20N20, 39B32). MR2282869 (2008a:43004).

[154] Parnami, J. C.; Singh, H.; Vasudeva, H. L., *On an exponential-cosine functional equation.* Period. Math. Hungar. **19** (1988), no. 4, 287–297. 39B20 (39B05 39B70 47D05). MR0984778.

[155] Parnami, J. C.; Vasudeva, H. L. *On Jensen's functional equation.* Aequationes Math. **43** (1992), no. 2-3, 211–218. 39B52. MR1158729 (93e:39013).

[156] Penney, R. C.; Rukhin, A. L. *d'Alembert's functional equation on groups.* Proc. Amer. Math. Soc. **77** (1979), no. 1, 73–80. 39B50 (39B30). MR0539634 (80h:39013).

[157] Pexider, Jan Vilém, *Notiz über Funktionaltheoreme.* Monatsh. Math. Phys. **14** (1903), 293–301. Fortschr. **34**, 621 (1903).

[158] Pontrjagin, Lev Semenovich: *"Topological groups."* Princeton University Press. 1946.

[159] Poulsen, Thomas A.; Stetkær, Henrik, *On the trigonometric subtraction and addition formulas.* Aequationes Math. **59** (2000), no. 1-2, 84–92. 39B52 (39B32). MR1741472 (2001f:39029).

[160] Prunesco, Mihai, *Concrete algebraic cohomology for the group* $(\mathbb{R},+)$ *or how to solve the functional equation* $f(x+y)-f(x)-f(y)=g(x,y)$. CUBO **9** (2007), no. 3, 39–45.

[161] Radó, Ferenc, *Remark 1.* Aequationes Math. **14** (1976), 228–229.

[162] Rassias, Michael Th., *Solution of a functional equation problem of Steven Butler.* Octogon Math. Mag. **12** (2004), 152–153.

[163] Rassias, Themistocles M.; Šimša, Jaromír: *"Finite sums decompositions in mathematical analysis."* Pure and Applied Mathematics (New York). A Wiley-Interscience Publication. John Wiley & Sons, Ltd., Chichester, 1995. vi+172 pp. ISBN: 0-471-94827-6. 26B40 (15A24 35C10 39B99 41A58). MR1339768 (96k:26006).

[164] Rickart, Charles E.: *"General theory of Banach algebras."* The University Series in Higher Mathematics. D. van Nostrand Co., Inc. Princeton, N.J.-Toronto-London-New York 1960. xi+394 pp. MR0115101 (22 #5903).

[165] Roukbi, Ahmed and Zeglami, Driss, *D'Alembert's functional equations on hypergroups.* Adv. Pure Appl. Math. **2** no. 2, (2011), 147–166. 39B52. MR2805073.

[166] Rudin, Walter: *"Functional analysis."* Tata McGraw-Hill Publishing Company Ltd. New Delhi. 1974.

[167] Sahoo, Prasanna K.; Kannappan, Palaniappan: *"Introduction to Functional Equations."* CRC Press, Boca Raton, FL, 2011. 39-02 (39B22 39B32 39B72 39B82). MR2757437 (2012b:39001).

[168] Šemrl, Peter, *Additive functions and two characterizations of inner-product spaces.* Glas. Mat. Ser. III **25**(45) (1990), no. 2, 309–317. 46C15 (39B22). MR1243735.

[169] Shapiro, Harold N., *A micronote on a functional equation.* Amer. Math. Monthly **80** (1973), 1041. 39A20. MR0326216 (48 #4560).

[170] Shin'ya, Hitoshi, *Spherical matrix functions and Banach representability for locally compact motion groups.* Japan. J. Math. (N.S.) **28** no. 2, (2002), 163–201. 43A65 (22D10). MR1947900 (2003k:43009).

[171] Shulman, Ekaterina, *Group representations and stability of functional equations.* J. London Math. Soc. (2) **54** (1996), no. 1, 111–120. 39B52 (22D12 39B72). MR1395071 (97h:39018).

[172] Shulman, Ekaterina, *A motion with a permanent segment-areal speed and the Levi-Cività functional equation.* Aequationes Math. **62** (2001), no. 3, 297–302. 53A17 (39B22 70F99). MR1869579 (2002h:53018).

[173] Shulman, Ekaterina, *Some extensions of the Levi-Civitá functional equation and richly periodic spaces of functions.* Aequationes Math. **81** (2011), no. 1-2, 109–120. 39B52 (22A25 46E10). MR2773093.

[174] Shulman, Ekaterina, *Subadditive set-functions on semigroups, applications to group representations and functional equations.* J. Funct. Anal. **263** (2012), no. 5, 1468–1484. 39B52 (22Axx 47Gxx). MR2943737.

[175] Sinopoulos, Pavlos, *Generalized sine equations. I.* Aequationes Math. **48** (1994), no. 2-3, 171–193. 39B52 (39B22 39B32). MR1295090.

[176] Sinopoulos, Pavlos, *Generalized sine equations. II.* Aequationes Math. **49** (1995), no. 1-2, 122–152. 39B22 (39B32 39B52). MR1309298.

[177] Sinopoulos, Pavlos, *Generalized sine equations. III.* Aequationes Math. **51** (1996), no. 3, 311–327. 39B32 (39B22 68Q40). MR1394736.

[178] Sinopoulos, Pavlos, *Wilson's functional equation for vector and matrix functions.* Proc. Amer. Math. Soc. **125** (1997), 1089-1094. 39B52. MR1363186 (97f:39034).

[179] Sinopoulos, Pavlos, *Contribution to the study of two functional equations.* Aequationes Math. **56** (1998), no. 1-2, 91–97. 39B52 (39B32). MR1628319 (99e:39093).

[180] Sinopoulos, Pavlos, *Functional equations on semigroups.* Aequationes Math. **59** no. 3, (2000), 255–261. 39B52. MR1761060.

[181] Sinopoulos, Pavlos, *Wilsons functional equation in dimension 3.* Aequationes Math. **66** no. 1-2, (2003), 164–179. 39B42 (39B52 39B72). MR2003463 (2004g:39044).

[182] Sinopoulos, Pavlos, *Applications of Wilsons functional equation.* Aequationes Math. **67** no. 1-2, (2004), 188–194. 39B42 (39B52 39B72). MR2049617 (2005a:39047).

[183] Stetkær, Henrik, *d'Alembert's equation and spherical functions.* Aequationes Math. **48** no. 2-3, (1994), 220–227. 39B52 (33C80 43A90). MR1295092.

[184] Stetkær, Henrik, *Functional Equations and Spherical Functions.* Preprint Series 1994 No. 18, Matematisk Institut, Aarhus University, Denmark.

[185] Stetkær, Henrik, *Wilson's functional equation on groups.* Aequationes Math. **49** no. 3, (1995), 252–275. 39B05 (33C50 39B52 43A90). MR1324352 (96j:39023).

[186] Stetkær, Henrik, *Functional equations on abelian groups with involution.* Aequationes Math. **54** no. 1-2, (1997), 144–172. 39B32. MR1466300.

[187] Stetkær, Henrik, *Functional equations on abelian groups with involution. II.* Aequationes Math. **55** (1998), no. 3, 227–240. 39B32. MR1615396 (99g:39031).

[188] Stetkær, Henrik, *D'Alembert's functional equations on metabelian groups.* Aequationes Math. **59** no. 3, (2000), 306–320. 39B52. MR1761065.

[189] Stetkær, Henrik, *D'Alembert's and Wilson's functional equations for vector and* 2×2 *matrix valued functions.* Math. Scand. **87** (2000), no. 1, 115–132. 39B52. MR1776968.

[190] Stetkær, Henrik, *On Multiplicative Maps.* Semigroup Forum **63** no. 3, (2001), 466–468. 20M15. MR1851826.

[191] Stetkær, Henrik, *On Jensen's functional equation on groups.* Aequationes Math. **66** no. 1-2, (2003), 100–118. 39B52 (39B32). MR2003458 (2004g:39048).

[192] Stetkær, Henrik, *d'Alembert's and Wilson's functional equations on step 2 nilpotent groups.* Aequationes Math. **67** (2004), no. 3, 241–262. 39B32 MR2162231 (2006h:39030).

[193] Stetkær, Henrik, *On a variant of Wilson's functional equation on groups.* Aequationes Math. **68** (2004), no. 3, 160–176. 39B32. MR2165413 (2006c:39041).

[194] Stetkær, Henrik, *Functional equations and matrix-valued spherical functions.* Aequationes Math. **69** (2005), no. 3, 271–292. 39B32 (43A90). MR2139609 (2006a:39043).

[195] Stetkær, Henrik, *Properties of d'Alembert functions.* Aequationes Math. **77** no. 3, (2009), 281–301. 39B52 (22E25). MR2520502

[196] Stromberg, Karl, *An elementary proof of Steinhaus's theorem.* Proc. Amer. Math. Soc. **36** (1972), 308. 28A70 (43A05). MR0308368 (46 #7482).

[197] Székelyhidi, László: *"Convolution Type Functional Equations on Topological Abelian Groups."* World Scientific Publishing Co., Inc., Teaneck, NJ, 1991. MR1113488 (92f:39017). 39B52 (43A99).

[198] Székelyhidi, László, *D'Alembert's functional equation on compact groups.* Banach J. Math. Anal. **1** (2007), no. 2, 221–226. MR2366105 (2008m:43007). 43A45 (39B52)

[199] Thielman, H. P., *On generalized Cauchy functional equations.* Amer. Math. Monthly **56** (1949), 452–457. 39.0X. MR0031643.

[200] Thurston, William P.: *"Three-Dimensional Geometry and Topology. Vol. 1."* Edited by Silvio Levy. Princeton Mathematical Series, 35. Princeton University Press, Princeton, NJ, 1997. 57M50 (53A35 57M25 57M60 57N10). MR1435975 (97m:57016).

[201] Varadarajan, V. S.: *"Lie groups, Lie algebras, and their representations."* Prentice-Hall Series in Modern Analysis. Prentice-Hall, Inc., Englewood Cliffs, N.J., 1974. 22-01 (17-02 22EXX). MR0376938 (51 #13113).

[202] Varadarajan, V. S.: *"Geometry of Quantum Theory."* Second edition. Springer-Verlag, New York, 1985. 81Bxx (81-02). MR0805158 (87a:81009).

[203] Vincze, Endre, *Eine allgemeinere Methode in der Theorie der Funktionalgleichungen. I.* (German) Publ. Math. Debrecen **9** (1962), 149–163. MR0142936 (26 #503)

[204] Vincze, Endre, *Eine allgemeinere Methode in der Theorie der Funktionalgleichungen. II.* (German) Publ. Math. Debrecen **9** (1962), 314–323. MR0152774 (27 #2749).

[205] Vincze, Endre, *Eine allgemeinere Methode in der Theorie der Funktionalgleichungen. III, IV.* (German) Publ. Math. Debrecen **10** (1963), 191–202; ibid. **10** 1963, 283–318. MR0166506 (29 #3781).

[206] Vleck, Edward B. Van, *A functional equation for the sine.* Ann. of Math. (2) **11** (1910), no. 4, 161–165. MR1502405.

[207] Vleck, Edward B. Van, *A functional equation for the sine. Additional note.* Ann. of Math. (2) **13** (1913), 154.

[208] Whitehead, John Henry Constantine, *A certain exact sequence.* Ann. of Math. (2) **52**, (1950). 51–110. 56.0X MR0035997 (12,43c).

[209] Wilson, W. Harold, *On certain related functional equations.* Bull. Amer. Math. Soc. **26** (1919–20), 300–312. Fortschr.: **47**, 320 (1919–20).

[210] Wilson, W. H., *Two general functional equations.* Bull. Amer. Math. Soc. **31**, no. 7, (1925), 330–334. MR1561055.

[211] Yale, Paul B., *Automorphisms of the Complex Numbers.* Math. Mag. **39** (1966), 135–141. Zbl 0156.27502.

[212] Yang, Dilian, *The quadratic functional equation on groups.* Publ. Math. Debrecen **66** (2005), no. 3-4, 327–348. 39B52 (20F99). MR2137773 (2005k:39035).

[213] Yang, Dilian, *Factorization of cosine functions on compact connected groups.* Math. Z. **254** (2006), no. 4, 655–674. MR2253463 (2007g:39036). 39B52 (39B62)

[214] Yang, Dilian, *Functional equations and Fourier analysis.* Canad. Math. Bull. 56 (2013), no. 1, 218–224. MR3009420.

[215] Yang, Dilian, *Cosine functions revisited.* Banach J. Math. Anal. 5 (2011), no. 2, 126–130. 39B52 (22C05 43A30). MR2805145.

[216] Yang, Dilian, *The symmetrized sine addition formula.* Aequationes Math. **82** no. 3, (2011), 299–318. 39B52. MR2855069.

[217] Zalcman, Lawrence, *Derivation Pairs on Algebras of Analytic Functions.* J. Funct. Anal. **5** (1970), 329–333. 46J05. MR0412819 (54 #940).

[218] Żelasko, Wiesław, *A characterization of multiplicative linear functionals in complex Banach algebras.* Studia Math. **30** (1968), 83–85. 46.50. MR0229042 (37 #4620).

Glossary

Page numbers refer to relevant definitions.

adj	The adjugate map, 310
$\langle A \rangle$	Subgroup generated by A, 297
$\widehat{\mathcal{A}}$	The non-zero, multiplicative and linear functionals, 296
$(ax+b)$-group	The $(ax+b)$-group, 304
$B^n(G,H)$	n-coboundaries, 352
$\mathbb{C}$	The complex numbers, 293
$C_b(X)$	The bounded functions in $C(X)$, 295
$C_c(X)$	The compactly supported functions in $C(X)$, 295
$C(F)$	Space of matrix-coefficients, 337
Cf	Cauchy difference of f, 16
C^2f	The second Cauchy difference of f, 16
χ	A character, 30
χ_π	Character of π, 341
$C^\infty(X)$	The smooth functions in $C(X)$, 295
$C_c^\infty(X)$	The smooth functions in $C_c(X)$, 295
$\mathbb{C}^n$	n-dimensional complex space, viewed as column vectors, 293
$\mathcal{C}^n(G)$	The nth term in the descending central series, 298
$C^n(G,H)$	Cochains, 352
$[A,B]$	Subgroup generated by commutators from A and B, 297
$[x,y]$	The commutator between x and y, 297
$[G,G]$	The commutator subgroup, 297

$S^{\text{cont}}(G,H)$	Continuous maps in $S(G,H)$, 215
$C(\pi)$	Space of matrix-coefficients, 341
$\mathbb{C}^*$	$\{z \in \mathbb{C} \mid z \neq 0\}$, 293
$C(X)$	The continuous complex-valued functions on X, 295
d	$d(x) = 2g(x)^2 - g(x^2)$, 318
Δ	Function associated to g, 318
$\widetilde{\Delta}$	Function associated to g, 318
det	Determinant, 295
$d^{(n)}$	Coboundary homomorphism, 352
e	Neutral element, 308
ev_x	Point evaluation at x, 339
$\check{f}$	f composed with the group inversion, 295
f_e	The even part of f, 295
f_o	The odd part of f, 295
$\overline{f}$	The complex conjugate of f, 295
G	A group, 297
g_0	Special d'Alembert function, 141
$\langle G^2 \rangle$	The subgroup generated by the squares, 297
G/H	Coset space with respect to the subgroup H, 307
$\text{GL}(n,\mathbb{C})$	The invertible matrices in $M(n \times n,\mathbb{C})$, 303
$\text{GL}(n,\mathbb{R})$	The invertible matrices in $M(n \times n,\mathbb{R})$, 303
$\text{GL}(V)$	The bijections in $\mathcal{L}(V)$, 295
g_x	$g_x(y) := g(xy) - g(x)g(y)$, 318
$\mathbb{H}$	The division algebra of the quaternions, 303
$H^2(G,H)$	The second cohomology group, 352
H_3	The Heisenberg group, 302
H_{2n+1}	The $(2n+1)$-dimensional Heisenberg group, 302
$\Vert - \Vert_{HS}$	The Hilbert-Schmidt norm, 294
$\text{Hom}(G,H)$	The homomorphisms from G to H, 299
$\text{Hom}^{\text{cont}}(G,H)$	Continuous maps in $\text{Hom}(G,H)$, 215
I	Identity matrix or operator, 301

id	The identity map, 294
$\mathbf{1}_X$	The indicator function of the set X, 295
L	The left regular representation, 344
log	The natural logarithm, 295
L'	Left regular antirepresentation, 344
$\mathcal{L}(V)$	The linear maps from V to V, 295
$\langle \lambda, x \rangle$	$= \lambda_1 x_1 + \cdots + \lambda_n x_n$, 293
M	A monoid, 308
$M(n \times m, \mathbb{C})$	The complex $n \times m$ matrices, 294
$M(n \times m, \mathbb{R})$	The real $n \times m$ matrices, 294
M^*	The adjoint of the matrix M, 294
M^t	The transpose of the matrix M, 294
μ	Usually multiplicative map, 308
$\mathbb{N}$	The natural numbers, 293
$\mathbb{N}_0$	$\mathbb{N} \cup \{0\}$, 293
$\circ$	Composition of maps, 294
$O(n)$	The orthogonal group, 303
P_3	P_3 groups are special nilpotent groups, 298
$\pi\|_W$	Subrepresentation, 341
$\bigoplus$	Direct sum, 296
$\mathbb{Q}$	The rational numbers, 293
$Q(G, H)$	Normalized solutions of the quadratic equation, 229
$Q_\tau(M, H)$	Normalized solutions of the quadratic equation, 229
Q_8	The quaternion group, 304
R	The right regular representation, 344
$\mathbb{R}$	The real numbers, 293
$\mathbb{R}^n$	n-dimensional euclidean space viewed as column vectors, 293
$\mathbb{R}^+$	$\{t \in \mathbb{R} \mid t > 0\}$, 293
$\mathbb{R}^*$	$\{t \in \mathbb{R} \mid t \neq 0\}$, 293
S	A semigroup, 308
$\times_s$	Semidirect product, 305

$S(G, H)$	Normalized solutions of Jensen's equation, 200
sgn	The sign, 295
$\mathrm{sl}(2, \mathbb{C})$	Traceless matrices, 126
$\mathrm{SL}(n, \mathbb{C})$	The matrices in $\mathrm{GL}(n, \mathbb{C})$ with determinant 1, 303
$\mathrm{SL}(n, \mathbb{R})$	The matrices in $\mathrm{GL}(n, \mathbb{R})$ with determinant 1, 303
$\mathrm{SO}(n)$	The special orthogonal group, 303
$\mathrm{SU}(n)$	The special unitary group, 303
$\langle x \rangle$	The subgroup generated by x, 297
$\mathbb{T}$	$\{z \in \mathbb{C} \mid \|z\| = 1\}$, 293
τ	An involution, 309
$f_1 \otimes f_2$	Tensor product, 295
T_n	Chebyshev polynomials, 139
tr	The trace, 295
$U(n)$	The unitary group, 303
V^*	The dual space of V, 296
$\langle v^*, v \rangle$	Value of v^* at v, 296
$W(g)$	Solutions of the symmetrized sine addition formula, 123
$\mathrm{Wil}(g)$	Wilson functions, 178
Y^X	The set of maps from X to Y, 294
$\mathbb{Z}$	The integers, 293
$Z(f)$	Elements central with respect to f, 315
$Z(G)$	The center of the group G, 297
$\mathbb{Z}_N$	The integers modulo N, 293
$Z^n(G, H)$	n-cocycles, 352
$Z(S)$	The center of the semigroup S, 308

Index

U

V

W

Y